Advances in Industrial Control

For other titles published in this series, go to
www.springer.com/series/1412

Other titles published in this series:

Digital Controller Implementation and Fragility
Robert S.H. Istepanian and James F. Whidborne (Eds.)

Optimisation of Industrial Processes at Supervisory Level
Doris Sáez, Aldo Cipriano and Andrzej W. Ordys

Robust Control of Diesel Ship Propulsion
Nikolaos Xiros

Hydraulic Servo-systems
Mohieddine Mali and Andreas Kroll

Model-based Fault Diagnosis in Dynamic Systems Using Identification Techniques
Silvio Simani, Cesare Fantuzzi and Ron J. Patton

Strategies for Feedback Linearisation
Freddy Garces, Victor M. Becerra, Chandrasekhar Kambhampati and Kevin Warwick

Robust Autonomous Guidance
Alberto Isidori, Lorenzo Marconi and Andrea Serrani

Dynamic Modelling of Gas Turbines
Gennady G. Kulikov and Haydn A. Thompson (Eds.)

Control of Fuel Cell Power Systems
Jay T. Pukrushpan, Anna G. Stefanopoulou and Huei Peng

Fuzzy Logic, Identification and Predictive Control
Jairo Espinosa, Joos Vandewalle and Vincent Wertz

Optimal Real-time Control of Sewer Networks
Magdalene Marinaki and Markos Papageorgiou

Process Modelling for Control
Benoît Codrons

Computational Intelligence in Time Series Forecasting
Ajoy K. Palit and Dobrivoje Popovic

Modelling and Control of Mini-Flying Machines
Pedro Castillo, Rogelio Lozano and Alejandro Dzul

Ship Motion Control
Tristan Perez

Hard Disk Drive Servo Systems (2nd Ed.)
Ben M. Chen, Tong H. Lee, Kemao Peng and Venkatakrishnan Venkataramanan

Measurement, Control, and Communication Using IEEE 1588
John C. Eidson

Piezoelectric Transducers for Vibration Control and Damping
S.O. Reza Moheimani and Andrew J. Fleming

Manufacturing Systems Control Design
Stjepan Bogdan, Frank L. Lewis, Zdenko Kovačić and José Mireles Jr.

Windup in Control
Peter Hippe

Nonlinear H_2/H_∞ Constrained Feedback Control
Murad Abu-Khalaf, Jie Huang and Frank L. Lewis

Practical Grey-box Process Identification
Torsten Bohlin

Control of Traffic Systems in Buildings
Sandor Markon, Hajime Kita, Hiroshi Kise and Thomas Bartz-Beielstein

Wind Turbine Control Systems
Fernando D. Bianchi, Hernán De Battista and Ricardo J. Mantz

Advanced Fuzzy Logic Technologies in Industrial Applications
Ying Bai, Hanqi Zhuang and Dali Wang (Eds.)

Practical PID Control
Antonio Visioli

(*continued after Index*)

Halim Alwi · Christopher Edwards · Chee Pin Tan

Fault Detection and Fault-Tolerant Control Using Sliding Modes

Springer

Dr. Halim Alwi
Department of Engineering
University of Leicester
University Road
LE1 7RH Leicester
UK
ha18@le.ac.uk

Prof. Christopher Edwards
Department of Engineering
University of Leicester
University Road
LE1 7RH Leicester
UK
chris.edwards@le.ac.uk

Dr. Chee Pin Tan
School of Engineering
Monash University Sunway Campus
Jalan Lagoon Selatan, Bandar Sunway
46150 Selangor
Malaysia
tan.chee.pin@eng.monash.edu.my

ISSN 1430-9491
ISBN 978-0-85729-649-8 e-ISBN 978-0-85729-650-4
DOI 10.1007/978-0-85729-650-4
Springer London Dordrecht Heidelberg New York

British Library Cataloguing in Publication Data
A catalogue record for this book is available from the British Library

Library of Congress Control Number: 2011930667

Cover design: VTeX UAB, Lithuania

Printed on acid-free paper

Springer is part of Springer Science+Business Media (www.springer.com)

Advances in Industrial Control

Professor (Emeritus) O.P. Malik
Department of Electrical and Computer Engineering
University of Calgary
2500, University Drive, NW
Calgary, Alberta
T2N 1N4
Canada

Professor K.-F. Man
Electronic Engineering Department
City University of Hong Kong
Tat Chee Avenue
Kowloon
Hong Kong

Professor G. Olsson
Department of Industrial Electrical Engineering and Automation
Lund Institute of Technology
Box 118
221 00 Lund
Sweden

Professor A. Ray
Department of Mechanical Engineering
Pennsylvania State University
0329 Reber Building
University Park
PA 16802
USA

Professor D.E. Seborg
Chemical Engineering
University of California Santa Barbara
3335 Engineering II
Santa Barbara
CA 93106
USA

Doctor K.K. Tan
Department of Electrical and Computer Engineering
National University of Singapore
4 Engineering Drive 3
Singapore 117576
Singapore

Professor I. Yamamoto
Department of Mechanical Systems and Environmental Engineering
Faculty of Environmental Engineering
The University of Kitakyushu
1-1, Hibikino, Wakamatsu-ku, Kitakyushu, Fukuoka, 808-0135
Japan

We dedicate this book to

Halim's wife Nor Mazuita Noor Azizuddin,
Chris' parents: with love and gratitude,
Chee Pin's wife Priscilla, and children Joseph & Joy-Anne.

Series Editors' Foreword

The series *Advances in Industrial Control* aims to report and encourage technology transfer in control engineering. The rapid development of control technology has an impact on all areas of the control discipline. New theory, new controllers, actuators, sensors, new industrial processes, computer methods, new applications, new philosophies..., new challenges. Much of this development work resides in industrial reports, feasibility study papers and the reports of advanced collaborative projects. The series offers an opportunity for researchers to present an extended exposition of such new work in all aspects of industrial control for wider and rapid dissemination.

In this *Advances in Industrial Control* monograph the authors, Halim Alwi, Christopher Edwards and Chee Pin Tan, remark that "In the last five decades, control system methodologies have evolved... into sophisticated and advanced electronic devices for controlling high performance and highly unstable systems... Some of these control methodologies... have found success in industry with a wide range of applications. Other control methodologies have not so readily been accepted by industry." One such technique that has not been so readily accepted is "sliding mode control" and it is this method that is at the heart of this new monograph. In fact, looking back over the *Advances in Industrial Control* monograph series (a series of over a hundred volumes from the series inception in 1991), we were surprised to find that this is the *very first* monograph on the sliding mode control method in the series!

In many ways, this monograph demonstrates the true theoretical and applications depth to which the sliding mode control paradigm has been developed today. It has three very strong themes: *control design, theoretical extensions* and *industrial applications*. For sliding mode *control*, Chap. 3 carefully builds up the reader's understanding of the design method. This is presented in structured steps using a number of simulation examples, which can easily be replicated and experienced by the reader. The phenomenon of "chattering" is exhibited and how easy it is to overcome is demonstrated. Of the extensions, there are many; however, there are two themes here. One theme is the development of technical tools like a sliding mode observer and its properties (Chaps. 4 and 5). The second theme is the use of sliding mode control in fault-tolerant control (FTC) and fault detection and isolation (FDI). There is an excellent overview of the FTC and FDI issues in Chap. 2 and then a number of Chaps. 6, 7, 8, 9, and 10 follow up these issues.

The third strong theme in the monograph is the application of the authors' schemes to computer simulations, test rigs and ultimately, an advanced flight simulator. The primary application area pursued in the monograph is the control of large passenger and cargo aircraft under extreme safety-critical conditions. These aircraft control problems supply a theme that is threaded through the monograph as it progresses. Most impressive are the two chapters at the end of the monograph describing a sliding mode control allocation scheme (Chap. 11) developed for testing on a high-fidelity 6 DOF research flight simulator, SIMONA (SImulation MOtion NAvigation) based at Delft University of Technology, The Netherlands, and a subsequent set of simulated recreations (Chap. 12) of the accident that occurred to the ELAL Flight 1862 known as the Bijlmereer Incident in which a cargo aircraft suffered the catastrophic loss of two engines from the same wing along with additional wing damage. The objective of the simulations was to demonstrate the potential of the new sliding mode control algorithm to assist in this type of situation. These simulation tests included trials using experienced pilots to obtain an independent assessment of the capabilities of the algorithm. This sort of research demonstrates industrial control engineering at its very best.

Other examples and demonstrations in the monograph include VTOL aircraft computer simulations, a laboratory-scale crane system and a d.c. motor system; all examples that give the monograph a real connection to problems of industrial control engineering.

As a first monograph on the methods of sliding mode control in the *Advances in Industrial Control* series, this is a very substantial and impressive contribution. Its mixture of the theoretical and the practical should appeal to a wide range of readers, from both the academic and industrial control engineering communities. The Editors are very pleased to have this monograph enter the series as it well demonstrates and offers the real prospects of advances in industrial control using a potentially undervalued control technique, that of the sliding mode method.

Industrial Control Centre
Glasgow
Scotland, UK
2010

M.J. Grimble
M.A. Johnson

Preface

In safety critical systems, there is an inherent requirement that, overall, some level of possibly degraded performance must be maintained even in the event of serious faults or failures occurring within the system. The ability to deal with situations in which faults and failures occur, was originally termed 'self repairing control'. However, it is now more commonly referred to by the moniker 'fault tolerant control'. The aerospace industry has often been the driver and focus of such research. As recent crashes in London and in Madrid demonstrate, malfunctions, however statistically unlikely, still occur in civil aviation contexts, and the prevention of significant loss of life depends almost solely on the correct judgement and skill of the pilot. Generally speaking fault tolerant control (FTC) schemes are classified as either passive or active. Passive schemes operate independently of any fault information and basically exploit the robustness of the underlying control paradigm. Such schemes are usually less complex, but are conservative, in order to cope with 'worst case' fault effects. Active fault tolerant controllers react to the occurrence of faults, typically by using information from a fault detection and isolation (FDI) scheme, and they invoke some form of reconfiguration. This represents a more flexible architecture. Early publications focussed on so-called projection methods whereby, if a particular fault was detected and identified, a corresponding control law from a pre-specified and pre-computed set of controllers was selected and switched online. Subsequent methods have tended to focus on online adaptation or online controller synthesis. Reconfiguration is usually necessary in the event of severe faults such as total failures in actuators/sensors. For example, if a sensor or actuator fails totally, no adaptation within that feedback loop can recover performance without modification to the choice of actuators and sensors coupled via the controller (i.e., reconfiguration). Fault tolerant control may be considered to be at the intersection of a number of research fields, and is essentially an open problem. Unsurprisingly many robust control paradigms have been used as the basis for fault tolerant controllers. The possibilities of exploiting the inherent robustness properties of sliding modes for fault tolerance have previously been explored for aerospace applications and the work in [128] argued that sliding mode control has the potential to become an alternative to reconfigurable control.

Observer-based methods are the most popular form of model-based fault detection filter. Typically (in linear observer schemes) the output estimation error formed as the difference between the measured plant output and the output of the observer, is scaled to form a residual. During fault-free operation, this residual should be 'zero' but should become 'large' and act as an alarm in the presence of a fault. A strand of work pioneered by the authors has been the development of sliding mode observers for fault estimation. This is achieved by appropriate scaling and filtering of the so-called 'equivalent output error injection', which represents the average value the nonlinear output error injection term has to take to maintain a sliding motion. This is a unique property of sliding mode observers and emanates from the fact that the introduction of a sliding motion forces the outputs of the observer to exactly track the plant measurements. Even in the presence of actuator faults, the sliding mode forces the outputs of the observer to perfectly track the measurements, and accurate estimation of the states is still possible. The fault reconstruction signal is not computed from a residual calculation based on the output estimation error (which will be zero during the sliding motion), but from the equivalent output error injection signal. Consequently accurate state estimation and fault estimation can be, in principle, achieved simultaneously from a single (sliding mode) observer. This is quite different to the situation in the case of traditional linear observer designs for FDI which require a trade-off between robustness with respect to the state estimation, and fault sensitivity for detection using output error based residuals. Robust state estimation, whilst retaining fault sensitivity, is a property unique to sliding mode observers.

The book will cover the theoretical development and implementation of sliding mode schemes for fault tolerant control. A key development in this book considers sliding mode control allocation schemes for fault tolerant control based on integral action and a model reference framework. Unlike many control allocation schemes in the literature, one of the main contributions described in this book is the use of actuator effectiveness levels to redistribute the control signals to the remaining healthy actuators when faults/failures occur. A rigorous stability analysis and design procedure is developed from a theoretical perspective for this scheme. A fixed control allocation structure is also rigorously analyzed in the situation when information on actuator effectiveness levels is not available. The proposed scheme shows that faults and even certain total actuator failures can be handled directly without reconfiguring the controller. The later chapters of the book present the results obtained from real-time hardware implementations of the controllers on the 6-DOF SIMONA flight simulator at Delft University as part of the GARTEUR AG16 programme.

Chapter 1 gives an overview of the recent developments in the area of fault detection and fault tolerance control. It is intended to provide motivation for the theoretical developments which follow in the subsequent chapters.

Chapter 2 begins with the definition of the terms fault and failure and briefly discusses the different types of faults and failures which can occur in actuators and sensors—with specific aircraft examples. The chapter introduces the concept of fault tolerant control and gives a general overview of the different FTC and FDI research fields. The main concepts and strategies behind some of the FTC and FDI schemes in the literature, as well as their advantages and drawbacks, are also discussed.

Chapter 3 gives a brief introduction to the concept of sliding mode control and examines its properties. This chapter also highlights the benefits of sliding modes when applied to the fields of FTC and FDI. A simple pendulum example is used to introduce the concept. The unit-vector approach for multi–input systems, sliding surface design and tracking requirements (integral action and model reference based tracking) are also discussed. Chapter 3 ends with some discussions on the benefits and motivation for sliding mode control in the fields of FTC and FDI.

Chapter 4 considers sliding modes applied to the problem of observer design. A historical development is outlined leading to the description of a specific class of sliding mode observer which will be used throughout the book. It will be shown how the unique properties associated with the so-called equivalent injection signal necessary to maintain sliding can be exploited to reconstruct actuator and sensor faults modelled as additive perturbations to the inputs and the outputs of the plant. Design methodologies based on Linear Matrix Inequalities (LMIs) are presented. These approaches exploit all the available degrees of freedom associated with the choice of the observer gains. The chapter describes sliding mode observers which can reconstruct faults and yet be robust to disturbances/uncertainties which may corrupt the quality of the reconstructions resulting from mismatches between the model about which the observer is designed and the real system. Initially, the design method is formulated for the case of actuator faults. A comparison is also made between the sliding mode observer schemes developed in the chapter and more traditional linear unknown input observers which are prevalent in the literature.

Chapter 5 examines the assumptions that must be made for the observer schemes described in Chap. 4 to be applicable. (These amount to relative degree one minimum phase limitations on the transfer function matrices relating the unknown fault signals to the measurements.) This chapter explores ways of obviating these limitations, at the expense of creating cascaded observer structures. The components of the cascade will be observer formulations taken from Chap. 4, and explicit constructive algorithms will be given to ensure the overall scheme can still accurately estimate actuator faults in the case where the relative degree between the faults and the measurements is greater than or equal to two. The advantages these schemes offer over traditional linear methods (particularly UIOs) will be demonstrated.

Chapter 6 will focus specifically on sensor faults. Different formulations will be considered in which the measured output signals are filtered to yield 'fictitious systems' in which sensor faults appear as 'actuator faults'. Consequently, the actuator fault reconstruction ideas from the previous chapters can then be applied to the fictitious system to reconstruct the sensor fault. The results will also be extended to the case of unstable plants which result in nonminimum phase configurations post-filtering.

Chapter 7 considers the real-time implementation of the sensor fault reconstruction schemes (for FDI and FTC) from Chap. 6 on a laboratory crane and a small DC motor rig. These rigs provide cheap, safe and practical demonstrators for the ideas presented in Chap. 6. The data collection and (subsequent) controller implementation has been achieved using MATLAB® and dSPACE®. Estimates of the sensor faults, obtained from online sliding mode FDI schemes have been used to correct the measured outputs from the sensors. The 'virtual sensors' have been used in the control algorithm to form the output tracking error signal which is processed to generate the fault tolerant control signal.

Chapter 8 presents a new sliding mode scheme for reconfigurable control. The controller is based on a state-feedback scheme where the nonlinear unit-vector term is allowed to adaptively increase when the onset of a fault is detected. The scheme is applied to a benchmark aircraft problem. In comparison to other fault tolerant controllers which have been previously implemented on this model, the controllers proposed in this book are simple and yet are shown to work across the entire 'up and away' flight envelope. Excellent rejection of a certain class of actuator faults is shown. However, the proposed controller cannot directly cope with the total failure of an actuator. In the second half of the chapter, the use of sensor fault reconstruction methods to correct faulty measurements prior to the control law calculations, hence effecting fault tolerant control, is demonstrated. Here, a formal closed-loop analysis is made of the resulting schemes. An example of such a method applied to a benchmark aircraft problem is described.

Chapter 9 proposes an online sliding mode control allocation scheme for fault tolerant control. The effectiveness level of the actuators is used by the control allocation scheme to redistribute the control signals to the remaining actuators when a fault or failure occurs. The chapter provides an analysis of the sliding mode control allocation scheme and determines the nonlinear gain required to maintain sliding. The allocation scheme shows that faults and even certain total actuator failures can be handled directly without reconfiguring the controller.

Chapter 10 describes an adaptive model reference sliding mode fault tolerant control scheme with online control allocation. As in Chap. 9, the control allocation scheme uses the effectiveness level of the actuators to redistribute the control signals to the remaining actuators when a fault or failure occurs. Meanwhile, the adaptive nonlinear gain and reference model provide online tuning for the controller. This chapter provides a rigorous stability analysis for the model reference scheme. The scheme has been tested on a linearisation of the ADMIRE aircraft model to convey the ideas associated with the proposed scheme and shows that various faults and even total actuator failures can be handled.

Chapter 11 describes the implementation of the sliding mode allocation schemes from Chap. 9 on the 6-DOF research flight simulator SIMONA at Delft University of Technology, the Netherlands. The controller from Chap. 9 is implemented in 'C' and runs on the 'flight control' computer associated with SIMONA. Real-time implementation issues are discussed and a range of fault scenarios from the GARTEUR AG16 benchmark are tested and discussed.

Chapter 12 presents the ELAL flight 1862 (Bijlmermeer incident) scenario—which is one of the case studies of GARTEUR AG16. The results presented in this chapter demonstrate the outcome of the 'flight testing' campaign and the GARTEUR AG16 final workshop at Delft University of Technology in November 2007. The results represent the successful real-time implementation of a sliding mode controller on SIMONA with experienced test pilots flying and evaluating the controller.

Finally, Chap. 13 makes some concluding remarks and offers suggestions for future work.

Leicester, UK — Halim Alwi
Leicester, UK — Christopher Edwards
Bandar Sunway, Malaysia — Chee Pin Tan

Acknowledgements

Whilst the research presented in this book is almost totally the efforts of the three authors, the contribution of Prof. Jan Albert (Bob) Mulder (Control and Simulation Division, Faculty of Aerospace Engineering) and Ir. Olaf Stroosma (International Research Institute for Simulation, Motion and Navigation (SIMONA)) from Delft University of Technology, The Netherlands, to Chaps. 11 and 12 is gratefully acknowledged. Without the offer from Prof. Mulder to use the SIMONA simulator as part of Delft's contribution to the GARTEUR FM-AG16[1] programme, none of the implementation results described in Chaps. 11 and 12 would have been possible. The technical expertise of Ir. Stroosma in terms of interfacing the controller code with the SIMONA platform made the implementation (almost) a joy. We thank him for all his help.

In addition to the hardware platform used to test the controllers, the underlying benchmark aircraft model, which was used as a basis for the simulator, has been the result of many man-hours of development by different contributors over the years—most recently Hafid Smaili and Jan Breeman of NLR (National Aerospace Laboratory), The Netherlands and Dr. Andres Marcos, formally at the University of Minnesota, USA and now at Deimos Space, Madrid, Spain. Other contributors to the development of the benchmark model include Coen van der Linden and Dr. Thomas Lombaerts (Delft University of Technology), Prof. Gary Balas (University of Minnesota), David Breeds (QinetiQ) and Stuart Runham (DSTL). We would like to thank all those who were involved in the GARTEUR FM-AG16 action group on fault tolerant control. Their support and contributions to the discussions in the AG16 program are highly appreciated.

The authors would like to thank all those who kindly gave their approval to use the pictures and illustrations in this book. The illustrations remain the property of the copyright holders.

[1]The European Flight Mechanics Action Group FM-AG(16) on Fault Tolerant Control was established in 2004 and concluded in 2008. It represented a collaboration involving thirteen European partners from industry, universities and research establishments under the auspices of the Group for Aeronautical Research and Technology in Europe (GARTEUR) program.

The provision of two grants from the Engineering and Physical Sciences Research Council (EPSRC), which funded most of the work which is described in this book, is also gratefully acknowledged.

Contents

List of Notations

Acronyms

air, ail	inboard right and inboard left ailerons
aor, aol	outboard right and outboard left ailerons
cmd	command signal
ru, rl	upper and lower rudders
sp	spoiler
6-DOF	6 Degree of Freedom
ATC	Air Traffic Controller (airport control tower)
CA	Control Allocation
CFIT	Controlled Flight Into Terrain
CG	Centre of Gravity
DFDR	Digital Flight Data Recorder
DI	Dynamic Inversion
DME	Distance Measuring Equipment
EPR	Engine Pressure Ratio
FBW	Fly-By-Wire
FDI	Fault Detection and Isolation
FPA	Flight Path Angle
FTC	Fault Tolerant Control
GARTEUR	Group for Aeronautical Research and Technology in Europe
GS	Glide Slope
IAS	Indicated Airspeed
ILS	Instrument Landing System
IMM	Interactive Multiple Model
KIAS	Indicated Air Speed in Knots
KLM	Royal Dutch Airlines
LMI	Linear Matrix Inequality
LOC	Localizer Capture
LPV	Linear Parameter Varying
LTI/LTV	Linear Time Invariant/Varying
MAC	Mean Aerodynamic Chord

MCT	Maximum Continuous Thrust
MMST	Multiple Model Switching and Tuning
MPC	Model Predictive Control
MRAC	Model-reference Adaptive Control
NLR	National Aerospace Laboratory, the Netherlands
PIM	Pseudo-Inverse Method
ROV	Remote Operating Vehicle
SIMONA	SImulation, MOtion and NAvigation (flight simulator)
SMC	Sliding Mode Control
s.p.d.	symmetric positive definite
TAS	True Airspeed
VOR	VHF Omni-directional Radio Range

Mathematical notation

p, q, r	roll, pitch and yaw rate (rad/s)
V_{tas}	true air speed (m/s)
α, β	angle of attack and sideslip angle (rad)
ϕ, θ, ψ	roll, pitch and yaw angle (rad)
h_e, x_e, y_e	geometric earth position along the z, x and y axis (m)
$\|\cdot\|$	Euclidean norm or induced spectral norm
$\lambda(\cdot)$	eigenvalues
$\bar{\lambda}(\cdot), \underline{\lambda}(\cdot)$	largest and smallest eigenvalues
Γ	integral action design matrix
$\nu(t)$	virtual control input and pseudo control
A^{T}	transpose of matrix A
$\mathbb{C}$	field of complex numbers
$\mathbb{C}_-$	the set of strictly negative complex numbers
D, E	robust sliding mode observer gain
F, G	feedback and feedforward control matrix
$G(\mathrm{s})$	transfer function
G_l, G_n	sliding mode observer gain matrices
J	cost function
K	actuator fault/failure distribution matrix
$\mathcal{K}$	sliding mode design matrix
L	sliding mode observer design matrix
L_x	sliding mode control design matrix
M	actuator fault distribution matrix (observer)
N	sensor fault distribution matrix
$\mathcal{N}(A)$	null space of the matrix A
Q	uncertainty distribution matrix
$\mathbf{Q}, \mathbf{R}$	LQR/LMI weighting matrix
$\mathcal{R}$	rectangle
$\mathbb{R}$	field of real numbers
$\mathbb{R}_+$	the set of strictly positive real numbers
$\mathrm{Re}(\cdot)$	real part of a complex number
s	sliding mode switching function

s	Laplace variable
$\mathrm{sgn}(\cdot)$	signum function
S	sliding mode matrix
$\mathcal{S}$	sliding surface
V	Lyapunov function
$\mathcal{V}_r$	truncated ellipsoid
$\mathbf{V}$, $\mathbf{W}$	weighting matrix
W	actuator effectiveness distribution matrix
$\mathcal{W}$	allowable fault set
X	LMI variable

Chapter 1
Introduction

In the last five decades, control system methodologies have evolved from simple mechanical feedback structures into sophisticated and advanced electronic devices for controlling high performance and highly unstable systems which optimise cost and control effort. Some of these control methodologies, for example the 'three term' PID (Proportion, Integral and Derivative) controller [12] and the Kalman filter [237, 268], have found success in industry with a wide range of applications. Other sophisticated control methodologies have not so readily been accepted by industry.

Some of the strategies that have received a good deal of attention in the last couple of decades are multivariable robust and adaptive control methods (see for example [175, 216, 228, 297]). This is motivated by the need to optimise the performance of safety critical systems such as aircraft, chemical plants and nuclear power plants, which require the control systems to deal with wide changes in the operating conditions of the plant. However some unexpected scenarios or unusual events in the system mean the designed controller is sometimes simply 'overwhelmed' and a loss of performance and stability might occur. Examples of these unexpected scenarios are faults, failures or system 'damage', which are typically not considered in the controller design process.

The problem of achieving some level of performance and stability in the case when these unexpected scenarios occur, especially for safety critical systems (e.g., chemical and nuclear power plants) and expensive autonomous systems (e.g., satellites and underwater remote operating vehicles (ROV)) requires a different strategy rather than just having a robust or adaptive controller (which only guarantees stability and performance for perturbations in the nominal plant). An example of a system which requires such a control strategy is the problem of increasing the survivability of an aircraft when faults or failures to the actuators/sensors or structural damage occurs during a flight. In such a situation the aircraft requires some 'emergency' strategy to allow the pilot to safely land the aircraft. This challenge has motivated a strategy widely known in the literature as *fault tolerant control* (FTC).

Many different control paradigms have been applied to the problem of FTC. Examples of some of the existing control approaches can be found in Table 1.1, whilst Table 1.2 shows different systems that FTC has been applied to. In this book, the ad-

H. Alwi et al., *Fault Detection and Fault-Tolerant Control Using Sliding Modes*,
Advances in Industrial Control,
DOI 10.1007/978-0-85729-650-4_1,

Table 1.1 An example of existing approaches in FTC (adapted from [294])

Design approaches	References
Model-reference adaptive control	[142, 143, 230, 292]
Adaptive control	[14, 76, 143]
Multiple model switching and tuning	[10, 30, 41, 109, 111, 148, 195, 196, 253]
Interactive multiple model	[149, 170, 214, 284, 290]
Gain scheduling	[165]
Linear parameter varying systems	[16, 105, 182, 185, 215, 223, 224, 273]
Model predictive control	[176–178, 220]
Pseudo-inverse method	[106, 132, 187, 206, 284]
Control allocation	[29, 31, 37–39, 63, 89, 122, 127, 222]
Dynamic Inversion	[137, 138, 144, 172, 249, 250]
Robust control e.g., $\mathcal{H}_\infty$	[178, 228]
Sliding mode control	[85, 127, 256]

Table 1.2 An example of applications of FTC and FDI (adapted from [294])

Applications	References
Aircraft	[16, 40–44, 105, 185, 230, 233, 253]
Spacecraft	[68, 108, 137]
Automotive	[115, 154]
Engine and propulsion control	[40, 155, 202, 253]
Chemical/petrochemical plants	[176]
Robots	[198]

vantages of FTC will be demonstrated on aircraft systems as an example of a safety critical plant.

1.1 Motivation for Fault Tolerant Control Systems

The safety of aircraft passengers has been and will continue to be an important issue in the commercial aviation industry. Figures 1.1 and 1.2 represent some recent civil aviation safety statistics. Although the number of flights has doubled since 1980, the number of fatal accidents has been maintained over the years, and in fact decreased during the period from 1999–2003 [1]. This is contributed to by many factors, such as the stringent safety measures imposed on the aircraft and the implementation of important safety technology. Furthermore, all pilots undergo extensive training to help them to react to unforeseen difficulties which may arise during a flight. Figure 1.2 shows that 'controlled flight into terrain' (CFIT) and 'loss of control in flight' are the two most important occurrences and involve the most fatalities [1].

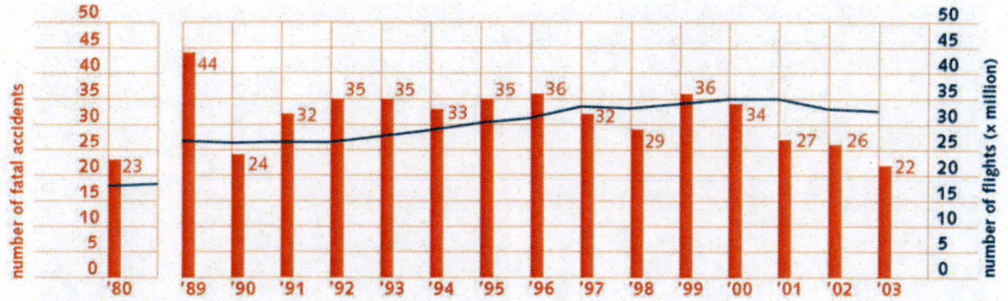

Fig. 1.1 Number of flights and fatal accidents (figure from [1])

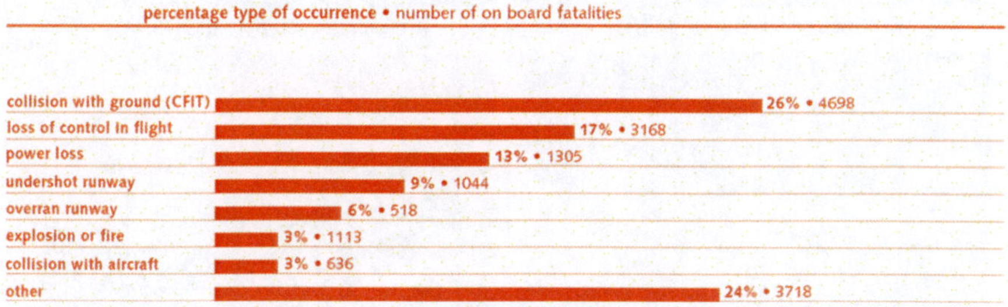

Fig. 1.2 Type of occurrences and fatalities (figure from [1])

Loss of control during flight is one of the motivating factors for fault tolerant control: the idea is to increase the 'flyability' of aircraft in the event of faults, failures or airframe damage. Learning from previous incidents, where pilots successfully landed crippled aircraft—such as Flight 232 in Sioux City, Iowa 1989,[1] the Kalita Air freighter in Detroit, Michigan, October 2004 (Fig. 1.4)[2] and the DHL freighter incident in Baghdad, November 2003 (Fig. 1.3)[3]—suggests that in many cases, the damaged aircraft is still 'flyable', with sufficient functionality to allow the pilot to safely land the aircraft.

It has been argued that with pilot skill and a fault tolerant control system, several accidents could have been avoided. For example, a recent report [40, 253] described a NASA experiment in which, by clever manipulation of thrust (in the event of total hydraulic loss), it was possible to land the 'crippled' plane. Pilot reviews and comments after the flight test indicate that fault tolerant control did help the pilot control the crippled plane when compared to pilot control alone [40, 253]. Although the work by NASA on propulsion controlled aircraft successfully handles total hydraulic loss, it is not sufficient to solve the general problem of fault tolerant control for aircraft, especially when other control surfaces are still functional or when dealing with structural damage and aerodynamic change (which for example occurred

[1] Flight 232 suffered tail engine failure causing total loss of hydraulics [41, 109].

[2] The freighter shed engine No. 1, but the crew managed to land safely.

[3] The A300B4 was hit by a missile and lost all hydraulics, but landed safely [41].

(a) (picture taken from [80])

(b) (picture: JetPhoto.net/DK (c))

(c) (picture: wikipedia)

(d) (picture: aviation-safety.net)

Fig. 1.3 DHL A300B4 emergency landing after being hit by missile in Baghdad, 2003

(a) (picture: pprune.org)

(b) (picture: airdisaster.com)

Fig. 1.4 Kalita Air emergency landing after losing one engine, 2004 in the ELAL flight 1862 Bijlmermeer incident [8]) or when dealing with control surface jams or runaways (for example flight 427, near Aliquippa, Pennsylvania in 1994 [9]).

1.2 Sliding Modes for FTC and FDI

Generally speaking, fault tolerant control (FTC) schemes are classified as either passive or active. Passive schemes operate independently of any fault information and basically exploit the robustness of the underlying controller. Such schemes are usually less complex, but in order to cope with 'worst case' fault effects, are conservative. Active fault tolerant controllers react to the occurrence of faults, typically by using information from a fault detection and isolation (FDI) scheme, and invoke some form of reconfiguration. This represents a more flexible architecture. Early publications focussed on so-called projection methods whereby if a particular fault was detected and identified, a corresponding control law from a pre-specified and pre-computed set of controllers, was selected and switched online. Subsequent methods have tended to focus on online adaption or online controller synthesis. Reconfiguration is usually necessary in the event of severe faults such as total failures in actuators/sensors. For example, if a sensor or actuator fails totally, no adaptation within that feedback loop can recover performance without modification to the choice of actuators and sensors coupled via the controller (i.e., reconfiguration). Fault tolerant control may be considered to be at the intersection of a number of research fields, and is essentially an open problem. Unsurprisingly many robust control paradigms have been used as the basis for fault tolerant controllers. These include LQR, adaptive, $\mathcal{H}_\infty$, QFT, model-following, neural networks, pseudo-inverse methods, Nonlinear Dynamic Inversion (NDI), multiple model approaches and MPC (see Table 1.1).

The possibilities of exploiting the inherent robustness properties of sliding modes for fault tolerance has previously been explored for aerospace applications and the work in [127] argued that sliding mode control has the potential to become an alternative to reconfigurable control. This is due to the inherent robustness properties of sliding modes to a certain class of uncertainty, including its ability to directly handle actuator faults without requiring the fault to be detected and without requiring controller reconfiguration. Despite its robustness property in handling actuator faults, sliding mode control (as with most other controllers) cannot handle total actuator failures. Some of the current research attempting to solve this problem has assumed that exact replication of the failed actuator is available [58]. However this is only applicable to a few over actuated systems. This is one aspect that will be explored in this book.

Observer-based methods are the most popular form of model-based fault detection filter. Typically (in linear observer schemes) the output estimation error, formed as the difference between the measured plant output and the output of the observer, is scaled to form a residual. During fault-free operation, this residual should be 'zero' but should become 'large' and act as an alarm in the presence of a fault. Many different design approaches have been employed: see for example [51, 69]. A strand of work by the authors has been the development of sliding mode observers for fault estimation [83]. This is achieved by appropriate scaling and filtering of the so-called 'equivalent output error injection', which represents the average value the nonlinear output error injection term has to take to maintain a sliding motion. This is a unique property of sliding mode observers and emanates from the fact that the introduction of a sliding motion forces the outputs of the observer to exactly track the plant measurements [83]. The results described in [86] indicate that sliding mode schemes perform better in several of the fault scenarios, and always at least as well as the other methods, when demonstrated on a well-known ship propulsion system benchmark [139]. Even in the presence of actuator faults, the sliding mode forces the outputs of the observer to perfectly track the measurements, and accurate estimation of the states is still possible. The fault reconstruction signal is not computed from a residual calculation based on the output estimation error (which will be zero during the sliding motion), but from the equivalent output error injection signal. Consequently accurate state estimation and fault estimation can be, in principle, achieved simultaneously from a single sliding mode observer. This is quite different from the situation in the case of traditional linear observer designs for FDI which require a trade-off between robustness with respect to the state estimation, and fault sensitivity for detection using output error based residuals. The exploitation of these unique properties of sliding mode observers for FDI will be covered in detail in this book.

Chapter 2
Fault Tolerant Control and Fault Detection and Isolation

When a fault occurs in a system, the main problem to be addressed is to raise an alarm, ideally diagnose what fault has occurred, and then decide how to deal with it. The problem of detecting a fault, finding the source/location and then taking appropriate action is the basis of fault tolerant control.

In this chapter, an introduction to fault tolerant control (FTC) and fault detection and isolation (FDI) will be presented. The chapter will start with some definitions and will describe different types of faults and failures which can occur in actuators and sensors. Later, different types of fault tolerant controllers and FDI schemes will be presented and discussed.

2.1 Faults and Failures

First, the terms fault and failure will be defined. The definition provided in this book is in compliance with the definitions given by the IFAC SAFEPROCESS technical committee (as given in [136]) which were developed to set a standard [51] in this area in order to avoid confusion among researchers. This will also enable the concept of fault tolerant control (FTC) to be specified in terms of faults and/or failures later in the chapter. The IFAC technical committee, as outlined in [136], makes the following definitions:

> ***fault***: an unpermitted deviation of at least one characteristic property or parameter of the system from the acceptable/usual/standard condition.
> ***failure***: a permanent interruption of a system's ability to perform a required function under specified operating conditions.

Clearly, a failure is a condition which is much more severe than a fault. When a fault occurs in an actuator for example, the actuator is still usable but may have a slower response or become less effective. But when a failure occurs, a totally different actuator is needed to be able to produce the desired effect.

In aircraft systems, there are some distinct types of actuator failure, the three most common are shown in Fig. 2.1. A *lock failure* is a failure condition when an actuator

H. Alwi et al., *Fault Detection and Fault-Tolerant Control Using Sliding Modes*, Advances in Industrial Control,
DOI 10.1007/978-0-85729-650-4_2,

Fig. 2.1 Type of fault and failures on actuator (figure adapted from [32])

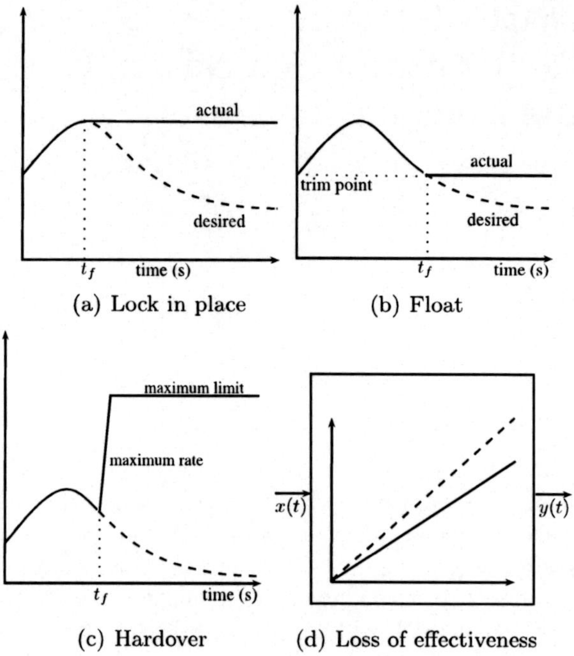

becomes stuck and immovable. This might be caused by a mechanical jam, due to lack of lubrication for example. This type of failure is considered in [53, 92, 105, 111, 292] and occurred in documented incidents such as flight 1080 (Lockheed L-1011, San Diego, 1977) [41] where one of the horizontal stabilisers jammed in the full trailing edge-up position; and flight 96 (DC-10, Windsor, Ontario, 1972) [41] where the rudder jammed with an offset.

A *float failure* is a failure condition whereby the control surface moves freely without providing any moment to the aircraft. An example of a float failure is the loss of hydraulic fluid. Examples of research considering float type failures are [42, 92, 105], and it has occurred in incidents such as Flight 123 (B-747, Japan, 1985) and DHL A300B4 (A300, Baghdad, 2003) [41] resulting from a total loss of hydraulics.

One of the most catastrophic types of failure is *runaway/hardover*. In a runaway situation the control surface will move at its maximum rate limit until it reaches its maximum position limit or its blowdown limit.[1] For example, a rudder runaway can occur when there is an electronic component failure which causes an uncommanded large signal to be sent to the actuators causing the rudder to be deflected at its maximum rate to its maximum deflection at low speed (or its blowdown limit at high

[1] A blowdown limit is an aerodynamic limit of the control surface deflection at a specified speed which overpowers the movement of the actuator. The blowdown limits might not be the maximum physical deflection of the control surface. Any deflection above the blowdown limit can cause structural damage [240] as it imposes the maximum physical and structural limit the control surface and the surrounding structure can have.

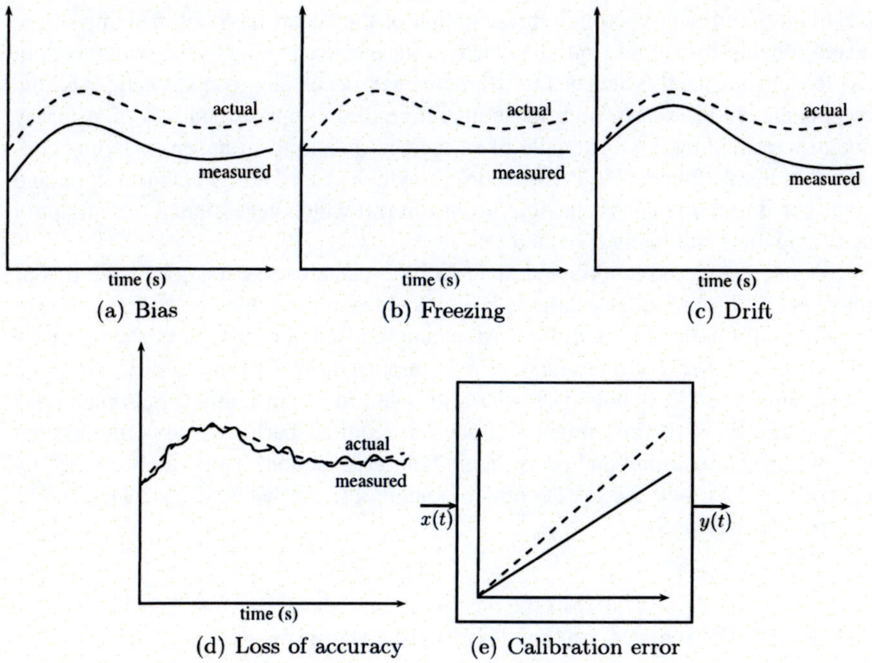

Fig. 2.2 Type of fault and failures on sensor (figure adapted from [32])

speed). This type of failure is considered in [234] and occurred in incidents such as Flight 85 (B-747, Anchorage, Alaska, 2002) [41] (which suffered a lower rudder runaway to full left deflection, causing the airplane to roll excessively) and flight 427 (B-737, Near Aliquippa, Pennsylvania, 1994) [9] (which suffered from a rudder runaway to its blowdown limits).

Note that the above faults and failures are related to the aircraft's control surfaces. Another type of fault that occurs in aircraft, is *structural damage*. Structural damage may change the operating conditions of the aircraft due to changes in the aerodynamic coefficients or a change in the centre of gravity. This constitutes a change to the dynamics of the system. Examples of failures that cause structural damage are wing battle damage [30], detachment of control surfaces, for example the rudder (flight 961, A310, Varadero, Cuba, 2005) [3] or engines (flight 1862, B-747, Amsterdam, 1992) [235], or detachments of some body parts of the aircraft e.g., the vertical fin/stabiliser (Flight 123, B-747, Japan, 1985) [41, 109] and (flight 587, A300, New York, 2001) [41], wing (DHL A300B4, A300, Baghdad, 2003) [41], fuselage skin or cargo doors (flight 981, DC-10, Paris, 1974) [41].

Figure 2.2 describes some typical sensor faults in aircraft. Bias is a constant offset/error between the actual and measured signals. Sensor drift is a condition whereby the measurement errors increase over time (and might be due to loss of sensitivity of the sensor). Loss of accuracy occurs when the measurements never reflect the true values of the quantities being measured. Freezing of sensor signals result in the sensor providing a constant value instead of the true value. Finally, a calibration error is a wrong representation of the actual physical meaning of the quantity being measured from the electrical or electronic signals that emerge from the sensor unit itself.[2] Sensor faults/failures can occur due to malfunctions in the components in the sensor unit, loose mounting of the sensors and loss of accuracy due to wear and tear. An example of an incident resulting from sensor failures occurred in flight 124 (B-777, Perth, 2005) [2] which caused a flight control upset and contributed to the violent behaviour of the aircraft which necessitated the auto pilot and navigation unit being switched off.

It is interesting to mention that faults/failures can also be categorised in terms of time [247]. Abrupt faults or failures exhibit sudden and unexpected changes and are usually easily noticed by the pilot. An example of an abrupt failure is an actuator jam, or a hardover. Incipient faults, for example a slow drift in a sensor, are more subtle and the effect is not so obvious. However, incipient faults if left unattended for a long period of time might degrade the required performance of the system and might lead to abrupt and catastrophic failures. Incipient faults can be caused by operational wear and tear as the effect is negligible but becomes gradually worse before it fails abruptly.

2.2 Fault Tolerant Control: General Overview

In the literature, most of the motivation and research work in fault tolerant control involves solving problems encountered in safety critical systems such as aircraft. Applications can also be found in other systems, for example robots [169], space systems [252] and underwater remotely operated vehicles (ROV) [198]. Patton in [206] stated that

> ...Research into fault tolerant control is largely motivated by the control problems encountered in aircraft system design. The goal is to provide a 'self-repairing' capability to enable the pilot to land the aircraft safely in the event of serious fault ...

Zhang and Jiang [291] define

> ...fault tolerant control systems (FTCS) as control systems that possess the ability to accommodate system component failures automatically. They are capable of maintaining overall system stability and acceptable performance in the event of such failures. FTCS were also known as self-repairing, reconfigurable, restructurable, or self designing control systems ...

FTC is a complex combination of three major research fields [206], FDI, robust control, and reconfigurable control (see Fig. 2.3). Patton [206] also discusses the relationship between these fields of research. A typical active fault tolerant control systems (AFTCS) architecture is shown in Fig. 2.4. For most FTC schemes, when a fault/failure occurs either in an actuator or sensor, the FDI scheme will detect and locate the source of the fault. This information is then passed to a mechanism to

[2] Sensors, most of the time, provide measurements in terms of current or voltage and therefore require transformation to represent the actual physical meaning of the quantities being measured.

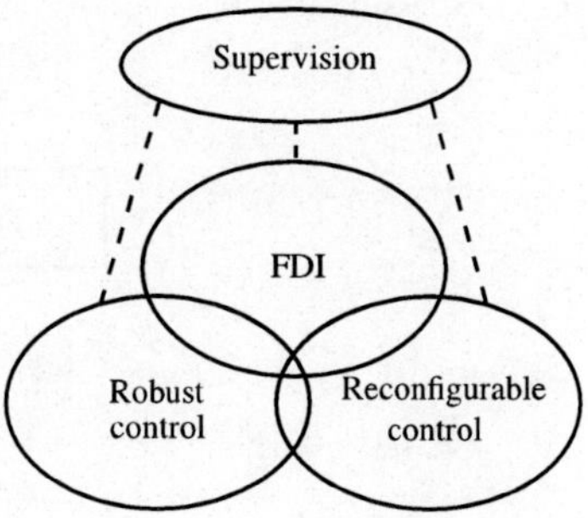

Fig. 2.3 Scattered areas of fault tolerant control research (figure adapted from [206])

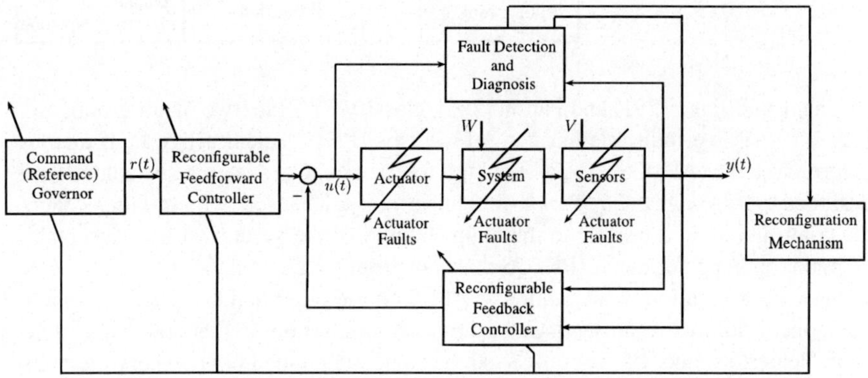

Fig. 2.4 General structure of active fault tolerant control systems (AFTCS) (figure adapted from [294])

initiate reconfiguration. The reconfigurable controller will try to adapt to the fault, therefore providing stability and some level of performance. Both the FDI and the reconfigurable controller need to be robust against uncertainties and disturbances.

Robust control is closely related to passive fault tolerant control systems (PFTCS) [206]. The controller is designed to be robust against disturbances and uncertainty during the design stage. This enables the controller to counteract the effect of a fault without requiring reconfiguration or FDI. In some robust methodologies, fault tolerant capability is limited, and importantly total, actuator failures cannot be handled. Some widely referred to surveys on FTC and FDI are [132, 143, 178, 206, 291, 293] and [51, 136]. Also there are recent publications (books and edited monographs) such as [24, 25, 47, 75, 80, 179, 199] in the field of FTC and [26, 69, 134, 135, 159] for FDI.

Zhang and Jiang [291] gives a good bibliographical review of reconfigurable fault tolerant control systems. The paper also proposes a classification of reconfiguration methods which is based on the mathematical tools used, the design approach and the way of achieving reconfiguration. It also provides a bibliographical classification based on the design approaches and the different applications. Open problems and current research topics in active fault tolerant control systems (AFTCS) are also discussed.

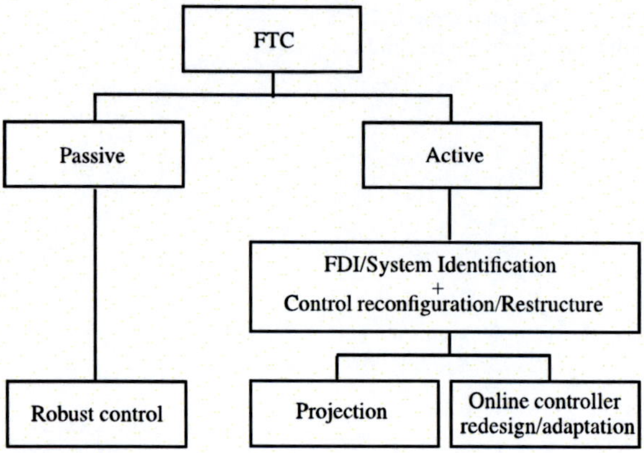

Fig. 2.5 Classification of FTC by [206]

Zhang and Jiang [291] and Patton [206], classify FTC into two major groups (see Fig. 2.5): passive fault tolerant control systems (PFTCS) and active fault tolerant control systems (AFTCS). In passive fault tolerant control systems, the controller is designed to be *robust against faults* and uncertainty. Therefore when a faults occurs, the controller should be able to maintain stability of the system with an acceptable degradation in performance. PFTCS does not require FDI and does not require controller reconfiguration or adaptation. AFTCS on the other hand responds to system component failures in an 'active' way by reconfiguration so that stability and acceptable performance of the entire system can be maintained [291]. Therefore, most AFTCS require FDI to provide the fault or failure information before reconfiguration can be undertaken.

Other surveys on reconfigurable control appear in [132] and [143]. The report [132], gives insight into many methods used for reconfigurable control for flight applications, while [143] gives a survey on reconfiguration methods used specifically for FTC in flight control applications. Table 2.1 presents a brief comparison of the FTC methods [143]. Note that in Table 2.1, the expression 'fault model' refers to the assumption that the faulty system is available and used in the design process. 'Actuator constraints' refers to the ability of the controller to handle actuator limits.

2.3 Redundancy

Redundancy can be categorised into two types; direct and analytical. In direct redundancy, actual physical hardware redundancy is available. In terms of sensors, two or three sensors that measure the same quantity is called double and triple redundancy. In normal operation, only one sensor is sufficient, however, two or three sensors are required to ensure reliable measurements in the case of faults. A voting system is a typical way to decide which channels are working correctly and which are faulty [112]. This hardware redundancy concept can also be extended to the actuators.

In terms of analytical redundancy, instead of having multiple sensors that measure the same signal, an observer that provides an estimate of the signals of inter-

Table 2.1 Current FTC methods comparison (table adapted from [262])

Method	Actuator failures	Structural failures	Robust	Adaptive	FDI	Fault model assumed	Actuator constraints	Linear model	Nonlinear model
Multiple model switching and tuning (MMST)		•		•	•			•	
Interactive multiple model (IMM)		•		•	•		∘	•	
Propulsion controlled aircraft (PCA)	•		∘			•		•	•
Control allocation (CA)	•					•	∘	•	
Feedback linearisation	•	•		•	•				•
Sliding mode control (SMC)	∘[a]	•	•[b]				•		•
Eigenstructure assignment (EA)		•				•		•	
Pseudo-inverse method (PIM)		•				•		•	
Model-reference adaptive control (MRAC)		•		•	•			•	
Model predictive control (MPC)	•	•	∘	∘	•	•	•	•	•

• means that the method has the property

∘ implies that an author suggested that the approach could be modified to incorporate the property [143]

[a]SMC can handle partial loss of effectiveness of actuators but not complete loss

[b]SMC assumes robust control can handle all forms of structural failures

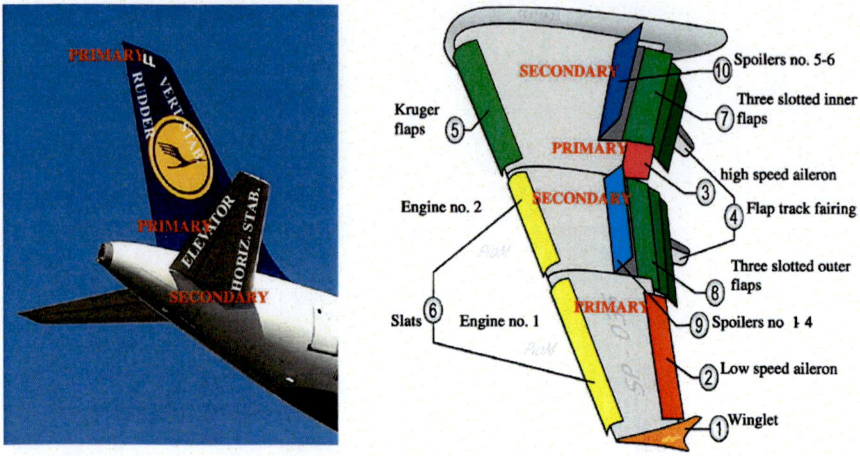

Fig. 2.6 Large transport aircraft: typical control surfaces (figures: Arpingstone and Piotr Jaworski via Wikipedia)

est provides analytical redundancy. There is no actual additional hardware implemented, instead some algorithm or mathematical model or observer runs in the control computer. This is desirable in many systems especially in aircraft and unmanned air vehicles (UAVs), since analytical redundancy eliminates the requirements for extra hardware therefore reducing weight and cost.

The development of new safety critical systems such as the re-entry vehicle [137, 138] allows the possibility of building in redundancy during the design process [206]. For many systems, however, the challenge is to use the existing available sensors and actuators to deal with faults/failures. In large transport aircraft, redundancy is already available in abundance. Even though it is not meant for the purpose of FTC, the use of these extra control surfaces provides the possibility of using them to obtain the same effect as the original control surface e.g., horizontal stabilisers can be used if elevators fail.

In large passenger transport aircraft, sensors are typically triple redundant [34, 35]. In view of the aerospace industry's attempts to reduce the 'carbon footprint' left by aircraft, many manufacturers have tried to reduce the consumption of fuel by designing high efficiency engines, and also by reducing weight by eliminating hardware redundancies, replacing them with analytical ones (observers to estimate the aircraft states). This is also beneficial in the development of cheap, robust and maintenance-free UAVs. Due to the low production cost, there is no requirement for repair, and instead, the whole unit is replaced.

In aircraft, a control surface, for example the rudder, can have three different hydraulic actuators running from three separate lines to three independent hydraulic pumps [35]. This means most control surfaces will have triple redundancy. In terms of the control surface itself, there exist secondary control surfaces that can be used in an emergency in an unconventional way to achieve the same effect as the primary control surface (see Fig. 2.6). In large passenger transport aircraft for example, the spoilers which are typically deployed to reduce speed can also be used differentially

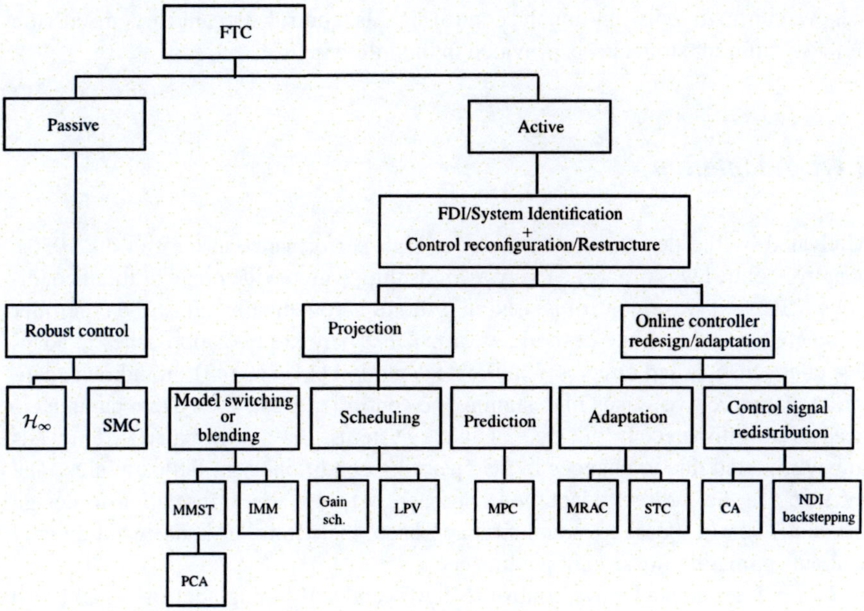

Fig. 2.7 Classification of FTC—how FTC is achieved

to create roll which normally is achieved by using ailerons; also engines can be used differentially to create yaw, which is typically achieved by using the rudder; and finally the horizontal stabiliser (see Fig. 2.6), which is normally used to set the angle of attack, can also replace elevators for pitch modulation.

2.4 Fault Tolerant Control

Figure 2.7 gives a general overview of how FTC is achieved. The top level of the tree diagram is based on the one proposed in [206]. The lower level is based on the different approaches for achieving FTC discussed above.

Passive FTC is usually based on robust control ideas and therefore handles faults/failures without requiring information from an FDI scheme. Active FTC (AFTC) in general requires some information on the faults/failures that occur and therefore typically FDI is required. AFTC can be divided into two sub-groups: projection type FTC; and online reconfiguration/adaptation. In projection based FTC, controllers are designed a priori for all possible faults/failures that might occur in the system. The projected controller will only be active when the corresponding fault/failure occurs. Projection based FTC is sub-divided into three categories which are model switching or blending, scheduling and prediction. AFTC is based on reconfiguration or online adaption. Here, two further sub-components have been proposed: FTC which is achieved through adaptive control; and FTC which can be achieved through redistributing the control signals (control allocation). A discussion of these different strategies is provided in the following subsections.

2.4.1 Adaptation

Motivated by the design of autopilots for high performance aircraft in the 1950s, adaptive control was proposed as a way of dealing with a wide range of flight conditions [230]. Adaptive control is used in order to automatically adjust the controller parameters to achieve the desired performance. There are two approaches in adaptive control: so-called direct and indirect adaptation [14, 76, 143]. In indirect adaptation, there are two stages in designing the controller. First, the system parameters need to be estimated. In the case of linear systems, the matrix pair (A, B) needs to be estimated due to changes in the operating conditions e.g., faults/failures. The next step in the indirect adaptation approach is to use this information to design the controller. In the direct adaptation approach, the controller is designed directly without estimating the system parameters.

Model-reference adaptive control (MRAC) and self tuning control (STC) [230] are two popular methodologies. In self tuning control, online parameter estimation is required for the controller adaptation. Meanwhile, in the MRAC, the unknown parameters are not perfectly estimated, but rather are tuned and adjusted so that the output of the plant follows the desired trajectory (the output of the reference model) by making the tracking error converge to zero.

2.4.2 Switching or Blending

The idea of using multiple models for reconfigurable control was introduced in the early 1990s [196]. Multiple model schemes have been motivated by the problem of coping with changes in operating conditions and varying flight envelopes. Most early classical control methods were based on linear methods, and multiple model schemes seemed an ideal extension to solve the problem of changing operating conditions. When implementing on a real system, usually linear controllers need to adapt to changes in operating conditions since the controller is only guaranteed to be stable near the linearisation condition. Therefore using multiple model schemes is one way to ensure that the controller can be designed so that stability and performance can be guaranteed for a wide flight envelope.

From an FTC point of view, the 'bank' of controllers acts as a backup and 'offline' dormant controllers are only activated when faults for which a particular controller is designed, occurs. This method depends on the FDI scheme providing the correct information on the type and location of the faults/failures to enable the correct controller to be switched on. The 'bank' of models must contain all the possible faults and failure modes. An FDI scheme can be created by comparing the current plant states and the outputs of all models in the bank [30]. Essentially, the model with the smallest error is the 'nearest' model to the actual plant and therefore its associated controller can be switched on. More elaborate descriptions of the switching rule are available in [195, 196]. These papers discuss the stability of the switching schemes as well as the performance after a switch has occurred. In [196], a review of the most recent development in MMST has been presented.

The switching between models and controllers sometimes introduces undesired transients. Therefore bumpless transfer methods [48, 81] are sometimes needed to reduce the effect. Another disadvantage is that some faults that occur are not predicted a priori. For example, in several flight incidents, unthinkable failures have occurred e.g., the Bijlmermeer incident in Amsterdam [233], where two engines detached from the right wing and caused unforeseeable effects on the aerodynamics of the aircraft due to the damaged airframe. A significant disadvantage of this method is its dependency on the robustness of the FDI scheme to identify the correct model and controller pair to be activated. Another disadvantage highlighted in [143], is the scheme's inability to handle multiple faults/failures. The survey in [143], gives a brief introduction to MMST. More detailed descriptions can be found in [195, 196]. The application of multiple model ideas in terms of FTC for aircraft systems can be found in [30, 111] and recently in [10].

Even though MMST can be used to tackle the problem of varying operating conditions, in some cases, to obtain a linear model that exactly matches the varying plant is hard to achieve; since hundreds (if not thousands) of linear models and controllers are needed to match every possible flight condition including faults/failures. In the 'Interacting Multiple Model' (IMM) [214, 290] approach, the idea is to obtain a set of linear models based on a few carefully chosen flight conditions and to design linear controllers at these selected operating conditions. When the operating conditions change (or faults/failures occur), an estimated plant output or control input is obtained by blending the predetermined models.

The main assumption used in IMM is that every possible flight condition including faults/failures can be modelled as a convex combination of the predetermined linear models. The second step is to obtain a control signal based on a blend of predefined controllers [290] or online control law calculations using the probability weight provided by the IMM estimator.

In the first step of the IMM scheme, [290] and [214] proposed the use of a bank of Kalman filters to calculate the probability of the individual faults/failures. This probability is also used to obtain a weighted average of each predefined linear model to estimate the state of the plant. In the second step, a bank of controllers is predesigned based on the anticipated faults/failures that might occur [214, 290]. The idea is that during faults/failures, the eigenvalues of the closed-loop system need to be as close as possible to the nominal no fault conditions. The reconfiguration of the controller comes from the online use of the probability weighted average to determine the blending ratio for the control input from the predefined/projected controllers when a fault or failure occurs [214, 290].

In [290], it has been shown that system faults[3] can be handled. In comparison to the MMST, the IMM has the ability to cope with non-anticipated faults/failures [214, 290]. One problem of IMM schemes is finding the right balance of blending/probability weights to get the best model match. The IMM method is also heavily dependent on the FDI scheme to correctly identify the faults/failures. Details of IMM schemes can be found in [149]. In [290], an integrated IMM approach is discussed where both FDI and FTC are integrated. The application of IMM to an Eagle-Eye UAV can be found in [214].

2.4.3 Prediction

Unlike many other control paradigms which came ostensibly from the academic community, the development of predictive control/model predictive control (MPC) was initiated in the process industry. This is due to the fact that the concept and the mathematical description is easy to understand by most control engineers in industry. Therefore it is no surprise that (other than classical PID control), MPC is the most widely used and implemented method in the process control industry [176].

The original idea for MPC is to allow the production process to run as close as possible to the process limits (both physical and safety) without violating any of the limits, in order to maximise production and therefore profit. The main benefit of MPC is its ability to handle limits and constraints. This is the main motivation for the study of MPC for flight control and especially FTC. Examples of MPC in the field of flight control and FTC can be found in [177, 178] respectively. During faults/failures, especially to the actuators, the remaining actuators will be driven to their limits [177]. MPC has the ability to handle this situation by including these limits in the optimisation process which is used to obtain the control signals. Structural damage can also be handled in MPC by modifying the internal/reference model [177].

MPC is an iterative control algorithm based on optimal control. The objective is to obtain predicted state trajectories in the future using the current states and the computed control signals. However only the first control signal from the optimisation is applied to the real actuators. Then the states are sampled again and the calculations are repeated. MPC is also known as receding horizon control [176, 178].

MPC in its most powerful form requires an online solution to the constrained optimisation problem [176]. However, with the current state of computer technology, online optimisation is still hard to achieve for systems requiring fast responses—such as aircraft. As in most FTC strategies, MPC is dependent on reliable FDI to provide information on the faulty system. In the case of actuator faults, the behaviour of the faulty actuator is needed from the FDI scheme so that a new constraint

[3] In terms of linear methods, system faults are the ones that affect the A matrix i.e., airframe or wing damage.

can be included in the optimisation process. In terms of tuning for flight control systems, there is still a lack of transparency in the design process [178], which typically requires trial and error and experience. The major benefit of MPC is that it can handle actuator constraints, and this has provided motivation for the study of MPC in flight control and FTC [178].

2.4.4 Control Signal Redistribution

The idea of the pseudo-inverse method (PIM) is to design a controller such that the poles of the system subject to a fault/failure condition will be as close as possible to the nominal closed-loop poles. The following equations give insight into the PIM method. Consider a linear system given by

$$\dot{x}(t) = Ax(t) + Bu(t) \tag{2.1}$$

Assume that a state feedback gain F has been designed, and the control law is defined as

$$u(t) = Fx(t) \tag{2.2}$$

and therefore the closed-loop system is given by

$$\dot{x}(t) = (A + BF)x(t) \tag{2.3}$$

During faults/failures, the closed-loop faulty system can be represented by

$$\dot{x}_f(t) = (A_f + B_f F_f)x_f(t) \tag{2.4}$$

The idea is to obtain a gain matrix F_f so that the faulty system closed-loop performance will be as close as possible to the nominal (2.3) one. The plant matrices A and B and the gain F is assumed to be known a priori. The faulty system (A_f, B_f) can be obtained from online system identification or from FDI: then in principle, F_f can be obtained online. For a non-square B_f matrix, the pseudo-inverse of B_f provides some degrees of freedom. In [187], these degrees of freedom were used to redistribute the control commands in order to improve the closed-loop system stability [206].

Even though the concept is quite simple and easy to understand, the PIM has several drawbacks which hindered its further progress. As argued in [132, 143, 206, 284], the main drawback is the lack of a stability analysis. The other drawbacks highlighted by [206] and [284] are associated with the assumption that the state measurements are always available. Meanwhile [284] highlights the problem of lack of robustness when the system pair (A_f, B_f) from the system identification is not perfectly known.

Some suggestions are given in [106, 132, 206] for improving the PIM method. In [106], the concept of modifying the PIM (MPIM) is discussed. It is based on the combination of PIM with the theory of robust stability of systems with structured uncertainty [132, 206]. In [132], a bank of pre-computed F_f matrices, which ensure a stable closed-loop system for all possible faults, was suggested. In [284], a robust 'control mixer' which relates to the IMM method was proposed. It is also interesting to point out the resemblance between the PIM approach and model following methods [206], where the closed-loop system is forced to follow pre-specified desired closed-loop dynamics.

In its early development, the idea of redistributing the control signals to the remaining healthy actuators was called 'restructuring' [206]. An early example is given in [131], where a 'restructuring controller' utilising a 'control mixer concept' is used to redistribute the control signals. Due to some drawbacks, this restructuring controller was not explored in the 90's. It has re-emerged in recent years as *control allocation* partly because of the development of high performance, highly redundant aircraft (such as [31, 37, 225, 269]) and improvements in computational power (which is necessary in order to solve online optimisation problems [20, 29, 77, 89, 144]).

Control allocation (CA) has the capability of redistributing the control command signals to the actuators especially during faults/failures. One major difference between CA and PIM is that in CA, the controller is designed based on a 'virtual control' signal and the CA element will map the virtual control to the actual control demand to the actuators. The benefit here is that the controller design is independent of the CA unit. Therefore, CA can be used in conjunction with any other controller design paradigm. Papers such as [122, 222] represent some of the recent work in this area.

CA has the capability of managing the actuator redundancy that exists in passenger aircraft [35] and modern fighter aircraft [97]. Not only is CA beneficial for FTC (see for example [38, 63]), it has also been used for different control strategies i.e., optimally using the actuators to reduce drag and increase efficiency. There is extensive literature on CA which discusses different algorithms, approaches and applications. Reference [89] discusses two (broadly) linked approaches (linear and quadratic programming) based on finding the 'best solution' to a system of linear equations. The work in [122] compares control allocation with optimal control design for distributing the control effort amongst redundant actuators. The authors in [39] demonstrate that feedback control systems with redundant actuators can be reduced to a feedback control system without redundancy using a special case of CA known as 'daisy chaining'. In this approach, a subset of the actuators, regarded as the primary actuators are used first, then secondary actuators are used if the primary actuators reach saturation. Other CA approaches which take into account actuator limits (using constrained optimisation) are discussed in [29, 31], while [121] discusses frequency weighted CA.

From an FTC point of view, the benefits of CA are that the controller structure does not have to be reconfigured in the case of faults and it can deal directly with total actuator failures without requiring reconfiguration/accommodation: the CA scheme automatically redistributes the control signal. As in MPC, another major benefit of CA is that actuator limitations can be handled by including the actuator constraint in the optimisation process. One of the drawbacks of CA is that, for linear systems, the pure factorisation of the input distribution matrix is a very strong

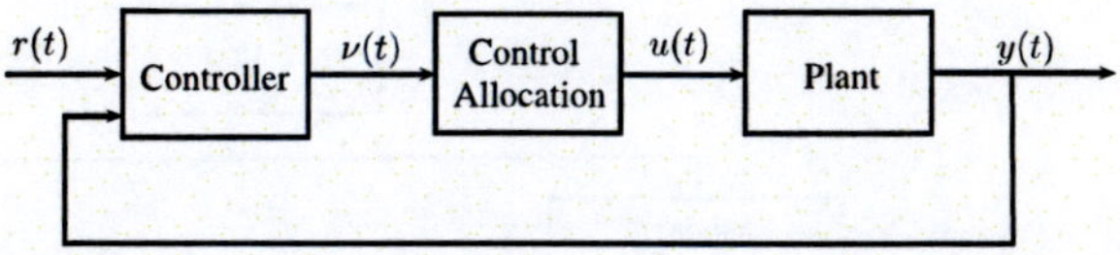

Fig. 2.8 Control allocation strategy

requirement and therefore some approximations have been made [38, 63, 122, 127]. In the case of optimal control surface deflection, linear or quadratic programming is required which is difficult to achieve online in real-time due to the requirements of high computational power during the optimisation process. There are only a few reported examples in the literature which have successfully implemented control allocation in real-time (see for example [63]).

CA occurs naturally in nonlinear methods like feedback linearisation and backstepping [20, 121, 225, 226]. It is based on separating the control law from the control allocation task (see Fig. 2.8). This is done by designing a controller to provide a 'virtual control' which will be mapped to the actual control signals sent to the actuators. Examples of the application of CA are given in [20, 37, 63, 236], while papers such as [75, 78, 122, 133, 144, 203, 222, 226, 263, 295] consider CA as an FTC strategy.

2.4.5 *Robust Control ($\mathcal{H}_\infty$ Control)*

Passive fault tolerant control relies on the robustness of the underlying control design paradigm. One of the most popular robust control methodologies developed during the 1980's was $\mathcal{H}_\infty$ control. It has become one of the most developed methods for multivariable control [178], with many applications ranging from industrial process control to aircraft control problems. Most robust control approaches do not require any information on faults and therefore work in nominal as well as in faulty conditions. The ability to deal with faults depends on the predesigned controller which is based on minimising the effect of uncertainty or disturbances on the system [178].

One disadvantage of $\mathcal{H}_\infty$ is the fact that in some cases, the controller is conservative in the nominal conditions in order to guarantee stability in the event of faults, the performance in the nominal condition is sometimes sacrificed for robustness. Another drawback is that the final controller is usually of a higher order than the system. In some cases model reduction is required to truncate the order of the controller (page 339 [178]). In the field of FTC, papers like [184] and the chapters in [178] describe some of the research in the area of flight control.

The theory of $\mathcal{H}_\infty$ control system design can be extended to Linear Parameter Varying (LPV) systems. Some general papers on Linear Parameter Varying (LPV) are [16, 204, 221, 273]. In the field of FTC, papers such as [105, 185] represent some of the research work in this area. Both of these papers have considered a LPV approach for dealing with faults/failures in a civil aircraft benchmark problem. The most recent LPV papers in the field of FTC are [200, 215, 223, 224].

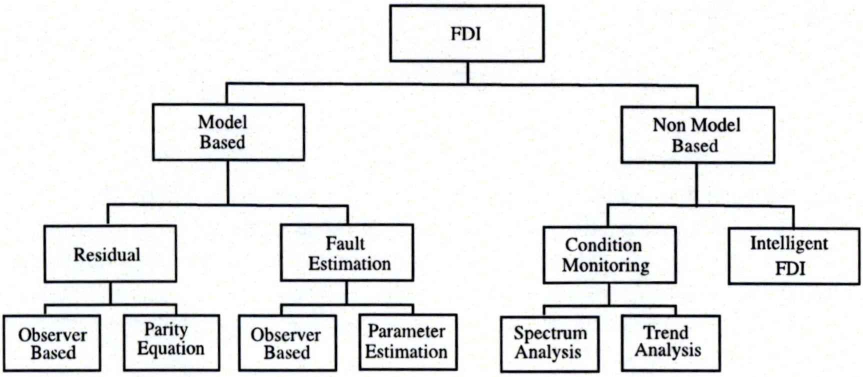

Fig. 2.9 FDI classification

2.5 Fault Detection and Isolation

In active FTC, FDI plays a vital role in providing information about faults/failures in the system to enable appropriate reconfiguration to take place. The main function of FDI is to detect a fault or failure and to find its location so that corrective action can be made to eliminate or minimise the effect on the overall system performance [247]. The IFAC technical committee, as stated in [136], makes the following definitions.

> ***fault detection***: determination of the faults present in a system and the time of detection.
> ***fault isolation***: determination of the kind, location and time of detection of a fault.
> ***fault identification***: determination of the size- and time-variant behaviour of a fault.

The interconnection of FDI with FTC is discussed in [206, 291, 293]. For most AFTC systems, the robustness of the FDI has a strong effect on the robustness of the FTC and this is discussed in [206, 289, 293]. Since the appearance of drive and fly-by-wire technology, there has been an increase in analytical redundancy. In analytical redundancy methods, the measured signals are compared to a mathematical model. The benefit of using analytical redundancy is clear: there is no need for redundant hardware to be installed, therefore reducing weight and cost. This is very useful for energy and weight reduction critical systems such as satellites and spacecraft.

There are many classifications of FDI in the literature [51, 136]. One obvious classification is model- and non-model-based FDI. In this book the emphasis will be on model-based FDI. Model-based FDI schemes can be grouped into two major categories; FDI using residual schemes and FDI which has the capability to estimate the faults.

Figure 2.9 represents a possible classification for FDI. The model-based residual classification is obtained from [51]. A brief description of a few key model-based FDI schemes are given below.

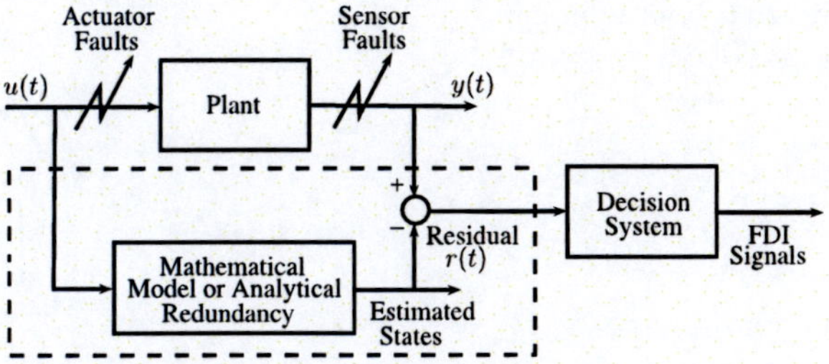

Fig. 2.10 Residual-based FDI

2.5.1 Residual-Based FDI

In residual-based FDI, signals from a mathematical model and hardware measurements are compared and the filtered difference forms a residual signal [51] (see Fig. 2.10). In nominal fault-free conditions, the residuals should be zero, and nonzero when faults/failures occur. This residual signal is usually applied with a threshold to avoid false alarms from disturbances or uncertainty. When the residual signal exceeds the threshold, a fault is said to occur. Usually in residual generation, a fault is detected and its location identified, but there is no further information on the fault.

A good deal of research has been focussed on residual-based FDI using different methods for various applications. In particular, [51] provides an excellent discussion on model-based residual FDI schemes covering all aspects including basic principles and robustness issues.

There are many benefits of using residual-based FDI. Most residual-based FDI systems are easy to understand and implement with many mature topic areas and examples of applications in the literature. For many systems, detection and isolation of the fault is sufficient to trigger the reconfiguration for FTC. For example multiple model controllers will switch on a particular controller when the designated failure occurs to the actuators or sensors based on the information about the location of the fault. However, for some FTC schemes, detecting and isolating the faults is not sufficient. Some FTC schemes require further information about the nature and behaviour of the fault.

2.5.2 Fault Identification and Reconstruction

Some FTC methods proposed in [275, 289, 296], require estimates of the actuator efficiency to allow the schemes to tolerate the faults/failures (see Fig. 2.11). In terms of sensor fault FTC, if the sensor fault can be estimated/reconstructed, this information can be used directly to correct the corrupted sensor measurements before they are used by the controller. This avoids reconfiguring or restructuring the controller. This is one aspect that will be considered in the later chapters of this book.

The Kalman filter is probably one of the most well known and used methodologies in industry. Conceived in the 1960's by Rudolf Kalman and made famous

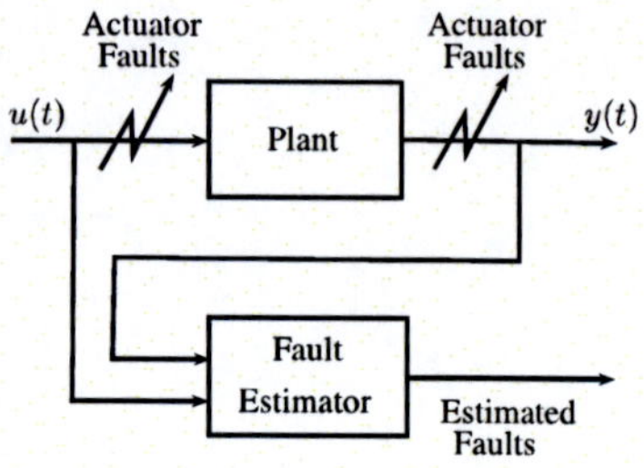

Fig. 2.11 Fault estimation based FDI

by its application in the NASA Apollo space program, the Kalman filter has found applications in many engineering systems (e.g., navigation, tracking targets such as aircraft and missiles using radar) as well as other fields such as economics.

A Kalman filter as summarised in [156], is an optimal estimator based on indirect, inaccurate and uncertain observations. It is recursive so that new measurements can be processed as they arrive. If all noise is Gaussian, the Kalman filter minimises the mean square error of the estimated parameters and therefore is optimal. Since its famous application in the Apollo space program, the Kalman filter has continued to be popular especially in industry for a number of reasons: (a) Kalman filters provide fairly accurate results in most applications due to its optimality and structure, and (b) Kalman filters have a recursive form and are suitable for online real-time digital processing and are easy to formulate and implement.

Standard Kalman filters act as an observer and therefore can be used to detect faults or failures by creating residual signals from comparing the actual and estimated outputs. The basic concept of the Kalman filter has been 'upgraded' to enable many applications, such as the extended Kalman filter for nonlinear systems [167] and for parameter estimation [107, 115] in which the parameters to be estimated are incorporated into the formulation as augmented states. Often this introduces bilinearities which can be overcome by the use of extended Kalman filters.

The Kalman filter can also be composed into a bank of Kalman filters [157, 158] or interacting multiple model Kalman filters (IMM-KF) [214, 290] in order to create a residual which can be used for fault detection. The IMM-KF uses the same IMM as used for controller reconfiguration which was discussed in Sect. 2.4.2. The Kalman filter also has been combined with the receding horizon (predictive control) method as shown in [162], which has the potential for fault diagnosis.

Another variant called the two stage Kalman filter [150, 151] has the ability not only to detect and isolate faults, but to estimate the effectiveness levels of actuators [140, 274, 275, 289]. This capability is a bonus for FTC schemes which depend on the effectiveness level of the actuator for reconfiguration [224, 296].

The early papers by Kalman can be found in [146, 147] whilst the most cited books and references are [13, 186, 237].

Using the same principles as those used for designing $\mathcal{H}_\infty$ controllers, an observer can be designed as a basis for a residual-based FDI scheme [185]. The idea is to allow the residual to be sensitive only to faults and robust against disturbances, modelling errors and noise [185]. This can be achieved by selecting the observer gains (using LMI formulations for example) which minimise the $\mathcal{H}_\infty$ norm between the uncertainty and the residual signal. Fault detection filters using $\mathcal{H}_\infty$ techniques are amongst the most popular and mature FDI schemes in the literature [130] with many applications in industry including aerospace [184, 185, 244]. Apart from the $\mathcal{H}_\infty$ optimisation technique, other frequency domain design approaches for model-based FDI, including μ synthesis, are discussed in [51] and [98].

Applications of $\mathcal{H}_\infty$ for robust detection of faults can be found in [184, 185]. In [184] an integrated design of both controller and observer is considered. The integrated design proposed in [185] gives some insight on designing a controller that is not only robust against actuator faults but also considers the robustness properties of the FDI scheme in the design of the controller.

The early $\mathcal{H}_\infty$ methods have recently been extended to linear parameter varying models. It has been claimed that even though there are various FDI approaches for LTI, LTV and bilinear systems, there are only a few available methods for LPV systems [27]. Therefore, the focus of the work in [27] was to introduce FDI based schemes for LPV systems using an extension of the approach called the fundamental problem of residual generation. Other recent LPV papers in the field of FDI are [11, 28, 114, 116, 153, 168, 200, 245, 267]. FDI based on LPV systems has inherent performance and stability guarantees for a range of operating conditions compared to multiple model or gain schedule based FDI.

2.5.3 Parameter Estimation

Parameter estimation schemes provide a means of updating the system's parameters online in real-time and for controller reconfiguration. Parameter estimation is one of many methodologies which have been applied to aircraft. Aircraft contain many parameters (especially aerodynamic coefficients) which change, based on the operating conditions. These parameters are typically pre-estimated offline through wind tunnel and flight test before being used for modelling or control design. However, during faults/failures (especially structural damage, such as wing damage or missing fuselage/skin), no accurate pre-estimate is available and therefore these aerodynamic coefficients need to be obtained online.

Examples of parameter estimation methods appear in [107, 115] which use Kalman filters, and [57, 172] which use the two step method. In the two step method (TSM), the original state-parameter estimation problem is decomposed into a state estimation one and a subsequent linear parameter identification sub-problem [57, 192]. Other sources of information on parameter estimation of aircraft systems can be found in [180, 181, 193, 201, 260].

In [115], parameter estimation based on an extended Kalman filter is used for FDI in an automotive engine. One of the most recent papers for aircraft FDI is [144]. This paper proposes the use of online parameter estimation provided by the two step method [172] (which identifies and estimates the current aircraft parameters which change due to structural damage). Here, not only are the changes to the aerodynamic coefficients used to detect faults/failures in the system, they are also used as part of the reconfiguration to achieve fault tolerance.

In most parameter estimation methods, in order to get good estimates, it may be necessary to introduce perturbation signals to make sure that all the plant modes are sufficiently excited [206]. For this reason, most parameter estimation methods work best in the presence of wind and gusts. However, in many practical applications, it is hard and not advisable to apply additional perturbation signals, especially when faults/failures or structural damage has occurred in the system.

2.5.4 Non-model-Based FDI (Intelligent FDI)

One of the main issues associated with model-based designs is the availability and quality of the model. Errors resulting from imperfect or inaccurate models will affect the performance of the fault diagnosis scheme [208, 210]. The use of robust model-based methods usually results in a design which is too conservative and insensitive to faults, too complicated or limited to certain classes of uncertainty [208]. Since the late 1990s there has been an increase in research on non-model-based FDI methods—especially those utilising artificial intelligence and 'soft computing' approaches such as neural networks, and fuzzy logic (see for example [26, 159, 160, 209, 212, 272]).

In [208], a combination of numerical (quantitative) and symbolic (qualitative) knowledge of the system in a single framework has been proposed. The idea was inspired by earlier work which uses observers for residual generation and fuzzy logic for decision making. The underlying concept is to structure the neural network in a fuzzy logic format which allows residual generation (through the rapid and correct training of the neural network to model the nonlinear dynamics of the system) and evaluation and diagnosis of the fault (through fuzzy logic). In [160], neuro-fuzzy modelling and diagnosis is considered with the addition of an adaptive threshold in the fault detection scheme, to achieve some level of robustness.

One of the benefits of using the intelligent approach, especially neural networks for FDI is its ability to model any nonlinear function [208]. In terms of FDI, neural networks have 'black box' characteristics and therefore the ability to learn from 'examples' and 'training', requiring little or no a priori information and knowledge of the system's structure [208]. Two major drawbacks of conventional neural networks are highlighted in [208]: namely, heuristic knowledge from an experienced expert cannot easily be incorporated, and the 'black box' characteristic means that its internal behaviour cannot be easily understood. Another drawback of neural networks is the lack of understanding of its internal behaviour, causing 'clearance problems'—especially for aircraft systems.

Recent research work can be found in [188, 213, 272] while application examples can be found in [5, 49, 189, 254, 287]. Examples of an intelligent approach for FDI in aircraft systems appear in papers such as [4, 238] and the references therein.

2.6 Summary

This chapter has presented a brief introduction to the fields of FTC and FDI. It includes definitions of terms regularly used in FTC and FDI such as faults and failures. This chapter also briefly discussed the possible types of faults and failures to actuators and sensors that can occur, together with a discussion on redundancy and its importance to FTC. Different methods to achieve FTC were discussed, ranging from robust control to control signal redistribution.

Chapter 3
First-Order Sliding Mode Concepts

The objective of this chapter is to introduce the concept of sliding modes and to examine the associated properties, in order to highlight its benefits in the fields of FTC and FDI. Sliding mode control (SMC) was conceived in the former USSR in the 1950s and spread into the West in the 1970s. SMC is a nonlinear type of control and a special case of Variable Structure Control. The controller design process is unique compared to other methods since the performance of the controller depends primarily on the design of the sliding surface. The idea is to drive the trajectory of the states towards the sliding surface and once reached, the states are forced to remain on the surface. SMC has an inherent robustness property to a certain type of uncertainty, which makes it a strong candidate for passive FTC to handle actuator faults.

3.1 Introduction

There are two stages involved in the design of sliding mode controllers. First to be designed is the sliding surface. Only then can the control law be designed so that sliding is achieved and maintained on the surface. Once sliding occurs, robustness to a certain class of uncertainty is guaranteed and the closed-loop system behaves as a reduced order motion independent of the control. Furthermore the closed-loop system performance depends on the choice of the sliding surface. A typical sliding mode control law consists of linear and nonlinear components. The nonlinear part of the control law determines the robustness property of the controller. This will be discussed in detail in the following sections.

3.1.1 Regular Form

In order to explain the concept of a sliding mode and its properties conveniently, the system can be transformed into a suitable canonical form. Consider the following linear time invariant (LTI) system:

$$\dot{x}(t) = Ax(t) + Bu(t) \tag{3.1}$$

H. Alwi et al., *Fault Detection and Fault-Tolerant Control Using Sliding Modes*, Advances in Industrial Control,
DOI 10.1007/978-0-85729-650-4_3,

where $A \in \mathbb{R}^{n\times n}$ and $B \in \mathbb{R}^{n\times m}$ with $1 \le m \le n$. Assume $\text{rank}(B) = m$, i.e., matrix B is full rank, and the pair (A, B) is controllable [85, 256]. Since $\text{rank}(B) = m$, there exists an orthogonal matrix $T_r \in \mathbb{R}^{n\times n}$ such that

$$T_r B = \begin{bmatrix} 0 \\ B_2 \end{bmatrix} \tag{3.2}$$

where $B_2 \in \mathbb{R}^{m\times m}$ and is nonsingular [85]. The orthogonal matrix T_r can be computed using 'QR' decomposition: details can be found in [85]. After the coordinate transformation $x(t) \mapsto T_r x(t) = z(t)$, the states can be partitioned as

$$z(t) = \begin{bmatrix} z_1(t) \\ z_2(t) \end{bmatrix} \tag{3.3}$$

where $z_1(t) \in \mathbb{R}^{n-m}$ and $z_2(t) \in \mathbb{R}^m$, so that (in the new coordinates), (3.1) can be written as

$$\begin{bmatrix} \dot{z}_1(t) \\ \dot{z}_2(t) \end{bmatrix} = \begin{bmatrix} A_{11} & A_{12} \\ A_{21} & A_{22} \end{bmatrix} \begin{bmatrix} z_1(t) \\ z_2(t) \end{bmatrix} + \begin{bmatrix} 0 \\ B_2 \end{bmatrix} u(t) \tag{3.4}$$

This representation is referred to as 'regular form' [85]. Define a linear combination of the states to be

$$s(t) = Sx(t) \tag{3.5}$$

where $S \in \mathbb{R}^{m\times n}$ is full rank, and let $\mathcal{S}$ be the hyperplane defined by

$$\mathcal{S} = \{x \in \mathbb{R}^n : Sx(t) = 0\} \tag{3.6}$$

Equation (3.5) is called the 'switching function' [85]. The matrix S can be partitioned as

$$S = [S_1 \quad S_2] T_r$$

where $S_1 \in \mathbb{R}^{m\times(n-m)}$ and $S_2 \in \mathbb{R}^{m\times m}$ then

$$\det(SB) = \det(S_2 B_2) = \det(S_2)\det(B_2)$$

Therefore a necessary and sufficient condition for the matrix SB to be nonsingular is that $\det(S_2) \neq 0$ since by construction $\det(B_2) \neq 0$. By design assume this to be the case. During ideal sliding on the surface $\mathcal{S}$, $Sz(t) = 0$ for all $t \ge t_s$, where t_s is the time when sliding commences, therefore

$$Sx(t) = S_1 z_1(t) + S_2 z_2(t) = 0 \Rightarrow z_2(t) = -\underbrace{S_2^{-1} S_1}_{\mathcal{K}} z_1(t) \tag{3.7}$$

Equation (3.7) implies that once $z_1(t)$ is known, the states $z_2(t)$ are completely determined. Therefore only the $\dot{z}_1(t)$ states i.e., the (top) partition from (3.4) needs to be considered. Partitioning (3.4) gives

$$\dot{z}_1(t) = A_{11} z_1(t) + A_{12} z_2(t) \tag{3.8}$$

and then substituting for $z_2(t)$ from (3.7) into (3.8) yields the following reduced order system

$$\dot{z}_1(t) = (A_{11} - A_{12}\mathcal{K}) z_1(t) \tag{3.9}$$

where $\mathcal{K} = S_2^{-1} S_1$. The choice of surface $\mathcal{S}$ from (3.6) clearly affects the dynamics in (3.9) through the design of $\mathcal{K}$ in (3.7). By analogy to 'classical' state feedback theory, it can be seen that this is the same as the problem of finding the state feedback matrix $\mathcal{K}$ for the system in (3.8), where the $z_2(t)$ plays the role of the 'control' signal.

The stability (and performance) of the system in (3.9) depends on the fictitious reduced order pair (A_{11}, A_{12}). Thus the design of $\mathcal{K}$ depends on the controllability (or otherwise) of the pair (A_{11}, A_{12}).

Proposition 3.1 *The matrix pair (A_{11}, A_{12}) is controllable, if and only if the pair (A, B) is controllable.*

Proof Because of the special structure of the regular form, and using the fact that $\det(B_2) \neq 0$, it follows that

$$\begin{aligned}\operatorname{rank}[sI - A \quad B] &= \operatorname{rank}\begin{bmatrix} sI - A_{11} & -A_{12} & 0 \\ -A_{21} & sI - A_{22} & B_2 \end{bmatrix} \\ &= \operatorname{rank}[sI - A_{11} \quad A_{12}] + m \quad \text{for all } s \in \mathbb{C}\end{aligned}$$

This implies

$$\operatorname{rank}[sI - A \quad B] = n \quad \Leftrightarrow \quad \operatorname{rank}[sI - A_{11} \quad A_{12}] = n - m$$

and from the Popov–Belevitch–Hautus (PBH) rank test, it follows that (A, B) is controllable if and only if the pair (A_{11}, A_{12}) is controllable. □

As a consequence of Proposition 3.1, if the original system is controllable, $\mathcal{K}$ can be designed using a 'classical' state feedback method, and once $\mathcal{K}$ is obtained, the surface $\mathcal{S}$ is determined.

The overall problem of designing a sliding mode controller is one of

- designing the matrix S to achieve the required performance and stable dynamics for the closed-loop system;
- designing a control law to ensure that the sliding surface is reached and subsequently maintained.

This design procedure is unique and differentiates SMC from other design methods.

Any 'classical' state feedback method can be used to compute $\mathcal{K}$, and then the matrix S in (3.5) can be obtained as

$$S = [S_2\mathcal{K} \quad S_2]T_r \tag{3.10}$$

The nonsingular matrix S_2 can be chosen arbitrarily, but for ease of computation, often it is chosen as $S_2 = I_m$. Notice S_2 has no effect on the dynamics of the sliding motion. In (3.7) and (3.10), S_2 acts only as a scaling factor for the switching function.

The following approaches have been suggested to obtain the matrix $\mathcal{K}$ and subsequently S:

- quadratic minimisation[1] [72, 257];
- robust eigenstructure assignment [73];
- direct eigenstructure assignment [72];
- LMI methods [56, 79].

Note that the control law $u(t)$ is not designed to directly specify any desired dynamics for the closed-loop system, but rather to ensure that the sliding surface is reached and motion on $\mathcal{S}$ is maintained. In the sliding mode literature, $u(t)$ is said to be designed in order for the 'reachability condition' to be satisfied [85]. The reachability condition specifies that the trajectory of the system states must always point towards the sliding surface. In the case of a single-input system, this can be expressed as

$$\lim_{s \to 0^+} \dot{s} < 0 \quad \text{and} \quad \lim_{s \to 0^-} \dot{s} > 0 \tag{3.11}$$

or more compactly as, near $s(t) = Sx(t) = 0$

$$s\dot{s} < 0 \tag{3.12}$$

Inequality (3.12) is referred to as the *reachability condition* [85, 256]. A more strict reachability condition to ensure that the sliding surface is reached despite the presence of uncertainty and in *finite time* is given by

$$s\dot{s} \le -\eta |s| \tag{3.13}$$

where η is a positive design scalar. Equation (3.13) is usually called the 'η-reachability condition' [85]. For multivariable systems, the natural extension of (3.13) is

$$s^{\mathrm{T}}\dot{s} \le -\eta \|s\| \tag{3.14}$$

where again η is a positive scalar.

3.1.2 Properties of the Sliding Mode

The following is a summary of the properties of the system in a sliding mode:

- the system behaves as a reduced order motion which (apparently) does not depend on the control signal $u(t)$;
- there are $n - m$ states that determine the dynamics of the closed-loop system;
- the sliding motion depends only on the choice of the sliding surface;
- the poles of the sliding motion are given by the invariant zeros of the system triple (A, B, S) [85, 88].

[1]This will be discussed later in Sect. 3.4.1.

A further property (which is probably the most important for FTC in terms of handling actuator faults) is discussed next. Consider the uncertain linear system

$$\dot{x}(t) = Ax(t) + Bu(t) + D\xi(t,x) \tag{3.15}$$

where $D \in \mathbb{R}^{n\times l}$ is known but the function $\xi : \mathbb{R}_+ \times \mathbb{R}^n \to \mathbb{R}^l$ is unknown and represents uncertainty. Suppose at time t_s the system states lie on the surface $\mathcal{S}$ and an ideal sliding motion takes place:

$$s(t) = \dot{s}(t) = 0 \quad \text{for all } t \geq t_s$$

which implies

$$\dot{s}(t) = S\dot{x}(t) = S\big(Ax(t) + Bu(t) + D\xi(t,x)\big) = 0 \tag{3.16}$$

Rearranging (3.16) with respect to $u(t)$ gives

$$u_{eq}(t) = -(SB)^{-1}\big(SAx(t) + SD\xi(t,x)\big) \quad \text{for all } t \geq t_s$$

The quantity $u_{eq}(t)$ is the so-called 'equivalent control' and is the (theoretical) average value the control signal must take to maintain a sliding motion on $\mathcal{S}$ [256]. Substituting $u_{eq}(t)$ into the system in (3.15) yields

$$\dot{x}(t) = \big(I_n - B(SB)^{-1}S\big)Ax(t) + \underbrace{\big(I_n - B(SB)^{-1}S\big)}_{P_s} D\xi(t,x)$$

It is easy to verify that the matrix P_s satisfies

$$SP_s = P_sB = 0 \tag{3.17}$$

Suppose the range space of D is contained within the range space of B, then it is possible to write $D = BR$ for some $R \in \mathbb{R}^{m\times l}$. Consequently

$$\dot{x}(t) = P_sAx(t) + P_sBR\xi(t,x)$$

and since $P_sB = 0$

$$\dot{x}(t) = P_sAx(t) \tag{3.18}$$

From (3.18) it can be seen that during the ideal sliding motion, the uncertainty (the signal $\xi(t,x)$) does not affect the reduced order sliding motion.

Definition 3.1 Any uncertainty which can be expressed as in (3.15), where the range space of D is contained within the range space of B, is described as *matched uncertainty*. Any uncertainty which does not lie within the range space of the input distribution matrix is described as *unmatched uncertainty* [85].

3.2 A Simple Example: Pendulum

This section will apply the above methods to a simple pendulum example to give some insight into the design of SMC systems and the characteristics of the sliding motion. Consider a typical pendulum consisting of a (weightless) shaft and a mass which is driven by a motor (torque) at the point of suspension (Fig. 3.1). The objective is to design a sliding mode controller so that the pendulum will return to its vertically downward equilibrium point when the pendulum is left to swing from a near-equilibrium initial condition.

Consider the following pendulum system (taken from [99]): The dynamic equation governing the motion is given by

$$T_q - mgl\sin\theta = I\ddot{\theta} \tag{3.19}$$

where θ represents the angular displacement from the vertical, T_q is the applied torque, m is the mass, g is the gravitational acceleration constant, l is the length of the shaft and I is the inertia. For a weightless shaft pendulum, the inertia is given by $I = ml^2$. Here it is assumed that $l = 1$ m, $m = 0.2$ kg, $g = 9.82$ m/s^2. Linearising (3.19) about the vertically downward equilibrium position yields the following state space model:

$$\begin{bmatrix} \dot{x}_1(t) \\ \dot{x}_2(t) \end{bmatrix} = \begin{bmatrix} 0 & 1 \\ -g & 0 \end{bmatrix} \begin{bmatrix} x_1(t) \\ x_2(t) \end{bmatrix} + \begin{bmatrix} 0 \\ 5 \end{bmatrix} u(t) \tag{3.20}$$

where $u(t)$ is the applied torque and the states $(x_1(t), x_2(t))$ represent $(\theta, \dot{\theta})$ which are angular displacement and angular velocity, respectively. Note that the input distribution matrix has only one nonzero contribution in the bottom row of the matrix and is already in regular form as described in (3.4). For other systems, state similarity transformations using an orthogonal transformation matrix T_r as in (3.2) can be employed to achieve this form [85].

The first step is to design the sliding surface matrix S. During an ideal sliding motion, $s(t) = Sx(t) = 0$ and from (3.7)

$$x_2(t) = -\mathcal{K}x_1(t) \tag{3.21}$$

where $x_1(t), x_2(t)$ and $\mathcal{K}$, as defined in (3.10), are scalars. Substituting (3.21) into the first row of (3.20), it follows that the reduced order sliding motion is given by

$$\dot{x}_1(t) = x_2(t) = -\mathcal{K}x_1(t) \tag{3.22}$$

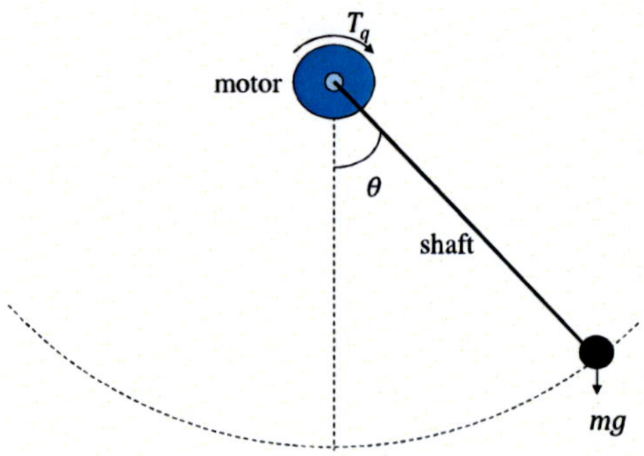

Fig. 3.1 Schematic of a pendulum

and the solution is $x_1(t) = x_1(t_s)e^{-\mathcal{K}(t-t_s)}$ where t_s is the time at which sliding occurs. If $\mathcal{K} = 2$, this gives a design with a settling time less than 3 s. From (3.10), choosing $S_2 = 1$ gives

$$S = [2 \quad 1] \tag{3.23}$$

Now a control law $u(t)$ needs to be developed to satisfy the reachability condition. Consider the following control law:

$$u(t) = -(SB)^{-1}\big(SAx(t) + \eta\,\mathrm{sgn}(s)\big) = \underbrace{-(SB)^{-1}SAx(t)}_{u_{eq}(t)} - (SB)^{-1}\eta\,\mathrm{sgn}\big(s(t)\big) \tag{3.24}$$

where η is a positive scalar and $\mathrm{sgn}(\cdot)$ represents the signum function. It follows that

$$\begin{aligned} \dot{s}(t) = S\dot{x}(t) &= S\big(Ax(t) + Bu(t)\big) \\ &= SAx(t) - (SB)(SB)^{-1}\big(SAx(t) + \eta\,\mathrm{sgn}\big(s(t)\big)\big) \\ &= -\eta\,\mathrm{sgn}\big(s(t)\big) \end{aligned}$$

Consequently

$$s\dot{s} = s\big(-\eta\,\mathrm{sgn}(s)\big) = -\eta|s|$$

Therefore the chosen control law $u(t)$ in (3.24) satisfies the η-reachability condition in (3.14). Using the matrix S obtained from the design in (3.23), and letting $\eta = 1$, the following control law is obtained from (3.24):

$$u(t) = [1.9640 \quad -0.4]x(t) - 0.2\,\mathrm{sgn}\big(s(t)\big)$$

Remark 3.1 The term $u_{eq}(t) = -(SB)^{-1}SAx(t)$ is sometimes called the 'nominal equivalent control'. It can be viewed as the control law required to maintain the ideal sliding motion (in the absence of uncertainty). However, it does not induce a sliding motion—the switching term $\mathrm{sgn}(s)$ is required for this purpose.

3.2.1 Simulations and Results

The following simulations are associated with an initial condition of 1 deg for the initial pendulum deflection angle θ. Figure 3.2 shows the results from the simulation and includes the states (angular velocity and deflection angle), the phase portrait, the input torque and the switching function $s(t)$. The deflection angle shows that the design requirement of a settling time less than 3 s with little or no overshoot is met. Since this particular example is a second-order system, it is possible to present a phase portrait [230]. The phase portrait shows the stability of the system since the trajectories converge to the origin. Here the sliding surface is indicated by the line with gradient -2. Define

$$\mathcal{L} := \big\{\big(x_1(t), x_2(t)\big) : x_2(t) = -2x_1(t)\big\}$$

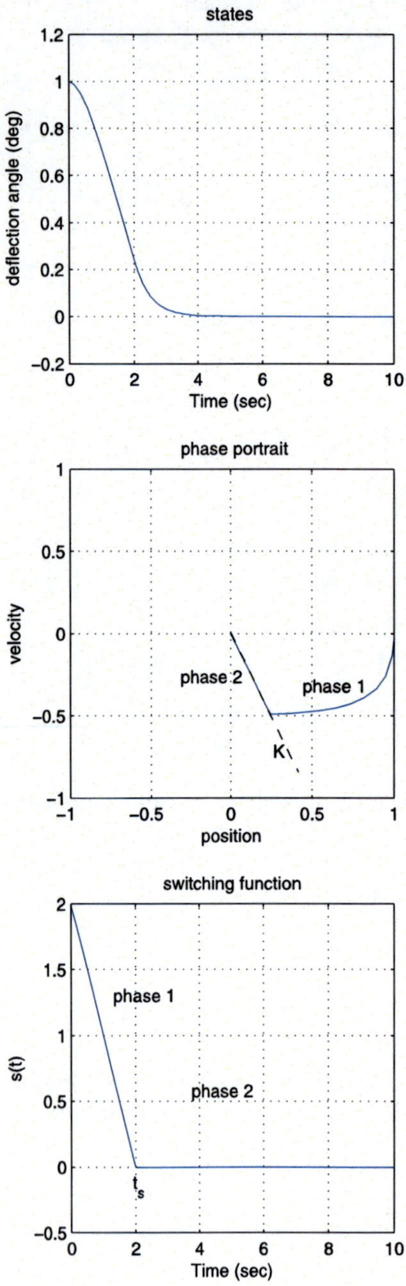

Fig. 3.2 Simulation results for the pendulum

to represent the sliding surface (line). During the sliding motion high frequency switching between the two different control structures takes place as the system trajectories repeatedly cross the line $\mathcal{L}$. This high frequency motion (associated with $\text{sgn}(s(t)) = 1$ and $\text{sgn}(s(t)) = -1$) is described as chattering. If infinite frequency switching was possible, the motion would be confined to the line $\mathcal{L}$. The motion when confined to the line $\mathcal{L}$ behaves like a first-order decay and the trajectories will slide along the line $\mathcal{L}$ to the origin. Such behaviour is described as an ideal sliding mode or an ideal sliding motion, and the line $\mathcal{L}$ is termed the sliding surface [85].

It is interesting to point out that the trajectory of the phase portrait can be classified into two phases. Phase 1 starts from the initial conditions. Due to the reachability conditions being met by the control law $u(t)$, sliding is induced and the trajectories move towards the sliding surface. Once the sliding surface is reached, Phase 2 begins, and the trajectory 'slides' along the surface towards the origin.

Even though the phase portrait shows the 'sliding' trajectory moving towards the origin, it does not, however, give any indication at which time sliding occurs (t_s). This information is available from the switching function plot. The plot shows that the sliding surface is reached in 2 s.

Note that the system's closed-loop poles given by $\lambda(A - B(SB)^{-1}SA)$ are $\{0, -2\}$. The pole at -2 is obtained from the choice of $\mathcal{K}$ and the other pole is zero. Generally, for an nth-order system with m inputs, the reduced order system during sliding has $n - m$ states and so will have $n - m$ nonzero closed-loop poles. The remaining m poles lie at the origin.

The input plot shows that the control action is highly discontinuous. This control law is not desirable for most systems due to the wear and tear that would occur on any mechanical components and to the actuators. It is therefore desirable for this discontinuity to be reduced or smoothed. A more practical control law design is introduced in the next section.

3.2.2 A Practical Control Law

Consider the following control law:

$$u(t) = -(SB)^{-1}(SA - \Phi S)x(t) - \rho(SB)^{-1}\frac{s(t)}{(|s(t)| + \delta)} \tag{3.25}$$

where Φ is a negative scalar. The quantity δ is a small positive scalar and ρ is a positive scalar. Note the difference between the controller in (3.25) and (3.24), is the introduction of the Φ term and the approximation of the $\text{sgn}(s)$ term.

The same choice of S as used in the example in Sect. 3.2 will be employed here. Let $\Phi = -6$ and $\rho = 1$ and $\delta = 0.001$. The control law in (3.25) becomes

$$u(t) = [-0.4360 \quad -1.6]x(t) - 0.2\frac{s(t)}{(|s(t)| + \delta)}$$

Figure 3.3 shows the results of the simulation. The phase portrait looks similar to the one in Fig. 3.2. The difference is mainly in the input and switching function

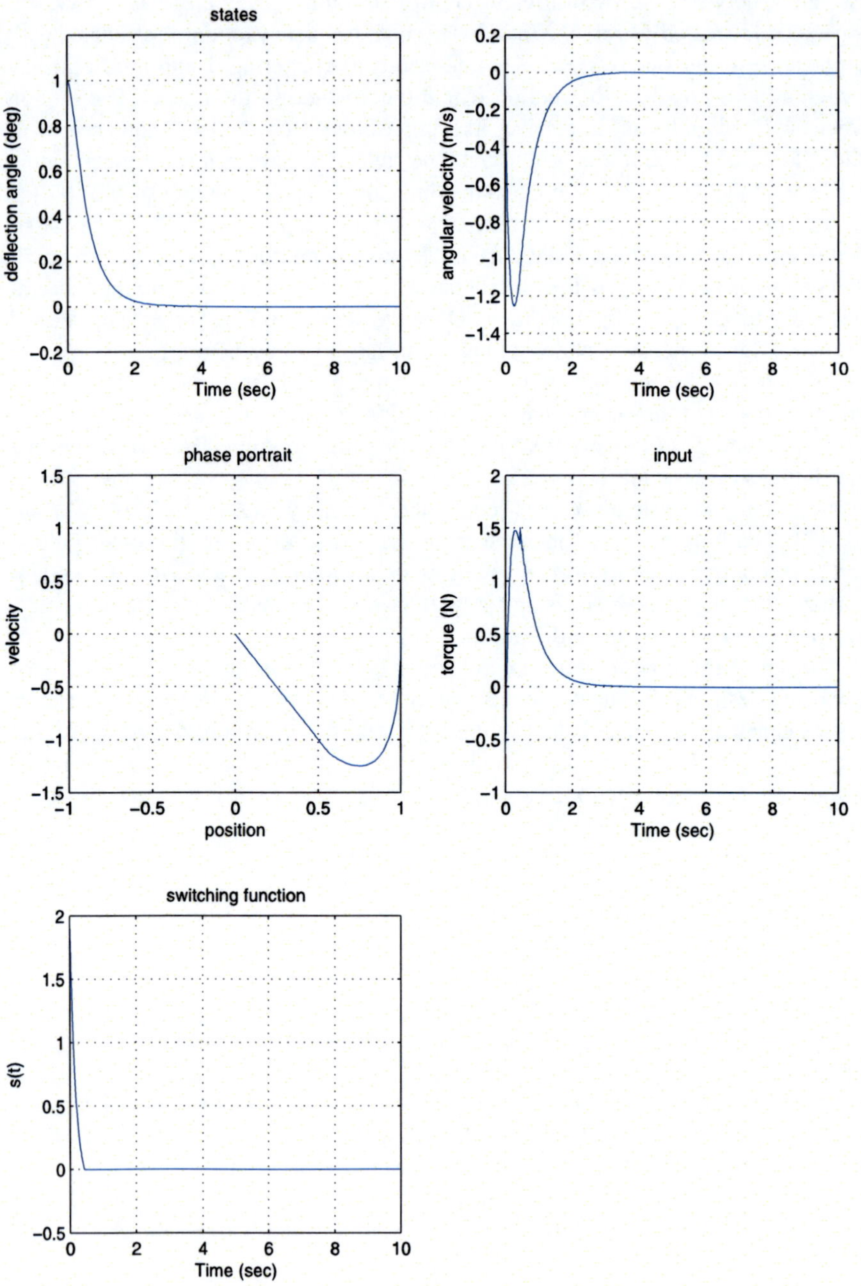

Fig. 3.3 Pendulum practical controller: simulation results

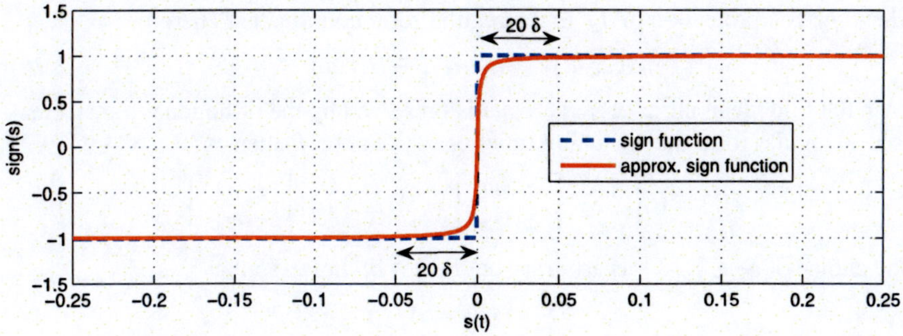

Fig. 3.4 An approximation of the signum function [85]

plots. The input plot shows no chattering, or high frequency switching, and a smooth signal is obtained. This is because of the approximation of the sgn(s) term (see Fig. 3.4). The δ term is chosen as a compromise between an ideal sliding motion and chatter prevention. A smaller δ will give an ideal siding motion but with chattering; a larger δ will give less chattering but yields a trajectory evolving further away from the sliding surface compared to ideal sliding. Figure 3.3 also shows that the extra degree of freedom Φ, has been used to moderate how quickly the sliding surface is attained. The difference between the previous controller design and the 'practical controller' design can be seen in Fig. 3.3. Sliding is now obtained in 0.46 s, whilst the previous design took almost 2 s.

3.3 Unit Vector Approach

In the previous section, a practical controller for a pendulum system was introduced. The pendulum system represented a single-input system. The most convenient control structure for multivariable systems, from a sliding mode perspective, is the 'unit vector' control structure attributed to Ryan and Corless [217]. This method will form the basis for the controller designs in this book.

Consider a system with matched uncertainty

$$\dot{x}(t) = Ax(t) + Bu(t) + f_m(t, x, u) \tag{3.26}$$

where $f_m(t, x, u)$ is in the range space of B, and is assumed to be unknown but bounded and satisfies

$$\|f_m(t, x, u)\| \leq k\|u(t)\| + \alpha(t, x) \tag{3.27}$$

As in Sect. 3.1.1, there exists an orthogonal transformation $x(t) \mapsto T_r x(t) = z(t)$ so that the system above can be transformed into the following 'regular form':

$$\begin{bmatrix} \dot{z}_1(t) \\ \dot{z}_2(t) \end{bmatrix} = \begin{bmatrix} A_{11} & A_{12} \\ A_{21} & A_{22} \end{bmatrix} \begin{bmatrix} z_1(t) \\ z_2(t) \end{bmatrix} + \begin{bmatrix} 0 \\ B_2 \end{bmatrix} u(t) + \begin{bmatrix} 0 \\ I_m \end{bmatrix} \bar{f}_m(t, z, u) \tag{3.28}$$

where $\bar{f}_m$ is a projection of f_m in the regular form coordinates. Therefore

$$\|\bar{f}_m(t,z,u)\| \leq k\|u(t)\| + \alpha(t,x) \tag{3.29}$$

since the Euclidean norm of $f_m(t,x,u)$ is preserved by the orthogonal transformation. In regular form the switching function $s(t)$ can be written as

$$s(t) = [S_1 \quad S_2]\begin{bmatrix} z_1(t) \\ z_2(t) \end{bmatrix} = S_2[\mathcal{K} \quad I_m]\begin{bmatrix} z_1(t) \\ z_2(t) \end{bmatrix}$$

The choice of $S_2 \in \mathbb{R}^{m\times m}$ is arbitrary but here it is chosen so that

$$S_2 B_2 = \Lambda$$

where Λ is a nonsingular diagonal matrix. Define another coordinate transformation $z(t) \mapsto T_s z(t)$ so that the system can be partitioned into:

$$\begin{bmatrix} z_1(t) \\ s(t) \end{bmatrix} = T_s \begin{bmatrix} z_1(t) \\ z_2(t) \end{bmatrix}$$

where the transformation matrix T_s is given by:

$$T_s = \begin{bmatrix} I & 0 \\ S_1 & S_2 \end{bmatrix}$$

Then the system in (3.28) can be written as:

$$\begin{bmatrix} \dot{z}_1(t) \\ \dot{s}(t) \end{bmatrix} = \begin{bmatrix} \bar{A}_{11} & A_{12}S_2^{-1} \\ S_2\bar{A}_{21} & S_2\bar{A}_{22}S_2^{-1} \end{bmatrix}\begin{bmatrix} z_1(t) \\ s(t) \end{bmatrix} + \begin{bmatrix} 0 \\ \Lambda \end{bmatrix} u(t) + \begin{bmatrix} 0 \\ S_2 \end{bmatrix} \bar{f}_m(t,z,u) \tag{3.30}$$

where $\bar{A}_{11} = A_{11} - A_{12}\mathcal{K}$, $\bar{A}_{21} = \mathcal{K}\bar{A}_{11} + A_{21} - A_{22}\mathcal{K}$ and $\bar{A}_{22} = \mathcal{K}A_{12} + A_{22}$. The Ryan and Corless [217] control law comprises linear and nonlinear components given by

$$u(t) = u_l(t) + u_n(t) \tag{3.31}$$

The linear component is defined as

$$u_l(t) = \Lambda^{-1}\big(-S_2\bar{A}_{21}z_1(t) - \big(S_2\bar{A}_{22}S_2^{-1} - \Phi\big)s(t)\big) \tag{3.32}$$

where $\Phi \in \mathbb{R}^{m\times m}$ is any stable design matrix and the nonlinear component

$$u_n(t) = -\rho(t,x)\Lambda^{-1}\frac{P_2 s(t)}{\|P_2 s(t)\|} \quad \text{if } s(t) \neq 0 \tag{3.33}$$

where $P_2 \in \mathbb{R}^{m\times m}$ is a symmetric positive definite matrix satisfying the Lyapunov equation

$$P_2\Phi + \Phi^{\mathrm{T}}P_2 = -I_m$$

The scalar function ρ depends on the magnitude of uncertainty and is any function satisfying:

$$\rho(t,x) \geq \frac{\|S_2\|(k\|u_l\| + \alpha(t, T_r x)) + \eta}{(1 - k\|\Lambda^{-1}\|\|S_2\|)} \tag{3.34}$$

where η is a positive scalar and k is a known constant with $k < \|B_2^{-1}\|^{-1}$.

3.3.1 Analysis of Stability for the Closed-Loop System

The problem of determining the stability of the closed-loop system under the influence of matched uncertainty becomes the problem of ensuring that sliding occurs despite the presence of uncertainty. This is due to the fact that when the controller induces an ideal sliding motion, the closed-loop system is stable by design. This section will show that the unit vector controller from (3.31) will still induce sliding in the presence of uncertainty.

Substituting the control law in (3.31) into system (3.30) gives

$$\dot{z}_1(t) = \bar{A}_{11} z_1(t) + A_{12} S_2^{-1} s(t) \tag{3.35}$$

$$\dot{s}(t) = \Phi s(t) - \rho(t,x) \frac{P_2 s}{\| P_2 s \|} + S_2 \bar{f}_m(t,z,u) \tag{3.36}$$

Consider a Lyapunov function $V(s) = s^{\mathrm{T}} P_2 s$ for (3.36). Differentiating the Lyapunov function yields

$$\begin{aligned} \dot{V} &= \dot{s}^{\mathrm{T}} P_2 s + s^{\mathrm{T}} P_2 \dot{s} \\ &= \left(\Phi s - \rho \frac{P_2 s}{\| P_2 s \|} + S_2 \bar{f}_m \right)^{\mathrm{T}} P_2 s + s^{\mathrm{T}} P_2 \left(\Phi s - \rho \frac{P_2 s}{\| P_2 s \|} + S_2 \bar{f}_m \right) \\ &= s^{\mathrm{T}} \big(\Phi^{\mathrm{T}} P_2 + P_2 \Phi \big) s - 2\rho \frac{1}{\| P_2 s \|} \big(s^{\mathrm{T}} P_2 P_2 s \big) + 2 s^{\mathrm{T}} P_2 S_2 \bar{f}_m \\ &= -s^{\mathrm{T}} s - 2\rho \| P_2 s \| + 2 s^{\mathrm{T}} P_2 S_2 \bar{f}_m \end{aligned} \tag{3.37}$$

since $s^{\mathrm{T}} P_2 P_2 s = \| P_2 s \|^2$ and $\Phi^{\mathrm{T}} P_2 + P_2 \Phi = -I$. Furthermore since

$$\big\| s^{\mathrm{T}} P_2 S_2 \bar{f}_m \big\| < \| P_2 s \| \| S_2 \| \| \bar{f}_m \|$$

from the Cauchy–Schwarz inequality,

$$\dot{V} \le -\| s \|^2 - 2 \| P_2 s \| \big(\rho - \| S_2 \| \| \bar{f}_m \| \big) \tag{3.38}$$

The idea is to represent ρ in (3.38) in terms of the uncertainty $\bar{f}_m$ using the definition of ρ given in (3.34). From (3.31) and (3.33), and using the triangle inequality property of norms

$$\big\| u(t) \big\| \le \big\| u_l(t) \big\| + \big\| u_n(t) \big\| \le \big\| u_l(t) \big\| + \rho \big\| \Lambda^{-1} \big\| \tag{3.39}$$

Equation (3.34) can be written as

$$\rho(t,x) \big(1 - k \big\| \Lambda^{-1} \big\| \| S_2 \| \big) \ge \| S_2 \| \big(k \| u_l \| + \alpha(t,x) \big) + \eta \tag{3.40}$$

Rearranging this equation yields

$$\begin{aligned} \rho(t,x) &\ge \| S_2 \| \big(k \| u_l \| + \alpha(t,x) \big) + \eta + \rho(t,x) k \| S_2 \| \big\| \Lambda^{-1} \big\| \\ &\ge \| S_2 \| \big(k \| u_l \| + \rho(t,x) k \big\| \Lambda^{-1} \big\| + \alpha(t,x) \big) + \eta \end{aligned}$$

Using (3.39) and (3.27), the above can be written as

$$\rho(t,x) \ge \| S_2 \| \big(k \| u \| + \alpha(t,x) \big) + \eta \ge \| S_2 \| \| \bar{f}_m \| + \eta \tag{3.41}$$

Substituting for (3.41) in (3.38) yields

$$\dot{V} \leq -\|s\|^2 - 2\|P_2 s\|\big(\|S_2\|\|\bar{f}_m\| + \eta\big) + 2\|P_2 s\|\|S_2\|\|\bar{f}_m\|$$
$$\leq -\|s\|^2 - 2\eta\|P_2 s\| \tag{3.42}$$

Equation (3.42) shows that the controller in the form (3.31), induces ideal sliding in the presence of matched uncertainty. This inequality will be used to show that sliding on $\mathcal{S}$ takes place in finite time. From the Rayleigh principle

$$\|P_2 s\|^2 = \big(P_2^{1/2}s\big)^{\mathrm{T}} P_2 \big(P_2^{1/2}s\big) \geq \lambda_{\min}(P_2)\big\|P_2^{1/2}s\big\|^2 = \lambda_{\min}(P_2)V(s) \tag{3.43}$$

which together with (3.42) gives

$$\dot{V} \leq -2\eta\sqrt{\lambda_{\min}(P_2)}\sqrt{V} \tag{3.44}$$

Integrating (3.44) implies that the time taken to reach the sliding surface $\mathcal{S}$ denoted by t_s satisfies

$$t_s \leq \eta^{-1}\sqrt{V(s_0)/\lambda_{\min}(P_2)} \tag{3.45}$$

where s_0 represents the initial value of $s(t)$ at $t = 0$ [85]. This is a natural multivariable analogue of the single-input case.

3.3.2 The Unit Vector Pseudo Sliding Term

In achieving this ideal sliding motion, discontinuous infinite frequency switching or chattering occurs. This is undesirable for some practical systems, especially for mechanical systems with actuators prone to wear and tear. It is therefore necessary that this discontinuity is 'smoothed' and an approximation to ideal sliding (sometimes called 'pseudo sliding') is achieved. Here the states of the system are only required to stay close to the sliding surface instead of on it. However, the total robustness property (invariance) to matched uncertainty is no longer guaranteed. On the other hand, if the approximation is close enough to the actual discontinuous term, a good approximation to ideal sliding can still be achieved. Therefore, there is a trade-off between robustness and reducing the chattering effect.

There are several methods used to achieve 'pseudo sliding'; but the one that will be used in this book is based on a method called 'fractional approximation' (or sigmoidal approximation) [85]. This is similar to the one in Fig. 3.4. Other approximation methods are discussed in [85], including the boundary layer approach and power law interpolation. The nonlinear term in the control law in (3.33) is replaced by

$$u_n = -\rho\Lambda^{-1}\frac{P_2 s}{\|P_2 s\| + \delta} \tag{3.46}$$

where $\delta > 0$ is a small positive scalar which determines the quality of the approximation. A very small δ will give a better approximation to the actual discontinuous function $\operatorname{sgn}(s)$; but to reduce chattering, a larger δ is needed.

3.4 Design of the Sliding Surface

The preceding sections have discussed the design of the control law. In the following section, the design of the switching surface, namely the matrix S in the switching function $s(t) = Sx(t)$, is discussed. The next section describes a design scheme based on a modification of the classical LQR design problem. This sliding surface design approach will be used extensively in this book. Details of other design approaches can be found in [85].

3.4.1 Quadratic Minimisation

Consider the problem of minimising the quadratic performance index

$$J = \frac{1}{2}\int_{t_s}^{\infty} \left(x(t)^{\mathrm{T}}\mathbf{Q}x(t)\right) dt \tag{3.47}$$

where $\mathbf{Q}$ is symmetric positive definite and t_s indicates the start of sliding. Consider a coordinate transformation $x(t) \mapsto T_r x(t) = z(t)$ so that the system is in regular form as in Sect. 3.1.1. In regular form, the matrix $\mathbf{Q}$ can be written as

$$T_r \mathbf{Q} T_r^{\mathrm{T}} = \begin{bmatrix} \mathbf{Q}_{11} & \mathbf{Q}_{12} \\ \mathbf{Q}_{21} & \mathbf{Q}_{22} \end{bmatrix} \tag{3.48}$$

where $\mathbf{Q}_{21} = \mathbf{Q}_{12}^{\mathrm{T}}$ and $\mathbf{Q}_{22} \in \mathbb{R}^{m\times m}$. Therefore, in regular form, the special 'cheap control' LQR problem associated with (3.47) involves minimising

$$J = \frac{1}{2}\int_{t_s}^{\infty} \left(z_1^{\mathrm{T}}\mathbf{Q}_{11}z_1 + 2z_1^{\mathrm{T}}\mathbf{Q}_{12}z_2 + z_2^{\mathrm{T}}\mathbf{Q}_{22}z_2\right) dt \tag{3.49}$$

Utkin and Young [257] proposed factorizing the last two terms of (3.49) to obtain

$$\begin{aligned} 2z_1^{\mathrm{T}}\mathbf{Q}_{12}z_2 + z_2^{\mathrm{T}}\mathbf{Q}_{22}z_2 = {} & \left(z_2 + \mathbf{Q}_{22}^{-1}\mathbf{Q}_{21}z_1\right)^{\mathrm{T}}\mathbf{Q}_{22}\left(z_2 + \mathbf{Q}_{22}^{-1}\mathbf{Q}_{21}z_1\right) \\ & - z_1^{\mathrm{T}}\left(\mathbf{Q}_{21}^{\mathrm{T}}\mathbf{Q}_{22}^{-1}\mathbf{Q}_{21}\right)z_1 \end{aligned}$$

Using the above, (3.49) can be written as

$$\begin{aligned} J = \frac{1}{2}\int_{t_s}^{\infty} & \left(z_1^{\mathrm{T}}\left(\mathbf{Q}_{11} - \mathbf{Q}_{12}\mathbf{Q}_{22}^{-1}\mathbf{Q}_{21}\right)z_1 \right. \\ & \left. + \left(z_2 + \mathbf{Q}_{22}^{-1}\mathbf{Q}_{21}z_1\right)^{\mathrm{T}}\mathbf{Q}_{22}\left(z_2 + \mathbf{Q}_{22}^{-1}\mathbf{Q}_{21}z_1\right)\right) dt \end{aligned} \tag{3.50}$$

Define

$$\hat{\mathbf{Q}} = \mathbf{Q}_{11} - \mathbf{Q}_{12}\mathbf{Q}_{22}^{-1}\mathbf{Q}_{21} \tag{3.51}$$

and introduce a pseudo control term as

$$v = z_2 + \mathbf{Q}_{22}^{-1}\mathbf{Q}_{21}z_1 \tag{3.52}$$

then (3.50) can be written as

$$J = \frac{1}{2}\int_{t_s}^{\infty} \left(z_1^{\mathrm{T}}\hat{\mathbf{Q}}z_1 + \upsilon^{\mathrm{T}}\mathbf{Q}_{22}\upsilon\right) dt \tag{3.53}$$

The minimisation of J is associated with the dynamical system in (3.8) which is given by:

$$\dot{z}_1(t) = A_{11}z_1(t) + A_{12}z_2(t) \tag{3.54}$$

Eliminating the z_2 term in (3.54) by using (3.52), the system in (3.54) becomes

$$\dot{z}_1(t) = \hat{A}z_1(t) + A_{12}\upsilon(t) \tag{3.55}$$

where $\hat{A} = A_{11} - A_{12}\mathbf{Q}_{22}^{-1}\mathbf{Q}_{21}$. This is now a classical LQR problem formulation. The 'optimal control law' is then given by

$$\upsilon(t) = -\left(\mathbf{Q}_{22}^{-1}A_{12}^{\mathrm{T}}P_1\right)z_1(t) \tag{3.56}$$

where P_1 satisfies

$$\hat{A}^{\mathrm{T}}P_1 + P_1\hat{A} - P_1A_{12}\mathbf{Q}_{22}^{-1}A_{12}^{\mathrm{T}}P_1 + \hat{\mathbf{Q}} = 0 \tag{3.57}$$

Recall that during sliding, $s(t) = 0$ and therefore

$$z_2(t) = -\mathcal{K}z_1(t) \tag{3.58}$$

where $\mathcal{K}$ is defined in (3.7). The manipulations resulting from solving for $z_2(t)$ from (3.52) and (3.56) yield

$$z_2(t) = -\mathbf{Q}_{22}^{-1}\left(A_{12}^{\mathrm{T}}P_1 + \mathbf{Q}_{21}\right)z_1(t) \tag{3.59}$$

and therefore the matrix $\mathcal{K}$ is defined as

$$\mathcal{K} = \mathbf{Q}_{22}^{-1}\left(A_{12}^{\mathrm{T}}P_1 + \mathbf{Q}_{21}\right) \tag{3.60}$$

Once the matrix $\mathcal{K}$ has been obtained, the matrix S can be calculated using (3.10) where S_2 can be arbitrarily chosen.

3.5 Design of a Controller with a Tracking Requirement

The control law discussions in this chapter have so far only considered state regulation. In this section, two methods for systems with a tracking requirement will be discussed in detail, one based on integral action and the other a model-reference approach.

3.5.1 Integral Action Approach

Consider a nominal linear system that is in regular form given by

$$\dot{x}(t) = Ax(t) + Bu(t) \tag{3.61}$$

where $A \in \mathbb{R}^{n\times n}$ and $B \in \mathbb{R}^{n\times m}$. In addition, identify some controlled outputs as

$$y(t) = Cx(t) \tag{3.62}$$

where $y(t) \in \mathbb{R}^m$. Consider additional states $x_r(t) \in \mathbb{R}^m$ defined as

$$\dot{x}_r(t) = r(t) - Cx(t) \tag{3.63}$$

where $r(t)$ is the 'filtered' reference signal given by

$$\dot{r}(t) = \Gamma\big(r(t) - R(t)\big) \tag{3.64}$$

where $\Gamma \in \mathbb{R}^{m\times m}$ is a stable design matrix and $R(t)$ is a piecewise constant demand vector. The signal $R(t)$ represents step changes in demand which is not differentiable at certain time instants. In (3.64), $r(t)$ represents a low pass filtered version of the signal $R(t)$ and so (3.64), can be viewed in classical terms as a pre-filtering of the demand signal to remove 'derivative kick'. The matrix Γ represents a useful design parameter for tailoring the closed-loop response to demand changes [85]. The analysis of the augmented tracking system is described below. Augment the nominal system with the new additional states $x_r(t)$ to obtain

$$\tilde{x}(t) = \begin{bmatrix} x_r(t) \\ x(t) \end{bmatrix} \tag{3.65}$$

Now the augmented system can be written as

$$\begin{bmatrix} \dot{x}_r(t) \\ \dot{x}(t) \end{bmatrix} = \begin{bmatrix} 0 & -C \\ 0 & A \end{bmatrix} \begin{bmatrix} x_r(t) \\ r(t) \end{bmatrix} + \begin{bmatrix} 0 \\ B \end{bmatrix} u(t) + \begin{bmatrix} I_m \\ 0 \end{bmatrix} r(t) \tag{3.66}$$

Assume the pair (A, B) is in regular form, then the state $\tilde{x}(t)$ can be partitioned as

$$\tilde{x}(t) = \begin{bmatrix} \tilde{x}_1(t) \\ \tilde{x}_2(t) \end{bmatrix} \tag{3.67}$$

where $\tilde{x}_1(t) \in \mathbb{R}^n$ and $\tilde{x}_2(t) \in \mathbb{R}^m$. In the new partition, the system can be written as:

$$\begin{bmatrix} \dot{\tilde{x}}_1(t) \\ \dot{\tilde{x}}_2(t) \end{bmatrix} = \begin{bmatrix} \tilde{A}_{11} & \tilde{A}_{12} \\ \tilde{A}_{21} & A_{22} \end{bmatrix} \begin{bmatrix} \tilde{x}_1(t) \\ \tilde{x}_2(t) \end{bmatrix} + \begin{bmatrix} 0 \\ B_2 \end{bmatrix} u(t) + \begin{bmatrix} B_r \\ 0 \end{bmatrix} r(t) \tag{3.68}$$

where the augmented and partitioned system matrix is given by

$$\tilde{A} = \begin{bmatrix} \tilde{A}_{11} & \tilde{A}_{12} \\ \tilde{A}_{21} & A_{22} \end{bmatrix} := \left[\begin{array}{cc|c} 0 & -C_1 & -C_2 \\ 0 & A_{11} & A_{12} \\ \hline 0 & A_{21} & A_{22} \end{array}\right] \quad \text{and} \quad B_r = \begin{bmatrix} I_m \\ 0 \end{bmatrix} \tag{3.69}$$

In expression (3.69), the matrix $[C_1 \; C_2]$ is a partition of the output distribution matrix C. The objective is to design a sliding surface of the form

$$\mathcal{S} = \big\{\tilde{x}(t) \in \mathbb{R}^{n+m} : S\tilde{x}(t) = 0\big\} \tag{3.70}$$

where $S \in \mathbb{R}^{m\times(n+m)}$ is designed to meet performance specifications for the closed-loop reduced order system. The matrix S can be partitioned as

$$S = [S_1 \quad S_2] \tag{3.71}$$

where $S_1 \in \mathbb{R}^{m\times n}$ and $S_2 \in \mathbb{R}^{m\times m}$. Assume that by choice of S_2, the matrix $\Lambda = S_2 B_2$ is a nonsingular diagonal design matrix. During an ideal sliding motion the reduced order system is governed by the top partition of (3.68), specifically

$$\dot{\tilde{x}}_1(t) = \tilde{A}_{11}\tilde{x}_1(t) + \tilde{A}_{12}\tilde{x}_2(t) + B_r r(t)$$

During the sliding motion $s(t) = S\tilde{x}(t) = 0$, and therefore

$$S_1\tilde{x}_1(t) + S_2\tilde{x}_2(t) = 0 \quad \Rightarrow \quad \tilde{x}_2(t) = -\mathcal{K}\tilde{x}_1(t) \tag{3.72}$$

where $\mathcal{K} = S_2^{-1}S_1$. Substituting (3.72) into the top partition of (3.68) yields the following:

$$\dot{\tilde{x}}_1(t) = (\tilde{A}_{11} - \tilde{A}_{12}\mathcal{K})\tilde{x}_1(t) + B_r r(t)$$

The design of the hyperplane gain matrix $\mathcal{K}$ is determined by the controllability of the pair $(\tilde{A}_{11}, \tilde{A}_{12})$.

Lemma 3.1 *If (A, B, C) is controllable and has no invariant zeros at the origin, then the matrix pair $(\tilde{A}_{11}, \tilde{A}_{12})$ is controllable.*

Proof Denote Rosenbrocks' system matrix by

$$P(\mathrm{s}) = \begin{bmatrix} \mathrm{s}I - A & B \\ -C & 0 \end{bmatrix}$$

The invariant zeros of the triple (A, B, C) are given by

$$\{\mathrm{s} \in \mathbb{C} : \det(P(\mathrm{s})) = 0\}$$

Therefore the system has zeros at the origin if and only if $\det(P(0)) = 0$. Because (A, B, C) is already in regular form and B_2 is nonsingular

$$\begin{aligned}\det(P(0)) = 0 &\Leftrightarrow \det\begin{bmatrix} -C & 0 \\ -A & B \end{bmatrix} = 0 \\ &\Leftrightarrow \det\begin{bmatrix} -C_1 & -C_2 & 0 \\ -A_{11} & -A_{12} & 0 \\ -A_{21} & -A_{22} & B_2 \end{bmatrix} = 0 \\ &\Leftrightarrow \det\begin{bmatrix} C_1 & C_2 \\ A_{11} & A_{12} \end{bmatrix} = 0\end{aligned}$$

Utilising the PBH rank test, the pair $(\tilde{A}_{11}, \tilde{A}_{12})$ is controllable if and only if

$$\operatorname{rank}\begin{bmatrix} \mathrm{s}I_m & C_1 & -C_2 \\ 0 & \mathrm{s}I - A_{11} & A_{12} \end{bmatrix} = n \quad \text{for all } \mathrm{s} \in \mathbb{C} \tag{3.73}$$

If $\mathrm{s} = 0$ then

$$\begin{aligned}\operatorname{rank}\begin{bmatrix} \mathrm{s}I_m & C_1 & -C_2 \\ 0 & \mathrm{s}I - A_{11} & A_{12} \end{bmatrix} = n &\Leftrightarrow \det\begin{bmatrix} C_1 & -C_2 \\ -A_{11} & A_{12} \end{bmatrix} \neq 0 \\ &\Leftrightarrow \det\begin{bmatrix} C_1 & C_2 \\ A_{11} & A_{12} \end{bmatrix} \neq 0 \\ &\Leftrightarrow (A, B, C) \text{ has no zeros at the origin}\end{aligned}$$

Otherwise $\mathrm{s} \neq 0$ and

$$\operatorname{rank}\begin{bmatrix} \mathrm{s}I_m & C_1 & C_2 \\ 0 & \mathrm{s}I - A_{11} & A_{12} \end{bmatrix} = n \Leftrightarrow \det[\mathrm{s}I - A_{11}\ A_{12}] = n - p$$

However, from Proposition 3.1, (A, B) is controllable if and only if (A_{11}, A_{12}) is controllable and therefore by assumption

$$\operatorname{rank}[\mathrm{s}I - A_{11} \quad A_{12}] = n - p \quad \text{for all s}$$

from the PBH rank test applied to (A_{11}, A_{12}). Therefore assertion (3.73) is true and $(\tilde{A}_{11}, \tilde{A}_{12})$ is controllable. □

If the above condition is satisfied, then the design methods described in Sect. 3.4.1 can be used for the augmented system above. The unit vector approach described earlier will be applied to the augmented system in (3.66) to induce a sliding motion. First transform the system using the coordinate change associated with the matrix

$$T_s = \begin{bmatrix} I_n & 0 \\ S_1 & S_2 \end{bmatrix} \tag{3.74}$$

The augmented and partitioned states become

$$\begin{bmatrix} \tilde{x}_1(t) \\ s(t) \end{bmatrix} := T_s \begin{bmatrix} \tilde{x}_1(t) \\ \tilde{x}_2(t) \end{bmatrix}$$

and the augmented system can be written as

$$\begin{bmatrix} \dot{\tilde{x}}_1(t) \\ \dot{s}(t) \end{bmatrix} = \begin{bmatrix} \bar{A}_{11} & \bar{A}_{12} \\ S_2\bar{A}_{21} & S_2\bar{A}_{22}S_2^{-1} \end{bmatrix} \begin{bmatrix} \tilde{x}_1(t) \\ s(t) \end{bmatrix} + \begin{bmatrix} 0 \\ \Lambda \end{bmatrix} u(t) + \begin{bmatrix} B_r \\ S_1 B_r \end{bmatrix} r(t) \tag{3.75}$$

where $\bar{A}_{11} = \tilde{A}_{11} - \tilde{A}_{12}\mathcal{K}$, $\bar{A}_{21} = \mathcal{K}\bar{A}_{11} + \tilde{A}_{21} - A_{22}\mathcal{K}$, $\bar{A}_{22} = \mathcal{K}\tilde{A}_{12} + A_{22}$, and $\bar{A}_{12} = \tilde{A}_{12}S_2^{-1}$. As in Sect. 3.3, the proposed controller is given by $u(t) = u_l(t) + u_n(t)$ where

$$u_l(t) = \Lambda^{-1}\big(-S_2\bar{A}_{21}\tilde{x}_1(t) + \big(\Phi - S_2\bar{A}_{22}S_2^{-1}\big)s(t) - S_1 B_r r(t)\big) \tag{3.76}$$

and

$$u_n(t) = -\rho\Lambda^{-1}\frac{\bar{P}_2 s}{\|\bar{P}_2 s\|} \quad \text{if } s(t) \neq 0 \tag{3.77}$$

where $\bar{P}_2$ is a symmetric positive definite matrix satisfying

$$\bar{P}_2\Phi + \Phi^{\mathrm{T}}\bar{P}_2 = -I \tag{3.78}$$

and $\Phi \in \mathbb{R}^{m\times m}$ is any stable design matrix. In the original coordinates, (3.76) can be written as

$$u_l(t) = L_x\tilde{x}(t) + L_r r(t) \tag{3.79}$$

with

$$L_x = -\Lambda^{-1}(S\tilde{A} - \Phi S) \tag{3.80}$$

$$L_r = -\Lambda^{-1}S_1 B_r \tag{3.81}$$

In the next section, a model-reference approach for tracking control is discussed.

3.5.2 Model-Reference Approach

In the model-reference approach, the idea is to compare the output of the nominal plant with an ideal model [17, 285, 286]. The information obtained from the tracking error between the nominal plant and the ideal model is taken as the state variable for design purposes.

Consider a nominal plant given by

$$\dot{x}(t) = Ax(t) + Bu(t) \tag{3.82}$$

and suppose the ideal model is given by

$$\dot{x}_m(t) = A_m x_m(t) + B_m r(t) \tag{3.83}$$

where $r(t) \in \mathbb{R}^r$ is an input vector representing the reference signal. Define the tracking error as

$$e(t) = x(t) - x_m(t) \tag{3.84}$$

The objective of the model-reference method is to design a controller so that the tracking error tends to zero, and therefore the nominal plant is said to have perfect tracking. Suppose the reference model pair (A_m, B_m) is obtained via the following equations:

$$A_m = A + BF \tag{3.85}$$

and

$$B_m = BG \tag{3.86}$$

One possible controller structure is

$$u(t) = u_{smc}(t) + Fx(t) + Gr(t) \tag{3.87}$$

where $u_{smc}(t)$ is a sliding mode controller based on the state error $e(t)$. Taking the derivative of the error given in (3.84) yields

$$\dot{e}(t) = \dot{x}(t) - \dot{x}_m(t) = Ax(t) - A_m x_m(t) + Bu(t) - B_m r(t) \tag{3.88}$$

Adding and subtracting $A_m x(t)$ to (3.88) gives

$$\dot{e}(t) = A_m e(t) + (A - A_m)x(t) + Bu(t) - B_m r(t) \tag{3.89}$$

Based on (3.89), a controller can be designed to eliminate the $x(t)$ and $r(t)$ terms from the right hand side of the equation. First define an error switching function

$$s(e) = Se(t) \tag{3.90}$$

which corresponds to the following hyperplane:

$$\mathcal{S}_e = \left\{ e(t) \in \mathbb{R}^n : Se(t) = 0 \right\} \tag{3.91}$$

During an ideal sliding motion,

$$s(e) = Se(t) = 0 \tag{3.92}$$

Differentiating this equation and substituting from (3.89) yields

$$\dot{s}(e) = S\dot{e}(t) = S\big(A_m e(t) + (A - A_m)x(t) + Bu - B_m r(t)\big) = 0 \tag{3.93}$$

Assuming that SB is nonsingular, the equivalent control is

$$u_{eq}(t) = -(SB)^{-1} S\big(A_m e(t) + (A - A_m)x(t) - B_m r(t)\big) \tag{3.94}$$

and the reduced order system is given by substituting for $u_{eq}(t)$ in (3.89) to obtain

$$\dot{e}(t) = \big(I - B(SB)^{-1}S\big)\big(A_m e(t) + (A - A_m)x(t) - B_m r(t)\big) \tag{3.95}$$

Substituting (3.85) and (3.86) in the above yields

$$\dot{e}(t) = \big(I - B(SB)^{-1}S\big)\big(A_m e(t) - BFx(t) - BGr(t)\big) \tag{3.96}$$

Using a similar argument to that in Sect. 3.1.2, the last two terms on the right hand side of (3.96) can be viewed as matched uncertainty. During the ideal sliding motion (3.96) reduces to

$$\dot{e}(t) = \big(I - B(SB)^{-1}S\big)A_m e(t) \tag{3.97}$$

$$= \big(I - B(SB)^{-1}S\big)Ae(t) \tag{3.98}$$

since $A_m = A + BF$ and $(I - B(SB)^{-1}S)B = 0$. Consequently if the pair (A, B) is controllable, then a hyperplane matrix S can be designed using any of the previously introduced methods, to make the tracking error $e(t) \to 0$ as $t \to \infty$.

Finally the controller component $u_{smc}(e)$ from (3.87) is defined as

$$u_{smc}(e) = u_l(t) + u_n(t) \tag{3.99}$$

where

$$u_l(t) = -(SB)^{-1}(SA_m - \Phi S)e(t) \tag{3.100}$$

$$u_n(t) = -\rho(SB)^{-1}\frac{P_2 s}{\|P_2 s\|} \quad \text{if } s(t) \neq 0 \tag{3.101}$$

and $P_2 \in \mathbb{R}^{m\times m}$ satisfies

$$P_2\Phi + \Phi^{\mathrm{T}} P_2 = -I_m \tag{3.102}$$

The scalar ρ depends on the magnitude of the uncertainty, and $\Phi \in \mathbb{R}^{m\times m}$ is a stable design matrix.

3.6 Sliding Modes for Fault Tolerant Control

As discussed in Sect. 3.1.2, during sliding, the trajectory of the closed-loop system is independent of the control input signal $u(t)$, and any uncertainty that occurs in these control input channels does not have any effect on the sliding motion and does not affect the system performance provided sliding can be maintained. If actuator faults in a linear system can be represented by

$$\dot{x}(t) = Ax(t) + Bu(t) + \underbrace{(-B)}_{D}\underbrace{K(t)u(t)}_{\xi(t,u,x)}$$

where $K(t) = \text{diag}(k_1(t), \ldots, k_m(t))$ and the scalars $k_i(t)$ satisfy $0 \le k_i(t) < 1$, this representation fits the definition of matched uncertainty given in Sect. 3.1.2 (see (3.15)). When $k_i(t) = 0$ the actuator is said to be working perfectly and when $k_i(t) > 0$ some degree of fault is present. When $k_i(t) = 1$ the actuator has failed completely. (This will be dealt with separately since sliding mode controllers per se cannot deal directly with total actuator failures.) Provided the modulation gain $\rho(\cdot)$ associated with the nonlinear control term is large enough to overcome the effect of matched uncertainty, sliding will always be guaranteed even in the presence of faults.

Despite its ability to handle actuator faults without requiring reconfiguring, SMC cannot deal directly with total actuator failures. During total actuator failures, some sort of reconfiguration or accommodation is needed. This is one motivation for the schemes described in the later chapters of this book.

In some systems, exact duplication of redundant actuators is available. This is considered in [58] for example. In this situation, the sliding mode controller can deal with total failures by simply channeling the control signals to the duplicate actuators without changing or reconfiguring the controller. This is simple in terms of design, since the same sliding mode controller output will be able to be used by many actuators. This is, however, restricted to systems with redundant actuators which are an exact duplicate of the originals. In many real engineering systems, this is simply not available. Thus, other tools are required to deal with total actuator failures.

This raises the question of whether other tools can be combined with sliding mode control to deal with total actuator failures. One potential candidate from the list of FTC methods in Chap. 2 is Control Allocation (CA). This will be one facet of the later chapters in this book. In safety critical systems such as large passenger transport aircraft, there already exists available redundancy. Learning from previous flight incidents and safe landings under extreme fault and failure conditions, this redundancy can be used unconventionally. In the case of an actuator fault/failure, CA has the ability to redistribute the control signals to the remaining functional actuators.

Even better, a careful combination of the robustness properties of SMC and the control reallocation capability of CA allows the possibility of creating simple robust controllers that deal with faults and failures without reconfiguration. This allows a single controller to work in almost all conditions. The strategy, the methods, and the theory on how SMC can deal directly with actuator failures, is one of the main contributions of this book.

3.7 Summary

In this chapter, the concept of sliding modes has been presented using a simple pendulum example. The properties of sliding mode controllers, especially robustness against matched uncertainty, have been presented. Some approaches for the design of the sliding surface (for closed-loop performance) and the control law (to ensure sliding is reached and maintained) have been introduced. Modifications of the control laws to include a tracking requirement (integral action and model-reference based tracking) have also been discussed. Finally, some advantages and drawbacks of sliding mode control and how it can be applied for FTC, have been considered. These ideas will be explored in the remainder of the book.

3.8 Notes and References

Many of the concepts in this chapter are based on the book by Edwards and Spurgeon [85], but have been reproduced here to set up the ideas and the notation which will be used in later chapters. Other excellent introductions to sliding modes can be found in Zinober [299, 300], Perruquetti and Barbot [211], and Utkin, Guldner and Shi [255]. The classical treatise on sliding modes is Utkin [256].

The control structures discussed in this chapter are largely based on the 'unit vector' approach from Ryan and Corless [217] although the specific development stems from [85]. The analysis in Sect. 3.3, assumes that the uncertainty is matched. The inclusion of unmatched uncertainty and the subsequent analysis can be found in [85]. Section 3.4 describes a design scheme based on a modification of the classical LQR design problem. Other design approaches can be found in the following references: robust eigenstructure assignment [73, 85], direct eigenstructure assignment [72, 85], and Linear Matrix Inequality (LMI) methods [56, 79].

The robustness properties of sliding modes against actuator faults make it a suitable candidate for FTC. A few researchers [127, 225, 226, 263, 269] have already studied the potential of sliding mode control in the field of reconfigurable control and FTC. For example, Hess and Wells [127] suggested that sliding mode control has the potential to become an alternative to reconfigurable control due to its robustness properties.

Chapter 4
Sliding Mode Observers for Fault Detection

In the previous chapter, the fundamental ideas of sliding mode control were presented. In this chapter, the concept of a sliding mode observer will be introduced. A simple design structure will be considered initially, and the resulting observer properties will be discussed. More advanced LMI design methods will then be presented. Robust fault reconstruction will be discussed and a comparison is made between (linear) unknown input observers and the proposed sliding mode schemes.

4.1 Introduction

The primary purpose of an observer is to estimate the unmeasurable states of the system by using the measured outputs and inputs of the system. It is essentially a mathematical replica of the system, where the input of the system is injected into the observer and its output is compared with the system output. The difference between the system outputs and observer outputs, termed the *output estimation error*, is then fed back as a corrective term so that the observer states will converge to the system states. The earliest observer is attributed to Luenberger [174] where the output estimation error is fed back *linearly* into the observer. However, in the presence of unknown signals, the Luenberger observer is usually unable to force the output estimation error to zero and the observer states will also not converge to the system states. A sliding mode observer [82, 256], which feeds back the output estimation error via a nonlinear switching term, provides an attractive solution to this issue. Provided a bound on the magnitude of the disturbances is known, the sliding mode observer can force the output estimation error to converge to zero in *finite time* (as opposed to the linear observer which only converges asymptotically), while the observer states converge asymptotically to the system states. During the sliding motion, the *equivalent output error injection* (the analogue to the equivalent control) contains information about the unknown signals, and by suitably scaling the equivalent output error injection, an accurate estimate of the unknown signals can be obtained. The first sliding mode observer in the literature appeared in [256]. Walcott and Zak [266] improved on this design by including a linear feedback term

H. Alwi et al., *Fault Detection and Fault-Tolerant Control Using Sliding Modes*, Advances in Industrial Control,
DOI 10.1007/978-0-85729-650-4_4,

such that the *sliding patch* can be enlarged. Edwards and Spurgeon [82] modified the sliding surface of the Walcott–Zak observer and presented a systematic numerical design method for the observer. In addition, the work in [82] identified necessary and sufficient conditions for the existence of the observer in terms of the original system matrices, and hence the class of systems for which the observer is feasible, is known.

By modelling faults as unknown signals, sliding mode observers can be used to reconstruct and thus detect and isolate faults. The early work in terms of fault reconstruction using sliding mode observers was reported by Edwards et al. [83, 86]. However, in the presence of other disturbances (which could represent unmodelled dynamics, parametric uncertainties or external disturbances), the reconstruction methods described in [83, 86] will no longer be accurate. It is therefore crucial to make the fault reconstruction robust to these disturbances (and hence achieve *robust fault reconstruction*). Tan and Edwards [248] proposed a design method for the observer parameters such that the $\mathcal{L}_2$ gain from the disturbances to the fault reconstruction is minimised. This chapter discusses sliding mode observers and demonstrates their use for fault reconstruction.

4.2 The Utkin Observer

Consider the linear system described by

$$\dot{x}(t) = Ax(t) + Bu(t) \tag{4.1}$$

$$y(t) = Cx(t) \tag{4.2}$$

where $A \in \mathbb{R}^{n\times n}$, $B \in \mathbb{R}^{n\times m}$ and $C \in \mathbb{R}^{p\times n}$. Assume that the pair (A, C) is observable. Introduce a linear nonsingular change of coordinates associated with the matrix

$$T_c = \begin{bmatrix} N_c^{\mathrm{T}} \\ C \end{bmatrix} \tag{4.3}$$

where the columns of N_c span the null space of C. Applying the change of coordinates $x \mapsto T_c x$, the triple (A, B, C) has the form

$$T_c A T_c^{-1} = \begin{bmatrix} A_{11} & A_{12} \\ A_{21} & A_{22} \end{bmatrix}, \qquad T_c B = \begin{bmatrix} B_1 \\ B_2 \end{bmatrix}, \qquad C T_c^{-1} = [0 \quad I_p] \tag{4.4}$$

where $A_{11} \in \mathbb{R}^{(n-p)\times(n-p)}$ and $B_1 \in \mathbb{R}^{(n-p)\times m}$. Utkin [256] proposed an observer of the form

$$\dot{\hat{x}}(t) = A\hat{x}(t) + Bu(t) + G_n \nu \tag{4.5}$$

$$\hat{y}(t) = C\hat{x}(t) \tag{4.6}$$

where $(\hat{x}, \hat{y})$ are the estimates of (x, y), and ν is a nonlinear discontinuous term. Define $e(t) := \hat{x}(t) - x(t)$ and $e_y(t) := \hat{y}(t) - y(t)$ as the state estimation and output estimation errors, respectively. The term ν is defined component-wise as

$$\nu_i = \rho \operatorname{sgn}(e_{y,i}), \quad i = 1, 2, \dots, p \tag{4.7}$$

where ρ is a positive real scalar and $e_{y,i}$ represents the ith component of $e_y(t)$. From (4.7), the term ν has been designed to switch discontinuously about the sliding surface $\mathcal{S} = \{e : Ce = 0\}$ and to drive the trajectories of $e(t)$ to $\mathcal{S}$. Assume (in the coordinate system of (4.4)) the gain G_n has the structure

$$G_n = \begin{bmatrix} G_{n,1} \\ -I_p \end{bmatrix} \tag{4.8}$$

where $G_{n,1} \in \mathbb{R}^{(n-p)\times p}$. The matrix $G_{n,1}$ represents the design freedom in the observer. Using the definition of $e(t)$, the following error system is obtained from (4.1)–(4.2) and (4.5)–(4.6):

$$\dot{e}(t) = Ae(t) + G_n \nu \tag{4.9}$$

Due to the structure of C in (4.4), the state estimation error can be partitioned as $e = \operatorname{col}(e_1, e_y)$ where $e_1 \in \mathbb{R}^{n-p}$. Then the error system (4.9) can be partitioned as

$$\dot{e}_1(t) = A_{11}e_1(t) + A_{12}e_y(t) + G_{n,1}\nu \tag{4.10}$$

$$\dot{e}_y(t) = A_{21}e_1(t) + A_{22}e_y(t) - \nu \tag{4.11}$$

From the definition of ν, (4.11) becomes (component-wise)

$$\dot{e}_{y,i}(t) = A_{21,i}e_1(t) + A_{22,i}e_y(t) - \rho \ \operatorname{sgn}(e_{y,i}) \tag{4.12}$$

where $A_{21,i}$ and $A_{22,i}$ represent the ith rows of A_{21} and A_{22}, respectively. From (4.12), it is straightforward to show that

$$\begin{aligned} e_{y,i}\dot{e}_{y,i} &= e_{y,i}(A_{21,i}e_1 + A_{22,i}e_y) - \rho|e_{y,i}| \\ &< -|e_{y,i}|\big(\rho - \big|(A_{21,i}e_1 + A_{22,i}e_y)\big|\big) \end{aligned}$$

If the scalar ρ is large enough such that it satisfies

$$\rho > |A_{21,i}e_1 + A_{22,i}e_y| + \eta \tag{4.13}$$

where $\eta \in \mathbb{R}_+$, then

$$e_{y,i}\dot{e}_{y,i} < -\eta|e_{y,i}| \tag{4.14}$$

Note that (4.14) is in the same form as (3.13) and hence implies that $e_{y,i}$ will converge to zero in *finite time*. When every component of $e_y(t)$ has converged to zero then a sliding motion takes place.

When a sliding motion has been achieved, $e_y(t) = \dot{e}_y(t) = 0$, and hence the error system defined by (4.10)–(4.11) can be written as

$$\dot{e}_1(t) = A_{11}e_1(t) + G_{n,1}\nu_{eq} \tag{4.15}$$

$$0 = A_{21}e_1(t) - \nu_{eq} \tag{4.16}$$

where ν_{eq} is the so-called *equivalent output error injection* that is required to maintain the sliding motion (analogous to the equivalent control in Sect. 3.1.2). This is not the term ν that is applied to the system, but rather, the *averaged* injection applied to maintain the sliding motion ($e_y(t) = \dot{e}_y(t) = 0$). Recall that this concept of equivalent output error injection is valid only during the sliding motion, and hence (4.15)–(4.16) are valid only when sliding takes place on the surface $\mathcal{S}$.

The properties of the sliding motion will be investigated in the following subsections: in particular, the effect of the choice of $G_{n,1}$ will be described.

4.2.1 Properties of the Sliding Motion

This subsection will analyse the behaviour of the system during the sliding motion. Eliminating the term ν_{eq} from (4.15)–(4.16) yields the following expression

$$\dot{e}_1(t) = (A_{11} + G_{n,1}A_{21})e_1(t) \tag{4.17}$$

This represents the reduced order motion (of order $n - p$) that governs the sliding mode dynamics.

Lemma 4.1 *The pair* (A_{11}, A_{12}) *is observable if and only if* (A, C) *is observable.*

Proof From the Popov–Belevitch–Hautus (PBH) rank test [85], if the pair (A, C) is observable, then the matrix

$$P(\mathrm{s}) = \begin{bmatrix} \mathrm{s}I_n - A \\ C \end{bmatrix} \tag{4.18}$$

will have full column rank for all values of s. Partitioned into the coordinates of (4.4), the expression in (4.18) becomes

$$P(\mathrm{s}) = \begin{bmatrix} \mathrm{s}I_{n-p} - A_{11} & -A_{12} \\ -A_{21} & \mathrm{s}I_p - A_{22} \\ 0 & I_p \end{bmatrix} \tag{4.19}$$

For (4.19) to have full column rank, the following matrix pencil must have full column rank for all values of s:

$$\begin{bmatrix} \mathrm{s}I_{n-p} - A_{11} \\ -A_{21} \end{bmatrix} \tag{4.20}$$

From the PBH rank test, this is equivalent to the pair (A_{11}, A_{21}) being observable. □

Therefore if (A, C) is observable, then (A_{11}, A_{21}) will also be observable, and an appropriate matrix $G_{n,1}$ can always be chosen to ensure that the reduced order motion in (4.17) is stable.

Remark 4.1 The differences between a sliding mode observer and sliding mode controller are two-fold. Firstly, in a controller, the switching term enters the system via the input distribution matrix B which is fixed; whereas for an observer the switching term enters via the matrix G_n which contains design freedom. Secondly in an observer, the sliding surface is fixed as $e_y(t) = 0$ whereas the sliding surface for a controller has design freedom that depends on the desired performance of the closed-loop system.

4.2.2 An Example

Consider a second order state-space system described by (4.1) and (4.2) where

$$A = \begin{bmatrix} 0 & 1 \\ -2 & 0 \end{bmatrix}, \qquad B = \begin{bmatrix} 0 \\ 1 \end{bmatrix}, \qquad C = [1 \quad 1] \tag{4.21}$$

which represents a simple harmonic oscillator. For simplicity assume $u(t) = 0$. A suitable choice for the nonsingular matrix T_c from (4.3) is

$$T_c = \begin{bmatrix} 1 & -1 \\ 1 & 1 \end{bmatrix} \tag{4.22}$$

Following the change of coordinates $x \mapsto T_c x$, the system triple (A, B, C) becomes

$$T_c A T_c^{-1} = \begin{bmatrix} 0.5 & 1.5 \\ -1.5 & -0.5 \end{bmatrix}, \qquad T_c B = \begin{bmatrix} -1 \\ 1 \end{bmatrix}, \qquad C T_c^{-1} = [0 \quad 1] \tag{4.23}$$

Suppose the nonlinear gain from (4.8) is chosen as $G_{n,1} = 3$. This results in the sliding motion being governed by $A_{11} + G_{n,1} A_{21} = -4$, which is stable. In the original coordinates of (4.21), the nonlinear gain can be calculated as

$$G_n = T_c^{-1} \begin{bmatrix} G_{n,1} \\ -I_p \end{bmatrix} = \begin{bmatrix} 0.5 & 0.5 \\ -0.5 & 0.5 \end{bmatrix} \begin{bmatrix} 3 \\ -1 \end{bmatrix} = \begin{bmatrix} 1 \\ -2 \end{bmatrix} \tag{4.24}$$

The following simulation was performed with the system assumed to have initial conditions $x(0) = \text{col}(0.5, -0.8)$ and the observer having zero initial conditions. In all the simulations $\rho = 1$.

Figure 4.1 shows the system states $x(t)$ and the observer estimates $\hat{x}(t)$. It can be seen that at approximately 1.5 s, perfect tracking of the states takes place. Figure 4.2 shows the output estimation error $e_y(t)$ and the state estimation errors $e(t)$. It can be seen that at approximately 0.66 s, $e_y(t)$ becomes zero, and remains there, and hence a sliding motion has taken place on $\mathcal{S} = \{e : Ce = 0\}$. Then, from that

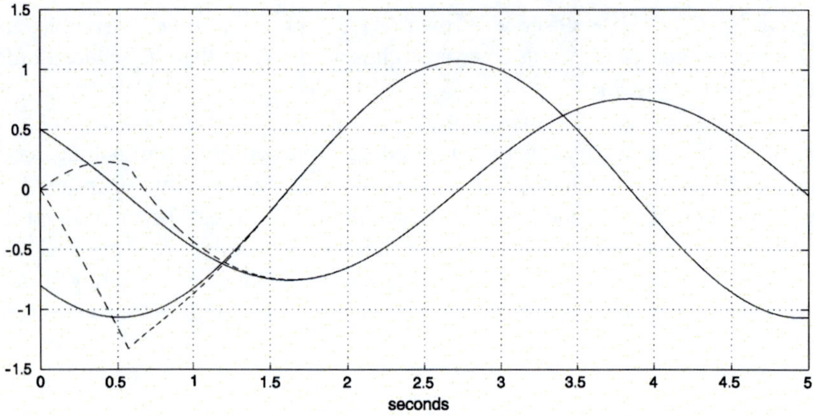

Fig. 4.1 System states $x(t)$ (*solid*) and the observer estimates $\hat{x}(t)$ (*dashed*)

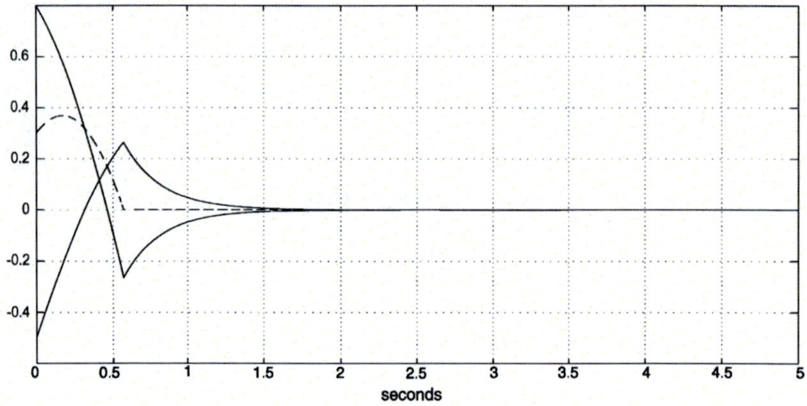

Fig. 4.2 The output estimation error $e_y(t)$ (*dashed*) and the components of the state estimation error $e(t)$ (*solid*)

time onwards, the errors (Fig. 4.2) evolve according to the dynamics of the *reduced order motion*. From Fig. 4.3, during the sliding motion, the nonlinear discontinuous switching term ν exhibits high frequency switching. Figure 4.4 shows the reduced order motion $e_1(t)$ which was obtained by multiplying $e(t)$ with the top $n-p$ rows of T_c. Figure 4.5 shows the equivalent output error injection signal ν_{eq} obtained from passing the term ν from Fig. 4.3 through a low-pass filter of time constant $\tau = 0.02$ s. Notice that the term ν_{eq} conforms to (4.16) once a sliding motion is taking place.

In the following simulations the same observer is used but the initial conditions of the states have been changed to 0.5 and -1.5, respectively. The initial conditions of the observer are once again set as zero. This situation represents effectively an increase in the initial conditions of $e_1(0)$ and $e_y(0)$.

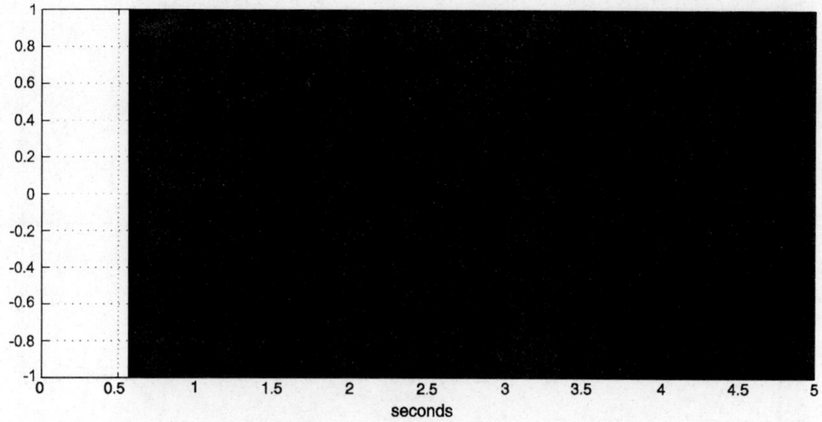

Fig. 4.3 The nonlinear injection switching term ν

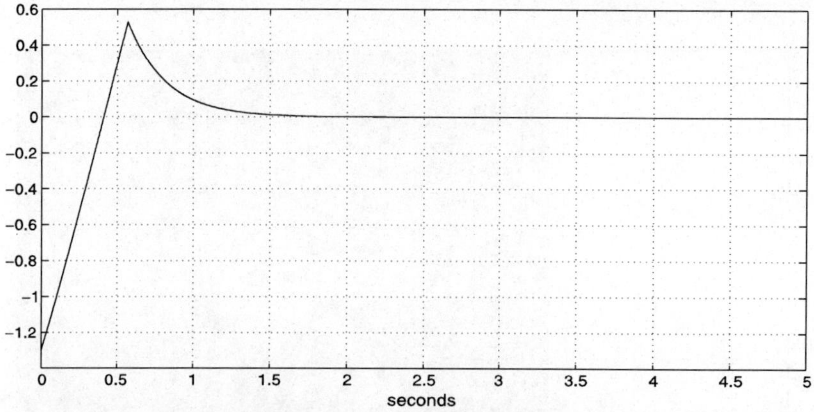

Fig. 4.4 The reduced order motion $e_1(t)$

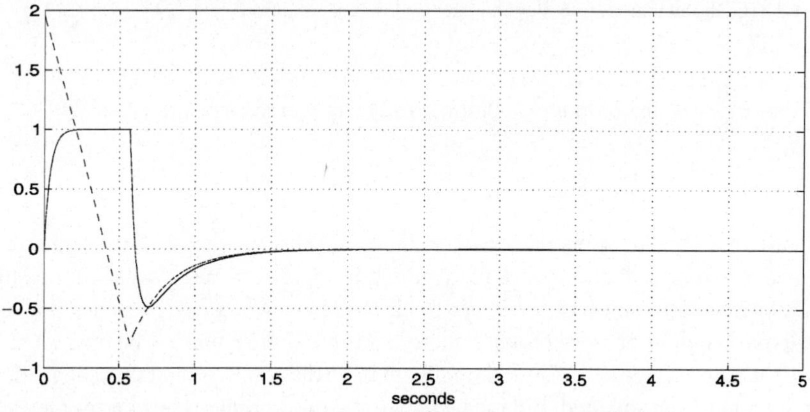

Fig. 4.5 The equivalent output error injection ν_{eq} (*solid*) and $A_{21}e_1(t)$ (*dashed*)

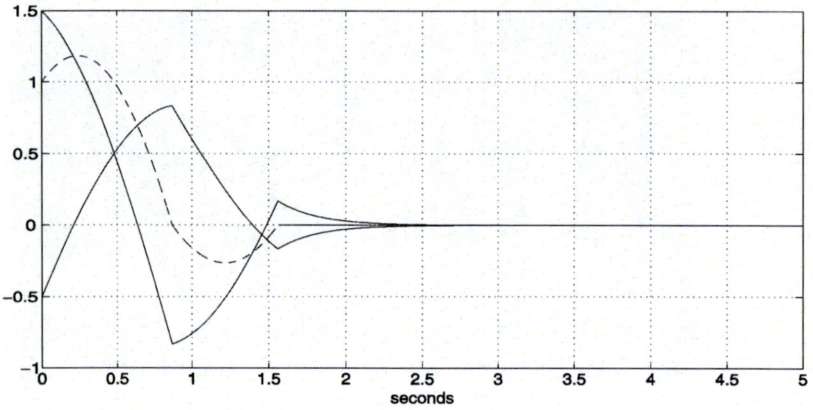

Fig. 4.6 The output estimation error $e_y(t)$ (*dashed*) and the components of the state estimation error $e(t)$ (*solid*)

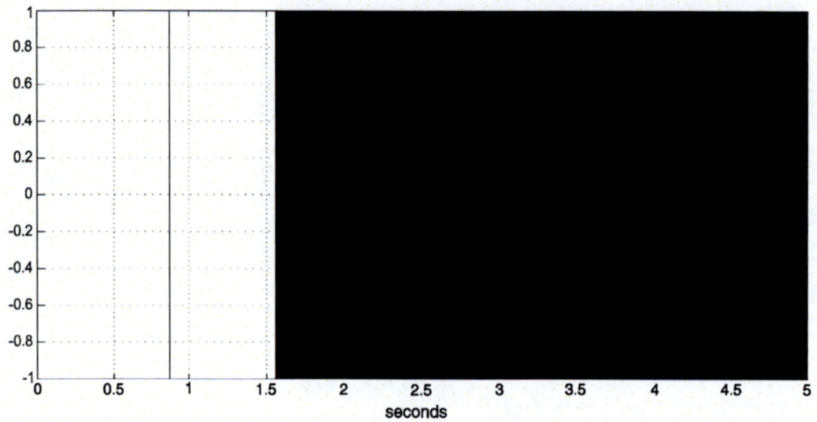

Fig. 4.7 The discontinuous term ν with larger error initial conditions

From Fig. 4.6, the output estimation error $e_y(t)$ pierces the surface

$$\mathcal{S} = \{e : Ce = 0\}$$

at approximately 0.87 s, but does not remain there. This is because of the large state estimation errors at that time instant, resulting in (4.13) not being satisfied; i.e., the reachability condition (4.14) does not hold. However, when $e_y(t)$ reaches 0 again at approximately 1.55 s, it remains there, and the sliding motion begins. At this point, the error vector $e(t)$ is much smaller than it was at 0.87 s, and the reachability condition has been satisfied. Figure 4.7 shows the discontinuous term ν for the case when the initial errors are large.

4.2.3 Disturbance Rejection Properties

Suppose (4.1) is now replaced by

$$\dot{x}(t) = Ax(t) + Bu(t) + Q\xi(t, x, u) \tag{4.25}$$

where $\xi(t, x, u) \in \mathbb{R}^h$ is a disturbance vector, and $Q \in \mathbb{R}^{n\times h}$ is its distribution matrix.

Suppose the gain G_n is designed such that it is matched to the disturbance matrix i.e., $Q = G_n X$ for some $X \in \mathbb{R}^{p\times h}$. Then in the coordinates of (4.4) and (4.8), the following condition will be satisfied

$$Q = \begin{bmatrix} G_{n,1}X \\ -X \end{bmatrix} \tag{4.26}$$

and the error system (4.10)–(4.11) becomes

$$\dot{e}_1(t) = A_{11}e_1(t) + A_{12}e_y(t) + G_{n,1}\nu - G_{n,1}X\xi(t, x, u) \tag{4.27}$$

$$\dot{e}_y(t) = A_{21}e_1(t) + A_{22}e_y(t) - \nu + X\xi(t, x, u) \tag{4.28}$$

From (4.28), it is straightforward to show that

$$\begin{aligned} e_{y,i}\dot{e}_{y,i} &= e_{y,i}(A_{21,i}e_1 + A_{22,i}e_y + X_i\xi) - \rho|e_{y,i}| \\ &< -|e_{y,i}|\big(\rho - |A_{21,i}e_1 + A_{22,i}e_y + X_i\xi|\big) \end{aligned}$$

where X_i is the ith row of X. If $\rho > |A_{21,i}e_1 + A_{22,i}e_y + X_i\xi| + \eta$ for a scalar $\eta > 0$, then the reachability condition in (4.14) is satisfied, and an ideal sliding motion takes place in finite time. When a sliding motion has been attained, (4.27)–(4.28) become

$$\dot{e}_1(t) = A_{11}e_1(t) + G_{n,1}\nu_{eq} - G_{n,1}X\xi(t, x, u) \tag{4.29}$$

$$0 = A_{21}e_1(t) - \nu_{eq} + X\xi(t, x, u) \tag{4.30}$$

Eliminating ν_{eq} from (4.29) and (4.30) yields the reduced order motion

$$\dot{e}_1(t) = (A_{11} + G_{n,1}A_{21})e_1(t) \tag{4.31}$$

which is independent of the disturbance $\xi(t, x, u)$. Notice that for the existence of an ideal sliding motion, the matching condition (4.26) is not required; a large enough ρ is sufficient to induce a sliding motion. The matching condition is only required for the reduced order motion (4.31) to be independent of $\xi(t, x, u)$. From (4.31), $e_1(t) \to 0$, and hence from (4.30), $\nu_{eq} \to X\xi(t, x, u)$. Hence the term ν_{eq} is able to provide information about the disturbance.

For the system in (4.21), consider the case when

$$Q = \begin{bmatrix} 1 \\ -2 \end{bmatrix}$$

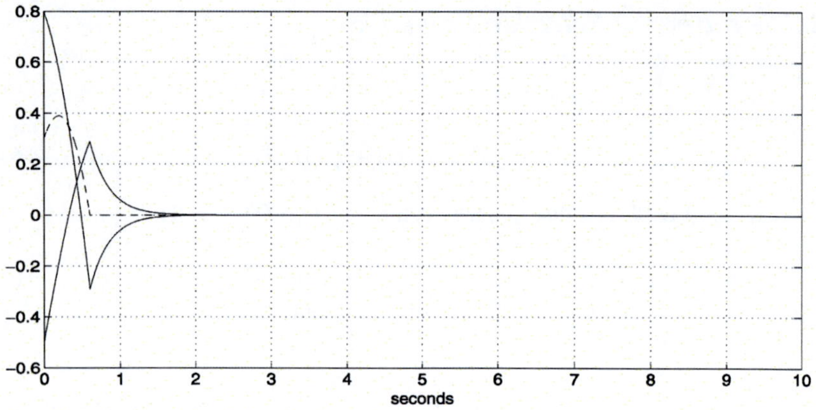

Fig. 4.8 The output estimation error $e_y(t)$ (*dashed*) and the components of the state estimation error $e(t)$ (*solid*)

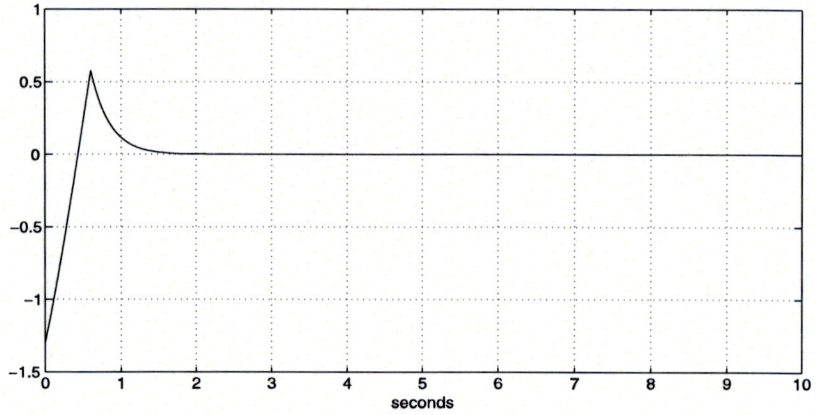

Fig. 4.9 The error vector associated with the sliding motion $e_1(t)$

and $\xi(t, x, u) = 0.2\sin(x_1(t))$. Notice from (4.24) that $G_n = Q$ and hence the matching condition from (4.26) is satisfied with $X = 1$. Assuming the same initial conditions as in Sect. 4.2.2, the following simulation results were obtained.

From Figs. 4.8 and 4.9, a sliding motion is achieved after approximately 0.66 s and the errors experience a first order decay, as before, unaffected by the disturbance. This disturbance rejection property is a major advantage of sliding mode observers over the nominal Luenberger observer. From Fig. 4.10, the effect of the disturbance $\xi(t, x, u)$ can be seen in the signal ν_{eq}. When the reduced order motion $e_1(t)$ has become small (at about 1.5 s), the signal ν_{eq} 'reproduces' the disturbance $\xi(t, x, u)$ (with a small delay due to the low-pass filter used to obtain ν_{eq}). Notice that the term ν was not designed with any a priori knowledge about $\xi(t, x, u)$, except that it is bounded. This feature of 'disturbance tracking' is essential to the fault reconstruction work in this book.

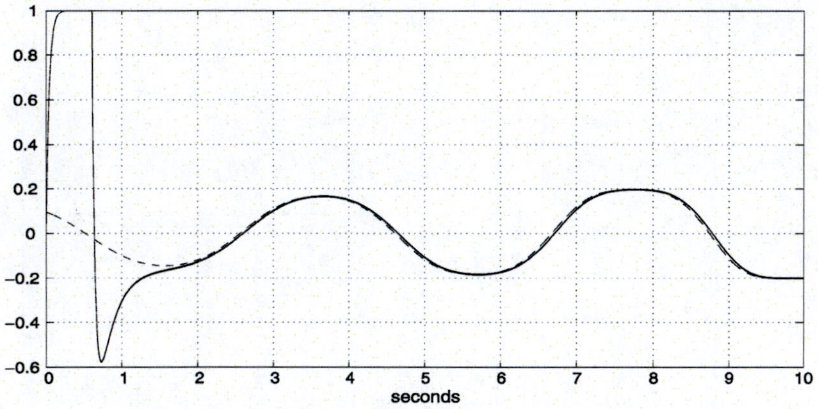

Fig. 4.10 The equivalent output error injection ν_{eq} (*solid*) and the disturbance $\xi(t, x, u)$ (*dashed*)

4.2.4 Pseudo-sliding by Smoothing the Discontinuity

From Fig. 4.3, the term ν is discontinuous with very high frequency switching. Systems with discontinuities often pose problems for simulation packages and generally cause an increase in the computational burden. Thus it is often useful to 'smooth' the discontinuity. (This is particularly true for sliding mode control systems as discussed earlier where high frequency switched control signals would represent an unacceptable input.) Recall that ν is defined component-wise by $\nu_i = \rho \ \mathrm{sgn}(e_{y,i})$, which can also be expressed as

$$\nu_i = \rho \frac{e_{y,i}}{|e_{y,i}|} \quad \text{if } e_{y,i}(t) \neq 0 \tag{4.32}$$

As in Sect. 3.3.2, a method to smooth ν is to approximate (4.32) by

$$\nu_i = \rho \frac{e_{y,i}}{|e_{y,i}| + \delta} \tag{4.33}$$

where δ is a small positive scalar. This results in a trade-off between ideal performance and maintaining a smooth output error injection signal. Repeating the simulation in Sect. 4.2.2, using ν as in (4.33) with $\delta = 0.0001$, the following figures were obtained. Figure 4.11 shows the smooth injection term ν from (4.33). Notice that its shape is similar to ν_{eq} from Fig. 4.5. From Fig. 4.12, it can be seen that the performance of the system is relatively unaffected (in comparison with Fig. 4.2).

Remark 4.2 The literature associated with the choice of smoothing coefficient δ in the unit vector approximation has been mainly associated with chatter avoidance in control schemes and its impact on the robustness of the closed-loop system [45, 64, 67]. For fault reconstruction, the objective is different because the emphasis is on the choice of δ and its impact on the quality of the reconstruction. Generally

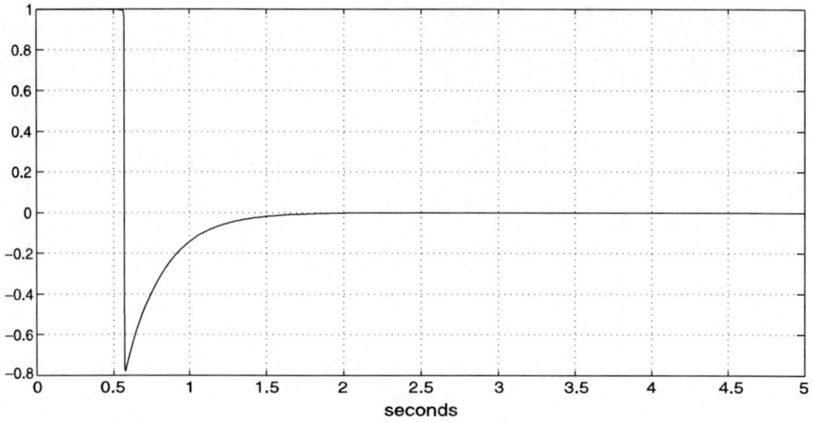

Fig. 4.11 The output error injection term ν after being smoothed

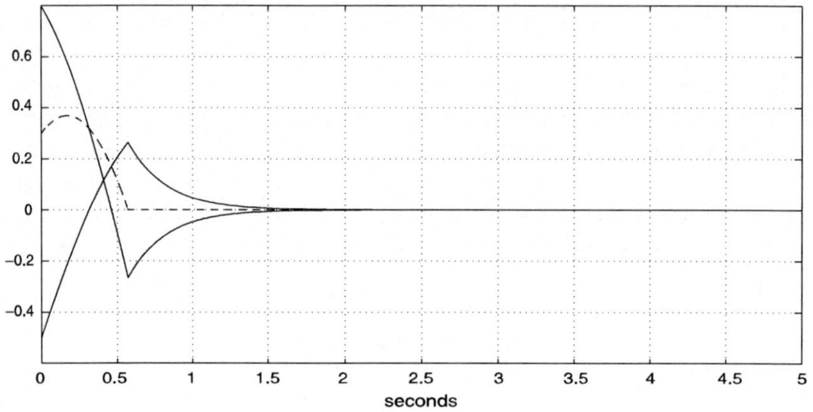

Fig. 4.12 The output estimation error $e_y(t)$ (*dashed*) and the components of the state estimation error $e(t)$ (*solid*)

speaking lower values of δ can be considered in observer problems than would be used in the controller case where chattering of the control signal is a key issue. It is argued in Sect. 3.3.2 that as a result of using the sigmoidally approximation a boundary layer of size 20δ is introduced (in the scalar case).

4.2.5 A Modification to Include a Linear Term

For the observer that has been discussed, the size of the parameter ρ dictates the size of the domain in which sliding takes place. However, for practical reasons, a very large value of ρ is not desirable and hence there is a trade-off.

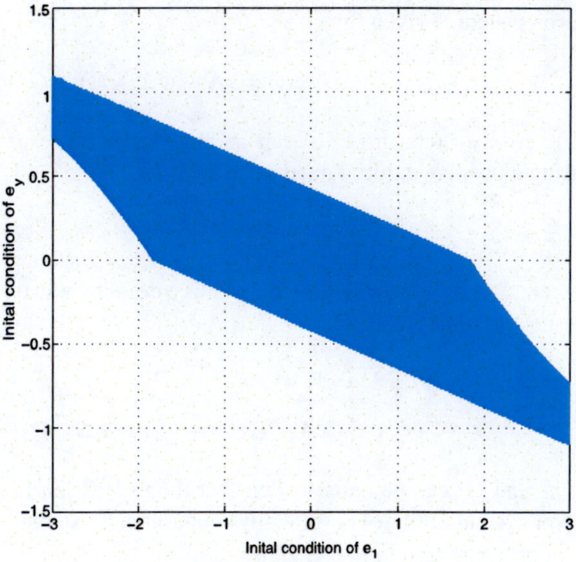

Fig. 4.13 The sliding region

For pedagogical purposes consider an unstable state-space system[1]

$$A = \begin{bmatrix} -2 & -3 \\ 1 & 3 \end{bmatrix}, \qquad B = \begin{bmatrix} 0 \\ 1 \end{bmatrix}, \qquad C = \begin{bmatrix} 0 & 1 \end{bmatrix} \tag{4.34}$$

To design an Utkin observer no change of coordinates is needed because the matrix C is already in the required structure of (4.4). Specifying $G_{n,1} = 0$ yields a reduced order motion pole of $A_{11} + G_{n,1}A_{21} = -2$. In the following simulations the gain ρ has been specified to be 1.

A series of simulations was carried out for different initial conditions ($e_1(0)$, $e_y(0)$), with both components ranging from -3 to $+3$. In Fig. 4.13, the 'shaded' area is the region in which the initial conditions $(e_1(0), e_y(0))$ must lie for a sliding motion to occur. Elsewhere the observer fails to provide converging state estimations. The shaded region is sometimes referred to as the *sliding patch* [229]. Of course the size of the shaded area can be enlarged by increasing the value of ρ, but for practical reasons that may be undesirable.

Instead consider the effect of adding an output error feedback term to the observer. Equation (4.5) can be modified to be

$$\dot{\hat{x}}(t) = A\hat{x}(t) - G_l e_y(t) + Bu(t) + G_n \nu \tag{4.35}$$

where $G_l \in \mathbb{R}^{n\times p}$. Slotine et al. [229] argue that an appropriate choice of the gain G_l will enlarge the sliding patch. From (4.1), (4.2) and (4.6), the state estimation

[1]In the following simulations full state feedback $u(t) = Fx(t)$ where $F = [1 \ -1]$ has been employed so that $\lambda(A + BF) = \{\pm 1.4142i\}$. The reason for this choice of closed-loop eigenvalues is that the states will be oscillatory, and the tracking of the states can be observed if desired.

error system is given by

$$\dot{e}(t) = (A - G_l C)e(t) + G_n \nu \tag{4.36}$$

The error system in (4.36) can be analysed with respect to quadratic stability by using a positive definite quadratic function

$$V = e^{\mathrm{T}} P e \tag{4.37}$$

where $P \in \mathbb{R}^{n \times n}$ is a symmetric positive definite matrix. Differentiating (4.37) with respect to time yields

$$\begin{aligned} \dot{V} &= \dot{e}^{\mathrm{T}} P e + e^{\mathrm{T}} P \dot{e} \\ &= e^{\mathrm{T}} \big(P(A - G_l C) + (A - G_l C)^{\mathrm{T}} P \big) e + 2 e^{\mathrm{T}} P G_n \nu \end{aligned} \tag{4.38}$$

If P and G_l can be chosen such that the expression in (4.38) is negative, then the error system in (4.36) is (globally) quadratically stable.

For the system in (4.34), choosing

$$G_l = \begin{bmatrix} -3 \\ 6 \end{bmatrix} \tag{4.39}$$

results in a closed-loop error system

$$\dot{e}_1(t) = -2e_1(t) \tag{4.40}$$

$$\dot{e}_y(t) = e_1(t) - 3e_y(t) - \mathrm{sgn}(e_y) \tag{4.41}$$

Consider a positive definite quadratic function as in (4.37) where the error vector e and matrix P are

$$e = \begin{bmatrix} e_1 \\ e_y \end{bmatrix}, \qquad P = \begin{bmatrix} \frac{1}{4} & 0 \\ 0 & 1 \end{bmatrix} \tag{4.42}$$

and hence the quadratic function from (4.37) is

$$V = \frac{1}{4} e_1^2 + e_y^2$$

Differentiating with respect to time yields

$$\begin{aligned} \dot{V} &= \frac{1}{2} e_1 \dot{e}_1 + 2 e_y \dot{e}_y \\ &= \frac{1}{2} e_1(-2e_1) + 2e_y \big(e_1 - 3e_y - \mathrm{sgn}(e_y) \big) \\ &= -e_1^2 - 6e_y^2 + 2e_1 e_y - 2|e_y| \\ &= -(e_1 - e_y)^2 - 5e_y^2 - 2|e_y| \end{aligned}$$

if $(e_1, e_y) \neq 0$. Hence global stability of this error system has been proven. When the magnitude of the error $e(t)$ becomes small enough, the reachability condition in (4.14) is satisfied and a sliding motion takes place. This example shows that in certain circumstances the introduction of a linear output error injection term can be beneficial.

4.3 The Edwards–Spurgeon Observer for Fault Reconstruction

Consider the following faulty dynamical system:

$$\dot{x}(t) = Ax(t) + Bu(t) + Mf(t) \tag{4.43}$$

$$y(t) = Cx(t) \tag{4.44}$$

where $x \in \mathbb{R}^n$ are the states, $y \in \mathbb{R}^p$ are the measurable outputs, and $u \in \mathbb{R}^m$ are the measurable inputs. The signal $f \in \mathbb{R}^q$ is the fault acting upon the system where $q < p$. It is unknown but assumed to be bounded so that

$$\|f(t)\| \leq \alpha \tag{4.45}$$

where α is known. Assume without loss of generality that the matrices C and M are full rank. The objective is to reconstruct the fault $f(t)$ based on the measured signals $u(t)$ and $y(t)$.

Two key lemmas will now be presented which underpin the rest of this chapter.

Lemma 4.2 *Let the triple (A, M, C) represent a linear system with $p > q$ and suppose* $\operatorname{rank}(CM) = \operatorname{rank}(M) = q$, *then there exists a change of coordinates $x \mapsto T_o x$ for the system* (4.43)–(4.44) *such that in the new coordinates the triple (A, M, C) of the transformed system has the following structure*:

$$A = \left[\begin{array}{c|c} A_{11} & A_{12} \\ \hline A_{211} & \multirow{2}{*}{A_{22}} \\ A_{212} & \end{array}\right], \qquad M = \begin{bmatrix} 0 \\ M_2 \end{bmatrix}, \qquad C = [0 \quad T] \tag{4.46}$$

where $A_{11} \in \mathbb{R}^{(n-p)\times(n-p)}$, $A_{211} \in \mathbb{R}^{(p-q)\times(n-p)}$, $T \in \mathbb{R}^{p\times p}$ is orthogonal and $M_2 \in \mathbb{R}^{p\times q}$ has the structure

$$M_2 = \begin{bmatrix} 0 \\ M_o \end{bmatrix} \tag{4.47}$$

where $M_o \in \mathbb{R}^{q\times q}$ is nonsingular.

Proof First introduce a coordinate transformation to make the last p states of the system the outputs. To achieve this, define

$$T_c = \begin{bmatrix} N_c^{\mathrm{T}} \\ C \end{bmatrix} \tag{4.48}$$

where $N_c \in \mathbb{R}^{n\times(n-p)}$ and the columns span the null space of C. The coordinate transformation $x \mapsto T_c x$ is nonsingular by construction and in the new coordinate system

$$C = [0 \quad I_p]$$

Suppose in the new coordinate system

$$M = \begin{bmatrix} M_1 \\ M_2 \end{bmatrix} \begin{matrix} \updownarrow n-p \\ \updownarrow p \end{matrix}$$

Then $CM = M_2$ and so by assumption $\operatorname{rank}(M_2) = q$. Hence the left pseudo-inverse $M_2^\dagger$ is well defined. Also there exists an orthogonal matrix $T \in \mathbb{R}^{p\times p}$ such that

$$T^{\mathrm{T}} M_2 = \begin{bmatrix} 0 \\ M_o \end{bmatrix} \tag{4.49}$$

where $M_o \in \mathbb{R}^{q\times q}$ is nonsingular. Consequently, the coordinate transformation $x \mapsto T_b x$ where

$$T_b = \begin{bmatrix} I_{n-p} & -M_1 M_2^\dagger \\ 0 & T^{\mathrm{T}} \end{bmatrix} \tag{4.50}$$

is nonsingular, and in the new coordinates the triple (A, M, C) is in the form

$$A = \begin{bmatrix} A_{11} & A_{12} \\ A_{21} & A_{22} \end{bmatrix}, \qquad M = \begin{bmatrix} 0 \\ M_2 \end{bmatrix}, \qquad C = [0 \quad T] \tag{4.51}$$

where $A_{11} \in \mathbb{R}^{(n-p)\times(n-p)}$ and the remaining sub-blocks in the system matrix are partitioned accordingly. □

Lemma 4.3 *The pair (A_{11}, A_{211}) is detectable if and only if the invariant zeros of (A, M, C) are Hurwitz.*

Proof From the PBH rank test [216], the unobservable modes of (A_{11}, A_{211}) are given by the values of s that make the following matrix pencil lose rank:

$$P_{\mathrm{obs}}(\mathrm{s}) = \begin{bmatrix} \mathrm{s}I - A_{11} \\ A_{211} \end{bmatrix}$$

The zeros of (A, M, C) are given by the values of s that make the Rosenbrock matrix $R_{\mathrm{obs}}(\mathrm{s})$ lose rank [216], where

$$R_{\mathrm{obs}}(\mathrm{s}) = \begin{bmatrix} \mathrm{s}I - A & -M \\ C & 0 \end{bmatrix} = \left[\begin{array}{cc|c} \mathrm{s}I - A_{11} & -A_{12} & 0 \\ -A_{211} & \star & 0 \\ -A_{212} & \star & -M_o \\ \hline 0 & T & 0 \end{array}\right]$$

and the $\star$ represent elements that do not play a role in the subsequent analysis. Since M_o and T are both square and invertible, then $R_{\mathrm{obs}}(\mathrm{s})$ loses rank if and only if

$P_{obs}(s)$ loses rank. Clearly the unobservable modes of (A_{11}, A_{211}) are the invariant zeros of (A, M, C), and hence the proof is complete. □

4.3.1 Observer Formulation and Stability Analysis

In this subsection, it will be assumed that for the system in (4.43)–(4.44)

A1. $\text{rank}(CM) = \text{rank}(M)$
A2. The invariant zeros (if any) of (A, M, C) are Hurwitz.

The following state observer will be considered for the system in (4.43)–(4.44)

$$\dot{\hat{x}}(t) = A\hat{x}(t) + Bu(t) - G_l e_y(t) + G_n \nu \tag{4.52}$$

$$\hat{y}(t) = C\hat{x}(t) \tag{4.53}$$

where $e_y(t) := \hat{y}(t) - y(t)$ is the output estimation error and both $G_l \in \mathbb{R}^{n\times p}$ and $G_n \in \mathbb{R}^{n\times p}$ are design matrices to be determined. Condition A1 implies that the coordinate structure of (4.46) can be attained and hence without loss of generality it can be assumed that the triple (A, M, C) is in the structure of (4.46). In the coordinates of (4.46), let

$$G_n = \begin{bmatrix} -L \\ I_p \end{bmatrix} T^{\mathrm{T}} P_o^{-1}, \quad L = [L^o \quad 0] \tag{4.54}$$

where $L^o \in \mathbb{R}^{(n-p)\times(p-q)}$ and $P_o = P_o^{\mathrm{T}} \in \mathbb{R}^{p\times p}$ are design matrices. The discontinuous vector ν is defined by

$$\nu(t) = -\rho \frac{e_y}{\|e_y\|} \quad \text{if } e_y(t) \neq 0 \tag{4.55}$$

where ρ is a positive scalar function dependent on the magnitude of f.

Defining $A_o = A - G_l C$ and the state estimation error as $e(t) := \hat{x}(t) - x(t)$, from (4.43) and (4.52), (4.44) and (4.53), the following error system can be obtained

$$\dot{e}(t) = A_o e(t) + G_n \nu - Mf(t) \tag{4.56}$$

Proposition 4.1 *If there exists a matrix G_l and a Lyapunov matrix P of the form*

$$P = \begin{bmatrix} P_1 & P_1 L \\ L^{\mathrm{T}} P_1 & T^{\mathrm{T}} P_o T + L^{\mathrm{T}} P_1 L \end{bmatrix} > 0 \tag{4.57}$$

where $P_1 \in \mathbb{R}^{(n-p)\times(n-p)}$ that satisfies

$$PA_o + A_o^{\mathrm{T}} P < 0 \tag{4.58}$$

and $\rho \geq \|P_o CM\|\alpha + \eta_o$ where $\eta_o > 0$, then the state estimation error $e(t)$ is asymptotically stable.

Proof Consider a positive definite Lyapunov function

$$V = e^{\mathrm{T}} P e \tag{4.59}$$

where P is given in (4.57). Differentiating (4.59) with respect to time,

$$\begin{aligned}\dot{V} &= \dot{e}^{\mathrm{T}} P e + e^{\mathrm{T}} P \dot{e} \\ &= e^{\mathrm{T}}(P A_o + A_o^{\mathrm{T}} P)e + 2e^{\mathrm{T}} P G_n \nu - 2e^{\mathrm{T}} P M f\end{aligned}$$

From the definitions of P, G_n and M in (4.57), (4.54) and (4.46), respectively, it is easy to prove that

$$P G_n = C^{\mathrm{T}} \quad \text{and} \quad P M = C^{\mathrm{T}} P_o C M \tag{4.60}$$

Using (4.58) and (4.60), $\dot{V}$ becomes

$$\dot{V} \le 2e^{\mathrm{T}} C^{\mathrm{T}} \nu - 2e^{\mathrm{T}} C^{\mathrm{T}} P_o C M f$$

From the definition of ν in (4.55) and using the bound of f in (4.45)

$$\begin{aligned}\dot{V} &< -2\rho\|e_y\| - 2e_y^{\mathrm{T}} P_o C M f \\ &\le -2\|e_y\|(\rho - \|P_o C M\|\alpha) \\ &\le -2\eta_o\|e_y\| \\ &< 0 \quad \text{for } e \neq 0\end{aligned}$$

which proves the state estimation error is quadratically stable. □

Corollary 4.1 *A stable sliding motion takes place on the surface*

$$\mathcal{S} = \{e : Ce = 0\} \tag{4.61}$$

in finite time and the sliding motion is governed by $\mathcal{A}_{11} = A_{11} + L^o A_{211}$.

Proof To prove a sliding motion is attained, it is convenient to firstly apply a change of coordinates $x \mapsto T_L x$ where

$$T_L = \begin{bmatrix} I_{n-p} & L \\ 0 & T \end{bmatrix} \tag{4.62}$$

such that the triple (A, M, C) in (4.46) is transformed to be

$$\mathcal{A} = \begin{bmatrix} \mathcal{A}_{11} & \mathcal{A}_{12} \\ \mathcal{A}_{21} & \mathcal{A}_{22} \end{bmatrix}, \qquad \mathcal{M} = \begin{bmatrix} 0 \\ \mathcal{M}_2 \end{bmatrix}, \qquad \mathcal{C} = [0 \quad I_p] \tag{4.63}$$

where $\mathcal{A}_{11} = A_{11} + L^o A_{211}$ (see the structure of L in (4.54)), $\mathcal{A}_{21} = T A_{21}$ and $\mathcal{M}_2 = T M_2$. Then the error system (4.56) can be partitioned as

$$\dot{e}_1(t) = \mathcal{A}_{11} e_1(t) + (\mathcal{A}_{12} - \mathcal{G}_{l,1}) e_y(t) \tag{4.64}$$

$$\dot{e}_y(t) = \mathcal{A}_{21}e_1(t) + (\mathcal{A}_{22} - \mathcal{G}_{l,2})e_y(t) + P_o^{-1}\nu - \mathcal{M}_2 f(t) \tag{4.65}$$

where

$$\begin{bmatrix} \mathcal{G}_{l,1} \\ \mathcal{G}_{l,2} \end{bmatrix} = T_L G_l \tag{4.66}$$

Introduce a Lyapunov function for the subsystem (4.65) as

$$V_s = e_y^{\mathrm{T}} P_o e_y$$

Differentiating V_s with respect to time and using (4.65), yields

$$\dot{V}_s = e_y^{\mathrm{T}}\big(P_o(\mathcal{A}_{22} - \mathcal{G}_{l,2}) + (\mathcal{A}_{22} - \mathcal{G}_{l,2})^{\mathrm{T}} P_o\big)e_y + 2e_y^{\mathrm{T}} P_o \mathcal{A}_{21} e_1 + 2e_y^{\mathrm{T}}\nu - 2e_y^{\mathrm{T}} P_o \mathcal{M}_2 f$$

It can be shown that

$$(T_L^{-1})^{\mathrm{T}} P T_L^{-1} = \begin{bmatrix} P_1 & 0 \\ 0 & P_o \end{bmatrix} \tag{4.67}$$

and

$$T_L A_o T_L^{-1} = \begin{bmatrix} \mathcal{A}_{11} & \mathcal{A}_{12} - \mathcal{G}_{l,1} \\ \mathcal{A}_{21} & \mathcal{A}_{22} - \mathcal{G}_{l,2} \end{bmatrix} \tag{4.68}$$

It follows that $(T_L^{\mathrm{T}})^{-1}(PA_o + A_o^{\mathrm{T}} P)T_L^{-1}$ can be expanded as

$$\begin{bmatrix} P_1\mathcal{A}_{11} + \mathcal{A}_{11}^{\mathrm{T}} P & \star \\ \star & P_o(\mathcal{A}_{22} - \mathcal{G}_{l,2}) + (\mathcal{A}_{22} - \mathcal{G}_{l,2})^{\mathrm{T}} P_o \end{bmatrix} < 0 \tag{4.69}$$

where the $\star$ are elements that do not play a role in the following analysis. Since (4.69) is symmetric, it is clear that $P_o(\mathcal{A}_{22} - \mathcal{G}_{l,2}) + (\mathcal{A}_{22} - \mathcal{G}_{l,2})^{\mathrm{T}} P_o < 0$ and hence

$$\begin{aligned} \dot{V}_s &< 2e_y^{\mathrm{T}} P_o \mathcal{A}_{21} e_1 + 2e_y^{\mathrm{T}}\nu - 2e_y^{\mathrm{T}} P_o \mathcal{M}_2 f \\ &\le 2\|e_y\|\|P_o\mathcal{A}_{21}e_1\| - 2\eta_o\|e_y\| \\ &= 2\|e_y\|\big(\|P_o\mathcal{A}_{21}e_1\| - \eta_o\big) \end{aligned} \tag{4.70}$$

Notice that

$$\|e_y\|^2 = \big(\sqrt{P_o}e_y\big)^{\mathrm{T}} P_o^{-1}\big(\sqrt{P_o}e_y\big) \ge \lambda_{\min}\big(P_o^{-1}\big)\|\sqrt{P_o}e_y\|^2 = \lambda_{\min}\big(P_o^{-1}\big)V_s \tag{4.71}$$

Define η as a scalar satisfying $0 < \eta < \eta_o$. Since from Proposition 4.1 the state estimation error is quadratically stable, in finite time $e_1(t)$ enters the domain

$$\Omega_\eta = \big\{e_1 : \|P_o\mathcal{A}_{21}e_1\| < \eta_o - \eta\big\}$$

and remains there. Inside the domain Ω_η inequality (4.70) becomes

$$\frac{dV_s}{dt} < -2\eta\|e_y\| < -2\eta\sqrt{\lambda_{\min}\big(P_o^{-1}\big)}\sqrt{V_s}$$

Integrating from the time when $e_1(t)$ enters Ω_η until the time when a sliding motion takes place,

$$\int_{V_s(t_\Omega)}^{0} \frac{1}{\sqrt{V_s}} dV_s < -2\eta\sqrt{\lambda_{\min}(P_o^{-1})} \int_{t_\Omega}^{t_s} dt$$

where $V_s(t_\Omega)$ is the initial condition of V_s at $t = t_\Omega$, the time at which $e_1(t)$ enters Ω_η, and t_s is the time at which the sliding motion begins. It can be shown that the time taken to attain a sliding motion t_s satisfies

$$t_s < \eta^{-1}\sqrt{\frac{V_s(t_\Omega)}{\lambda_{\min}(P_o^{-1})}} + t_\Omega$$

This proves that a sliding motion takes place on $\mathcal{S}$ in finite time.

When a sliding motion has been achieved, $e_y(t) = \dot{e}_y(t) = 0$ and from (4.64) the remaining dynamics $e_1(t)$ are governed by $\mathcal{A}_{11} = A_{11} + L^o A_{211}$. Since $P_1\mathcal{A}_{11} + \mathcal{A}_{11}^{\mathrm{T}} P_1 < 0$ and $P_1 > 0$, the matrix $\mathcal{A}_{11}$ is stable. □

4.3.2 Reconstruction of Actuator Faults

During the sliding motion, $e_y(t) = \dot{e}_y(t) = 0$. Recalling that the matrices $\mathcal{A}_{11} = A_{11} + L^o A_{211}$, $\mathcal{A}_{21} = T A_{21}$ and $\mathcal{M}_2 = T M_2$, (4.64)–(4.65) in the coordinates of (4.46) become

$$\dot{e}_1(t) = \left(A_{11} + L^o A_{211}\right) e_1(t) \tag{4.72}$$

$$0 = T A_{21} e_1(t) + P_o^{-1} \nu_{eq} - T M_2 f(t) \tag{4.73}$$

where ν_{eq} is the *equivalent output error injection* required to maintain a sliding motion. Define a measurable reconstruction signal:

$$\hat{f}(t) := \left(M_2^{\mathrm{T}} M_2\right)^{-1} M_2^{\mathrm{T}} T^{\mathrm{T}} P_o^{-1} \nu_{eq} \tag{4.74}$$

If L^o is chosen such that $A_{11} + L^o A_{211}$ is stable, then $e_1(t) \to 0$ and hence from (4.73) and (4.74)

$$\hat{f}(t) \to f(t) \tag{4.75}$$

In (4.47), since M_o is invertible, $(M_2^{\mathrm{T}} M_2)^{-1}$ is well defined. The reconstruction signal $\hat{f}(t)$ is computable online, since ν_{eq} is computable online by replacing (4.55) with

$$\nu_\delta = -\rho \frac{e_y}{\|e_y\| + \delta} \tag{4.76}$$

where δ is a small positive constant which governs the accuracy to which the equivalent injection is approximated.

4.4 LMI Design Methodologies

This section presents a systematic method to design a sliding mode observer of the structure given in (4.52)–(4.53) using Linear Matrix Inequalities (LMIs) [33]. From Proposition 4.1 the design of the observer requires finding matrices L^o, G_l, P_1 and P_o from (4.54) and (4.57) that satisfy the following inequality

$$P(A - G_l C) + (A - G_l C)^{\mathrm{T}} P < 0$$

where A and C have the structure of (4.46), and $P > 0$ has the structure in (4.57). A design method to achieve this can be summarised as:

Minimise trace(P^{-1}) subject to the inequalities

$$PA + A^{\mathrm{T}} P - C^{\mathrm{T}} \mathbf{V}^{-1} C + P\mathbf{W}P < 0, \quad P > 0 \tag{4.77}$$

where $\mathbf{W} \in \mathbb{R}^{n\times n}$ and $\mathbf{V} \in \mathbb{R}^{p\times p}$ are symmetric positive definite user-defined matrices. From the solution for P that is obtained, the observer gain G_l can then be directly calculated as

$$G_l = P^{-1} C^{\mathrm{T}} \mathbf{V}^{-1} \tag{4.78}$$

The motivation for the choice of the inequality posed in (4.77), and for minimising trace(P^{-1}) subject to (4.77), will be discussed here. In the absence of the fault $f(t)$ and as $\rho \to 0$, the observer tends to a linear formulation. Defining $\mathbf{X} := P^{-1}$, then from pre and post multiplying inequality (4.77) by $\mathbf{X}$, the following inequality can be obtained:

$$A\mathbf{X} + \mathbf{X}A^{\mathrm{T}} - \mathbf{X}C^{\mathrm{T}} \mathbf{V}^{-1} C\mathbf{X} + \mathbf{W} < 0 \tag{4.79}$$

The linear gain can now be calculated as $G_l = \mathbf{X}C^{\mathrm{T}}\mathbf{V}^{-1}$ and the objective becomes the minimisation of trace$(\mathbf{X})$ subject to (4.79).

The standard Linear Quadratic Gaussian (LQG) optimal observer design method as described in [175] uses the stabilising solution $\mathbf{X}_{\mathrm{are}}$ to the Algebraic Riccati Equation (ARE)

$$A\mathbf{X}_{\mathrm{are}} + \mathbf{X}_{\mathrm{are}} A^{\mathrm{T}} - \mathbf{X}_{\mathrm{are}} C^{\mathrm{T}} \mathbf{V}^{-1} C \mathbf{X}_{\mathrm{are}} + \mathbf{W} = 0 \tag{4.80}$$

to calculate the optimal observer gain $G_{l,are} := \mathbf{X}_{\mathrm{are}} C^{\mathrm{T}} \mathbf{V}^{-1}$. The associated optimal cost is given by trace$(\mathbf{X}_{\mathrm{are}})$.

Lemma 4.4 *Let* $\mathbf{X}$ *be any symmetric positive definite matrix satisfying* (4.79), *and let* $\mathbf{X}_{\mathrm{are}}$ *be the stabilising solution to the ARE* (4.80). *Then* $\mathbf{X} > \mathbf{X}_{\mathrm{are}}$ *and hence* trace$(\mathbf{X}) >$ trace$(\mathbf{X}_{\mathrm{are}})$.

Proof Inequality (4.79) can be expressed as

$$A\mathbf{X} + \mathbf{X}A^{\mathrm{T}} - \mathbf{X}C^{\mathrm{T}}\mathbf{V}^{-1}C\mathbf{X} + \mathbf{W} + \Delta = 0 \tag{4.81}$$

for some symmetric positive definite matrix Δ. Subtracting (4.80) from (4.81) and defining $\tilde{\mathbf{X}} = \mathbf{X} - \mathbf{X}_{\mathrm{are}}$ results in

$$A\tilde{\mathbf{X}} + \tilde{\mathbf{X}}A^{\mathrm{T}} - \mathbf{X}C^{\mathrm{T}}\mathbf{V}^{-1}C\mathbf{X} + \mathbf{X}_{\mathrm{are}}C^{\mathrm{T}}\mathbf{V}^{-1}C\mathbf{X}_{\mathrm{are}} + \Delta = 0 \tag{4.82}$$

Then substituting $\mathbf{X}_{\mathrm{are}} = \mathbf{X} - \tilde{\mathbf{X}}$ into (4.82) yields

$$\big(A - \mathbf{X}C^{\mathrm{T}}\mathbf{V}^{-1}C\big)\tilde{\mathbf{X}} + \tilde{\mathbf{X}}(A - \mathbf{X}C^{\mathrm{T}}\mathbf{V}^{-1}C)^{\mathrm{T}} + \Delta + \tilde{\mathbf{X}}C^{\mathrm{T}}\mathbf{V}^{-1}C\tilde{\mathbf{X}} = 0 \tag{4.83}$$

Since inequality (4.79) can be re-written as

$$\big(A - \mathbf{X}C^{\mathrm{T}}\mathbf{V}^{-1}C\big)\mathbf{X} + \mathbf{X}\big(A - \mathbf{X}C^{\mathrm{T}}\mathbf{V}^{-1}C\big)^{\mathrm{T}} + \mathbf{X}C^{\mathrm{T}}\mathbf{V}^{-1}C\mathbf{X} + \mathbf{W} < 0 \tag{4.84}$$

and $\mathbf{X} > 0$, it follows that $(A - \mathbf{X}C^{\mathrm{T}}\mathbf{V}^{-1}C)$ is stable. Therefore, as argued in Lemma 3 in [271], (4.83) implies $\tilde{\mathbf{X}} > 0$ and hence $\mathbf{X} > \mathbf{X}_{\mathrm{are}}$ as claimed. The fact that trace($\mathbf{X}$) > trace($\mathbf{X}_{\mathrm{are}}$) follows from the properties of the trace operator. □

From Lemma 4.4, the requirement of minimising trace($\mathbf{X}$) follows from the desire to approach the true minimal cost given by trace($\mathbf{X}_{\mathrm{are}}$). Of course a particular sub-optimal cost is enforced here by the requirement $P := \mathbf{X}^{-1}$ has the structure of (4.57).

In inequality (4.79), $\mathbf{W}$ is the performance weighting matrix for the observer, and $\mathbf{V}$ is the co-variance matrix of the system sensor noise. As in classical LQG theory the choice of $\mathbf{W}$ and $\mathbf{V}$ can be used to trade-off performance and noise amplification: increasing the 'magnitude' of $\mathbf{W}$ will generally increase the 'magnitude' of G_l whereas increasing the 'magnitude' of $\mathbf{V}$ will have the opposite effect on G_l; this gives the designer a systematic way to tune the gain G_l.

4.4.1 Software Implementation

By using the Schur complement [33], the matrix inequality in (4.77) is equivalent to

$$\begin{bmatrix} PA + A^{\mathrm{T}}P - C^{\mathrm{T}}\mathbf{V}^{-1}C & P \\ P & -\mathbf{W}^{-1} \end{bmatrix} < 0 \tag{4.85}$$

If $X \in \mathbb{R}^{n\times n}$ is symmetric positive definite, then (again using the Schur complement) the following inequality

$$\begin{bmatrix} -P & I_n \\ I_n & -X \end{bmatrix} < 0 \tag{4.86}$$

is equivalent to $X > P^{-1}$. Thus minimising trace(P^{-1}) subject to (4.77) can be implemented by minimising trace(X) subject to the LMIs (4.85) and (4.86). Writing P from (4.57) in terms of LMI variables

$$P = \begin{bmatrix} P_{11} & P_{12} \\ P_{12}^{\mathrm{T}} & P_{22} \end{bmatrix} > 0 \tag{4.87}$$

where $P_{11} \in \mathbb{R}^{(n-p)\times(n-p)}$ and $P_{12} := [P_{121}\ 0]$, $P_{121} \in \mathbb{R}^{(n-p)\times(p-q)}$ (due to the structure of L), then the elements of P in (4.57) can be calculated in terms of the LMI variables P_{11}, P_{121}, P_{22} by the following equations

$$P_1 = P_{11} \tag{4.88}$$

$$L^o = P_{11}^{-1} P_{121} \tag{4.89}$$

$$P_o = T\big(P_{22} - P_{12}^{\mathrm{T}} P_{11}^{-1} P_{12}\big) T^{\mathrm{T}} \tag{4.90}$$

It follows that the constrained minimisation problem represents a convex optimisation problem with regard to P_{11}, P_{121}, P_{22} and X. The approach can be formally stated as:

> Minimise trace(X) with respect to the variables P_{11}, P_{121}, P_{22} and X subject to the LMIs given in (4.85) and (4.86).

Standard LMI software, such as [104] can be used to synthesise numerically P and X, which will return values for P_{11}, P_{121}, P_{22} and X. Subsequently the observer parameters can be obtained: L^o from (4.89), P_o from (4.90), G_l from (4.78) and G_n from (4.54).

4.5 Robust Fault Reconstruction using Sliding Mode Observers

In Sect. 4.3, the design of the fault reconstruction schemes was based on an ideal linear system that did not consider quantities such as disturbances, parametric uncertainties or nonlinearities. This section presents an analysis and design method for the observer so that the fault reconstruction is robust to such quantities.

Consider a system that is subject to an actuator fault and also uncertainties and/or disturbances

$$\dot{x}(t) = Ax(t) + Bu(t) + Mf(t) + Q\xi(t, x, u) \tag{4.91}$$

$$y(t) = Cx(t) \tag{4.92}$$

where $A \in \mathbb{R}^{n\times n}$, $B \in \mathbb{R}^{n\times m}$, $C \in \mathbb{R}^{p\times n}$, $M \in \mathbb{R}^{n\times q}$ and $Q \in \mathbb{R}^{n\times h}$ where $n > p \ge q$. Assume that the matrices C and M are full rank and the function $f \in \mathbb{R}^q$

is unknown but bounded so that

$$\|f(t)\| \leq \alpha(t, u) \tag{4.93}$$

where $\alpha(t, u) \in \mathbb{R}_+ \times \mathbb{R}^m \times \mathbb{R}^q$ is a known function. The signal $\xi(t, x, u) \in \mathbb{R}^h$ encapsulates the uncertainty present in the system such as unmodelled dynamics, parametric uncertainties and external disturbances. It is assumed to be unknown but bounded subject to $\|\xi(t, x, u)\| < \beta$ where the scalar β is known.

From Sect. 4.3, an observer for the system (4.91)–(4.92) is

$$\dot{\hat{x}}(t) = A\hat{x}(t) + Bu(t) - G_l e_y(t) + G_n \nu \tag{4.94}$$

$$\hat{y}(t) = C\hat{x}(t) \tag{4.95}$$

where

$$\nu = -\rho \frac{e_y}{\|e_y\|} \quad \text{if } e_y \neq 0 \tag{4.96}$$

and $e_y(t) := \hat{y}(t) - y(t)$. Defining $e(t) := \hat{x}(t) - x(t)$ as the state estimation error, then from (4.91)–(4.92) and (4.94)–(4.95) the following error system can be obtained

$$\dot{e}(t) = A_o e(t) + G_n \nu - Mf(t) - Q\xi(t, x, u) \tag{4.97}$$

where

$$A_o = A - G_l C \tag{4.98}$$

Assume as in Sect. 4.3 that the conditions A1 and A2 have been satisfied and the triple (A, M, C) has the structure in Lemma 4.2. In this coordinate system, the gain G_n has the structure

$$G_n = \begin{bmatrix} -L \\ I_p \end{bmatrix} T^{\mathrm{T}} P_o^{-1} \tag{4.99}$$

where $L = [L^o \ 0]$, $L^o \in \mathbb{R}^{(n-p)\times(p-q)}$ and $P_o = P_o^{\mathrm{T}}$ are design matrices. The disturbance distribution matrix has the general structure

$$Q = \begin{bmatrix} Q_1 \\ Q_2 \end{bmatrix} \tag{4.100}$$

where $Q_1 \in \mathbb{R}^{(n-p)\times h}$. Suppose there exists a symmetric positive definite matrix $P \in \mathbb{R}^{n\times n}$ with the structure

$$P = \begin{bmatrix} P_1 & P_1 L \\ L^{\mathrm{T}} P_1 & T^{\mathrm{T}} P_o T + L^{\mathrm{T}} P_1 L \end{bmatrix} > 0 \tag{4.101}$$

where $P_1 \in \mathbb{R}^{(n-p)\times(n-p)}$ and $PA_o + A_o^{\mathrm{T}} P < 0$. Define two positive scalars

$$\mu_0 = -\lambda_{\max}\big(PA_o + A_o^{\mathrm{T}} P\big), \qquad \mu_1 = \|PQ\|$$

Proposition 4.2 *If the positive scalar gain function ρ in (4.96) satisfies*

$$\rho \geq \|P_o CM\|\alpha(t,u) + \eta_o \tag{4.102}$$

where η_o is a small positive scalar, then the state estimation error $e(t)$ in (4.97) is ultimately bounded with respect to the set

$$\Omega_\epsilon = \{e : \|e\| < 2\mu_1\beta/\mu_0 + \epsilon\}$$

where $\epsilon > 0$ is an arbitrarily small positive scalar.

Proof Define a Lyapunov function $V(e) = e^{\mathrm{T}}Pe$. The derivative along the estimation error state trajectory is:

$$\begin{aligned}\dot{V} &= e^{\mathrm{T}}(PA_o + A_o^{\mathrm{T}}P)e - 2e^{\mathrm{T}}PMf - 2e^{\mathrm{T}}PQ\xi + 2e^{\mathrm{T}}PG_n\nu \\ &\leq -\mu_0\|e\|^2 + 2\|e\|\mu_1\beta - 2e^{\mathrm{T}}PMf + 2e^{\mathrm{T}}PG_n\nu \end{aligned} \tag{4.103}$$

By using (4.60), inequality (4.103) can be written as

$$\begin{aligned}\dot{V} &\leq -\mu_0\|e\|^2 + 2\|e\|\mu_1\beta - 2e^{\mathrm{T}}C^{\mathrm{T}}P_oCMf + 2e^{\mathrm{T}}C^{\mathrm{T}}\nu \\ &= -\mu_0\|e\|^2 + 2\|e\|\mu_1\beta - 2e_y^{\mathrm{T}}P_oCMf + 2e_y^{\mathrm{T}}\nu\end{aligned}$$

From the Cauchy–Schwarz inequality, and the bound on ρ from (4.102),

$$\begin{aligned}\dot{V} &\leq -\mu_0\|e\|^2 + 2\|e\|\mu_1\beta - 2\|e_y\|(\rho - \|P_oCM\|\alpha(t,u)) \\ &\leq \|e\|(-\mu_0\|e\| + 2\mu_1\beta)\end{aligned} \tag{4.104}$$

which proves that the magnitude of $e(t)$ decreases when $\|e\| > 2\mu_1\beta/\mu_0$. This implies that in finite time $\|e(t)\|$ will be bounded with respect to Ω_ϵ. □

Proposition 4.2 will now be used to prove that for an appropriate choice of ρ, a sliding motion can be induced on the surface $\mathcal{S} = \{e : Ce = 0\}$. It is convenient to first introduce a new change of coordinates associated with the nonsingular matrix

$$T_L = \begin{bmatrix} I_{n-p} & L \\ 0 & T \end{bmatrix} \tag{4.105}$$

Applying the change of coordinates induced by T_L to the triple in Lemma 4.2 yields

$$\mathcal{A} = \begin{bmatrix} \mathcal{A}_{11} & \mathcal{A}_{12} \\ \mathcal{A}_{21} & \mathcal{A}_{22} \end{bmatrix}, \qquad \mathcal{M} = \begin{bmatrix} 0 \\ \mathcal{M}_2 \end{bmatrix}, \qquad \mathcal{C} = [0 \quad I_p] \tag{4.106}$$

where $\mathcal{A}_{11} = A_{11} + L^o A_{211}$ and $\mathcal{M}_2 \in \mathbb{R}^{p\times q}$. Since (A_{11}, A_{211}) is detectable, L^o can be chosen so that $\mathcal{A}_{11}$ is stable. In this coordinate system, the nonlinear gain from (4.99) will have the structure

$$\mathcal{G}_n = \begin{bmatrix} 0 \\ P_o^{-1} \end{bmatrix} \tag{4.107}$$

and the Lyapunov matrix P from (4.101) will have the block diagonal structure

$$\mathcal{P} = (T_L^{-1})^{\mathrm{T}} P T_L^{-1} = \begin{bmatrix} P_1 & 0 \\ 0 & P_o \end{bmatrix} \tag{4.108}$$

The uncertainty/disturbance distribution matrix will have the structure

$$\mathcal{Q} = T_L Q = \begin{bmatrix} \mathcal{Q}_1 \\ \mathcal{Q}_2 \end{bmatrix} = \begin{bmatrix} Q_1 + L Q_2 \\ T Q_2 \end{bmatrix} \tag{4.109}$$

and state estimation error in the new coordinate system is

$$\dot{e}_L(t) = \mathcal{A}_o e_L(t) + \mathcal{G}_n \nu - \mathcal{M} f(t) - \mathcal{Q}\xi(t, x, u) \tag{4.110}$$

where $\mathcal{A}_o = \mathcal{A} - \mathcal{G}_l \mathcal{C}$. Partitioning the state estimation error conformably with (4.106)–(4.109) yields

$$\dot{e}_1(t) = \mathcal{A}_{11} e_1(t) + (\mathcal{A}_{12} - \mathcal{G}_{l,1}) e_y(t) - \mathcal{Q}_1 \xi(t, x, u) \tag{4.111}$$

$$\begin{aligned} \dot{e}_y(t) = {} & \mathcal{A}_{21} e_1(t) + (\mathcal{A}_{22} - \mathcal{G}_{l,2}) e_y(t) + P_o^{-1} \nu - \mathcal{M}_2 f(t) \\ & - \mathcal{Q}_2 \xi(t, x, u) \end{aligned} \tag{4.112}$$

where $\mathcal{G}_{l,1}$ and $\mathcal{G}_{l,2}$ represent appropriate partitions of $\mathcal{G}_l$.

Proposition 4.3 *If the gain function $\rho(.)$ from (4.96) satisfies*

$$\rho \geq 2\|P_o \mathcal{A}_{21}\| \mu_1 \beta / \mu_0 + \|P_o \mathcal{Q}_2\| \beta + \|P_o \mathcal{M}_2\| \alpha(t, u) + \eta_o \tag{4.113}$$

where η_o is a positive scalar, then an ideal sliding motion takes place on the surface $\mathcal{S} = \{e : Ce = 0\}$ in finite time.

Proof Consider a Lyapunov function $V_s(e_y) = e_y^{\mathrm{T}} P_o e_y$. The derivative along the trajectory is

$$\begin{aligned} \dot{V}_s = {} & e_y^{\mathrm{T}} \big(P_o(\mathcal{A}_{22} - \mathcal{G}_{l,2}) + (\mathcal{A}_{22} - \mathcal{G}_{l,2})^{\mathrm{T}} P_o\big) e_y \\ & + 2 e_y^{\mathrm{T}} P_o \big(\mathcal{A}_{21} e_1 - \mathcal{M}_2 f - \mathcal{Q}_2 \xi + P_o^{-1} \nu\big) \end{aligned}$$

Using (4.67)–(4.69), the term $P_o(\mathcal{A}_{22} - \mathcal{G}_{l,2}) + (\mathcal{A}_{22} - \mathcal{G}_{l,2})^{\mathrm{T}} P_o < 0$ because $\mathcal{P}$ from (4.108) is a block diagonal Lyapunov matrix for $(\mathcal{A} - \mathcal{G}_l \mathcal{C})$. Therefore it follows:

$$\begin{aligned} \dot{V}_s & \leq 2 e_y^{\mathrm{T}} P_o (\mathcal{A}_{21} e_1 - \mathcal{M}_2 f - \mathcal{Q}_2 \xi) - 2\rho \|e_y\| \\ & \leq -2\|e_y\| \big(\rho - \|P_o \mathcal{A}_{21}\| \|e_1\| - \|P_o \mathcal{M}_2\| \alpha(t, u) - \|P_o \mathcal{Q}_2\| \beta\big) \end{aligned}$$

From Lemma 4.2, in finite time $e(t) \in \Omega_\epsilon$ which implies $\|e_1\| < 2\mu_1 \beta / \mu_0 + \epsilon$. Therefore from the definition of ρ in (4.113), and using the inequality (4.71), it follows that

$$\dot{V}_s \leq -2\eta_o \|e_y\| \leq -2\eta_o \eta \sqrt{V_s} \tag{4.114}$$

where $\eta := \sqrt{\lambda_{\min}(P_o^{-1})}$. Using the proof of Corollary 4.1, this proves that the output estimation error $e_y(t)$ will reach zero in finite time, and a sliding motion takes place. □

Remark 4.3 Since $\mathcal{M}_2 = CM$, it follows that the definition of ρ in (4.113) is consistent with the assumption on its size in (4.102).

4.5.1 Robust Actuator Fault Reconstruction

In this section the sliding mode observer described in (4.94)–(4.96) will be analysed with regard to its ability to robustly reconstruct the fault $f(t)$ despite the presence of the uncertainty $\xi(t, x, u)$.

Assume that the sliding mode observer described in (4.94)–(4.96) has been designed, and that a sliding motion has been achieved, then the output estimation error satisfies $e_y(t) = \dot{e}_y(t) = 0$ and (4.111) and (4.112) will become

$$\dot{e}_1(t) = \mathcal{A}_{11} e_1(t) - \mathcal{Q}_1 \xi(t, x, u) \tag{4.115}$$

$$0 = \mathcal{A}_{21} e_1(t) + P_o^{-1} \nu_{eq} - \mathcal{M}_2 f(t) - \mathcal{Q}_2 \xi(t, x, u) \tag{4.116}$$

where ν_{eq} is the equivalent output error injection required to maintain a sliding motion. In (4.115) and (4.116), the vector $\xi(t, x, u)$ will be treated as an unknown exogenous signal.

In the case when $\xi(t, x, u) \neq 0$, the attempted reconstruction of $f(t)$ from the equivalent injection signal will be corrupted by the exogenous signal $\xi(t, x, u)$. The objective here is to choose a scaling of the equivalent output error injection signal ν_{eq} and the gain L^o, to minimise the effect of the exogenous signal on the fault reconstruction. To this end define

$$W_{sc} := \begin{bmatrix} W_1 & M_o^{-1} \end{bmatrix} \tag{4.117}$$

where $W_1 \in \mathbb{R}^{q \times (p-q)}$ and represents design freedom and M_o is given in Lemma 4.2. Define a would-be reconstruction signal for $f(t)$ as

$$\hat{f} := W_{sc} T^{\mathrm{T}} P_o^{-1} \nu_{eq} \tag{4.118}$$

Rewriting (4.115) and (4.116) in terms of the coordinates of (4.51) in Lemma 4.2, and re-arranging yields

$$\dot{e}_1(t) = (A_{11} + L A_{21}) e_1(t) - (Q_1 + L Q_2) \xi(t, x, u) \tag{4.119}$$

$$P_o^{-1} \nu_{eq} = -T A_{21} e_1(t) + T M_2 f(t) + T Q_2 \xi(t, x, u) \tag{4.120}$$

Pre-multiplying (4.120) by $W_{sc} T^{\mathrm{T}}$ implies

$$\hat{f}(t) = -W_{sc} A_{21} e_1(t) + W_{sc} Q_2 \xi(t, x, u) + f(t) \tag{4.121}$$

and therefore

$$\hat{f}(t) = f(t) + \hat{G}(\mathrm{s})\xi(t,x,u) \tag{4.122}$$

where from (4.119) and (4.121) the transfer function matrix

$$\hat{G}(\mathrm{s}) = W_{sc}A_{21}\big(\mathrm{s}I_{n-p} - (A_{11} + LA_{21})\big)^{-1}(Q_1 + LQ_2) + W_{sc}Q_2 \tag{4.123}$$

Remark 4.4 For square systems (when $p = q$), it can be seen that L^o from (4.99) and W_1 from (4.117) do not exist (and $A_{21} = A_{212}$). Hence $\hat{G}(\mathrm{s})$ in (4.123) reduces to become $\hat{G}(\mathrm{s}) = M_o^{-1}A_{21}(\mathrm{s}I_{n-p} - A_{11})^{-1}Q_1 + M_o^{-1}Q_2$, which is made up of constant matrices. Consequently there is no design freedom to specify the effect of ξ on $\hat{f}$.

The objective now is to minimise the effect of ξ on the reconstruction $\hat{f}$, by choice of L^o and W_1. Using the Bounded Real Lemma [55, 104], the $\mathcal{L}_2$ gain of the transfer function $\hat{G}(\mathrm{s})$ from the exogenous signal ξ to $\hat{f}$ will not exceed $\gamma \in \mathbb{R}_+$ if the following inequality holds

$$\begin{bmatrix} \hat{P}(A_{11} + LA_{21}) + (A_{11} + LA_{21})^{\mathrm{T}}\hat{P} & -\hat{P}(Q_1 + LQ_2) & -(W_{sc}A_{21})^{\mathrm{T}} \\ -(Q_1 + LQ_2)^{\mathrm{T}}\hat{P} & -\gamma I_h & (W_{sc}Q_2)^{\mathrm{T}} \\ -W_{sc}A_{21} & W_{sc}Q_2 & -\gamma I_q \end{bmatrix} < 0 \tag{4.124}$$

where $\hat{P} \in \mathbb{R}^{(n-p)\times(n-p)}$ is symmetric positive definite. The objective is therefore to find $\hat{P}, L$ and W_{sc} to minimise γ subject to the inequality (4.124) and $\hat{P} > 0$. However, this must be done in conjunction with satisfying the requirements of obtaining a suitable sliding mode observer as expressed in Propositions 4.2 and 4.3: i.e., to find a value of G_l and P that has the structure in (4.101) such that $P(A - G_lC) + (A - G_lC)^{\mathrm{T}}P < 0$.

Theorem 4.1 *Define $D_1 \in \mathbb{R}^{p\times p}$, $\gamma_o \in \mathbb{R}_+$ to be user-defined quantities. Also define the following matrices that have $p + h$ columns*

$$B_d = [0 \quad Q], \qquad D_d = [D_1 \quad 0], \qquad H = [0 \quad W_{sc}Q_2] \tag{4.125}$$

Suppose there exists a Lyapunov matrix $P \in \mathbb{R}^{n\times n}$ with the structure

$$P = \begin{bmatrix} P_{11} & P_{12} \\ P_{12}^{\mathrm{T}} & P_{22} \end{bmatrix} > 0, \qquad P_{12} = [P_{121} \quad 0], \qquad P_{121} \in \mathbb{R}^{(n-p)\times(p-q)} \tag{4.126}$$

and W_1, E_2 and γ that satisfy the following inequalities:

$$\begin{bmatrix} PA + A^{\mathrm{T}}P - \gamma_o C^{\mathrm{T}}(D_d D_d^{\mathrm{T}})^{-1}C & -PB_d & E^{\mathrm{T}} \\ -B_d^{\mathrm{T}}P & -\gamma_o I_{p+h} & H^{\mathrm{T}} \\ E & H & -\gamma_o I_q \end{bmatrix} < 0 \tag{4.127}$$

$$\begin{bmatrix} P_{11}A_{11} + A_{11}^{\mathrm{T}}P_{11} + P_{12}A_{21} + A_{21}^{\mathrm{T}}P_{12}^{\mathrm{T}} & \star & \star \\ -(P_{11}Q_1 + P_{12}Q_2)^{\mathrm{T}} & -\gamma I_h & \star \\ -W_{sc}A_{21} & W_{sc}Q_2 & -\gamma I_q \end{bmatrix} < 0 \tag{4.128}$$

where the $\star$ *are terms that will make (4.128) symmetric and*

$$E = [-W_{sc}A_{21} \quad E_2] \tag{4.129}$$

If the observer gains are chosen as

$$L^o = P_{11}^{-1}P_{121} \tag{4.130}$$

$$G_l = \gamma_o P^{-1}C^{\mathrm{T}}\big(D_d D_d^{\mathrm{T}}\big)^{-1} \tag{4.131}$$

then $P(A - G_lC) + (A - G_lC)^{\mathrm{T}}P < 0$ *is satisfied and* $\|\hat{G}(s)\| < \gamma$.

Proof By using the Schur complement, inequality (4.127) can be written as

$$\begin{bmatrix} PA + A^{\mathrm{T}}P - \gamma_o C^{\mathrm{T}}(D_d D_d^{\mathrm{T}})^{-1}C & E^{\mathrm{T}} \\ E & -\gamma_o I_q \end{bmatrix} + \frac{1}{\gamma_o}\begin{bmatrix} -PB_d \\ H \end{bmatrix}\begin{bmatrix} -B_d^{\mathrm{T}}P & H^{\mathrm{T}} \end{bmatrix} < 0$$

which implies

$$\begin{bmatrix} PA + A^{\mathrm{T}}P - \gamma_o C^{\mathrm{T}}(D_d D_d^{\mathrm{T}})^{-1}C + \frac{1}{\gamma_o}PB_dB_d^{\mathrm{T}}P & E^{\mathrm{T}} - \frac{1}{\gamma_o}PB_dH^{\mathrm{T}} \\ E - \frac{1}{\gamma_o}HB_d^{\mathrm{T}}P & -\gamma_o I_q + \frac{1}{\gamma_o}HH^{\mathrm{T}} \end{bmatrix} < 0 \tag{4.132}$$

From (4.125), it is obvious that $B_d D_d^{\mathrm{T}} = 0$. By using

$$G_l = \gamma_o P^{-1}C^{\mathrm{T}}\big(D_d D_d^{\mathrm{T}}\big)^{-1}$$

it can be shown that

$$\Psi - PG_lC - (PG_lC)^{\mathrm{T}} = -\gamma_o C^{\mathrm{T}}\big(D_d D_d^{\mathrm{T}}\big)^{-1}C + \frac{1}{\gamma_o}PB_dB_d^{\mathrm{T}}P \tag{4.133}$$

where $\Psi := \frac{1}{\gamma_o}P(G_lD_dD_d^{\mathrm{T}}G_l^{\mathrm{T}} + B_dB_d^{\mathrm{T}})P$. Substituting (4.133) into the top left submatrix of (4.132) yields

$$\begin{bmatrix} P(A - G_lC) + (A - G_lC)^{\mathrm{T}}P + \Psi & E^{\mathrm{T}} - \frac{1}{\gamma_o}PB_dH^{\mathrm{T}} \\ E - \frac{1}{\gamma_o}HB_d^{\mathrm{T}}P & -\gamma_o I_q + \frac{1}{\gamma_o}HH^{\mathrm{T}} \end{bmatrix} < 0 \tag{4.134}$$

Since from (4.125) $D_dH^{\mathrm{T}} = 0$, it is clear that

$$\frac{1}{\gamma_o}\begin{bmatrix} P(G_lD_d - B_d) \\ H \end{bmatrix}\begin{bmatrix} (G_lD_d - B_d)^{\mathrm{T}}P & H^{\mathrm{T}} \end{bmatrix} = \begin{bmatrix} \Psi & -\frac{1}{\gamma_o}PB_dH^{\mathrm{T}} \\ -\frac{1}{\gamma_o}HB_d^{\mathrm{T}}P & \frac{1}{\gamma_o}HH^{\mathrm{T}} \end{bmatrix} \tag{4.135}$$

Substituting (4.135) into (4.134) and using the Schur complement results in

$$\begin{bmatrix} P(A-G_lC)+(A-G_lC)^{\mathrm{T}}P & E^{\mathrm{T}} & P(G_lD_d-B_d) \\ E & -\gamma_o I_q & H \\ (G_lD_d-B_d)^{\mathrm{T}}P & H^{\mathrm{T}} & -\gamma_o I_{q+h} \end{bmatrix} < 0 \tag{4.136}$$

Since (4.136) is symmetric and $P(A-G_lC)+(A-G_l)^{\mathrm{T}}P<0$, the first part of the theorem is proven.

Then, by substituting $P_{12}=P_{11}L$ it is clear that (4.128) will be in the same form as (4.124) by equating $\hat{P}=P_{11}$. Recalling that (4.124) describes the Bounded Real Lemma for $\hat{G}(s)$, the second part of the theorem is proven. □

By comparing P in (4.126) with (4.101), it can be seen that there is a one-to-one correspondence between the variables $(P_{11}, P_{121}, P_{22})$ and (P_1, L^o, P_o) since

$$P_1 = P_{11} \tag{4.137}$$

$$L^o = P_{11}^{-1}P_{121} \tag{4.138}$$

$$P_o = T\big(P_{22}-P_{12}^{\mathrm{T}}P_{11}^{-1}P_{12}\big)T^{\mathrm{T}} \tag{4.139}$$

Hence, the observer design can be summarised as:

> Minimise γ with respect to the variables $P_{11}, P_{121}, P_{22}, W_1$ and E_2 subject to (4.126), (4.127) and (4.128) where the scalar γ_o and D_1 are a priori user defined.

Then calculate $G_l=\gamma_o P^{-1}C^{\mathrm{T}}(D_dD_d^{\mathrm{T}})^{-1}$, $L=P_{11}^{-1}P_{12}$, P_o from (4.139) and G_n from (4.99).

Remark 4.5 It should be noted that the design method above is not unique; the only condition that needs to be satisfied is $P(A-G_lC)+(A-G_lC)^{\mathrm{T}}P<0$ where P has the structure in (4.101). The motivation for the observer design method above will be discussed in Remarks 4.6 and 4.7 below. The effect of the user-defined parameters γ_o and D_1 on the observer design will also be discussed.

Remark 4.6 Define

$$PA+A^{\mathrm{T}}P=\begin{bmatrix} X_{11} & X_{12} \\ X_{12}^{\mathrm{T}} & X_{22} \end{bmatrix}$$

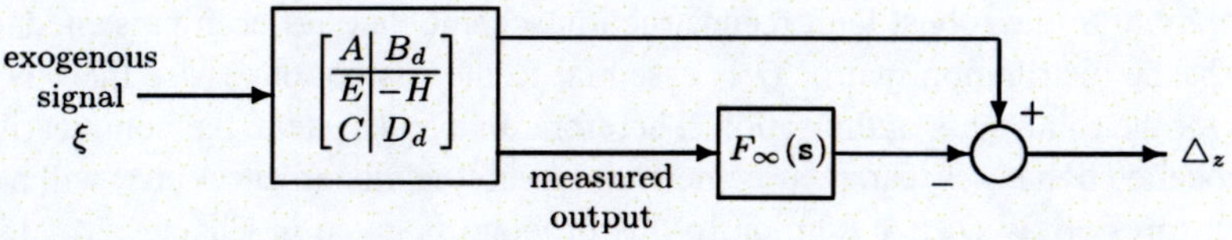

Fig. 4.14 The $\mathcal{H}_\infty$ filtering problem (notation taken from [297])

where $X_{11} \in \mathbb{R}^{(n-p)\times(n-p)}$. Hence $X_{11} = P_{11}A_{11} + A_{11}^{\mathrm{T}}P_{11} + P_{12}A_{21} + A_{21}^{\mathrm{T}}P_{12}^{\mathrm{T}}$ and using (4.129), inequality (4.127) can be written as

$$\begin{bmatrix} X_{11} & X_{12} & 0 & -(P_{11}Q_1 + P_{12}Q_2) & -(W_{sc}A_{21})^{\mathrm{T}} \\ X_{12}^{\mathrm{T}} & \star & 0 & \star & E_2^{\mathrm{T}} \\ 0 & 0 & -\gamma_o I_p & 0 & 0 \\ -(P_{11}Q_1 + P_{12}Q_2)^{\mathrm{T}} & \star & 0 & -\gamma_o I_h & (W_{sc}Q_2)^{\mathrm{T}} \\ -W_{sc}A_{21} & E_2 & 0 & W_{sc}Q_2 & -\gamma_o I_q \end{bmatrix} < 0$$

where the $\star$ are elements that make the matrix symmetric. A necessary condition for inequality (4.127) to hold is that

$$\begin{bmatrix} X_{11} & -(P_{11}Q_1 + P_{12}Q_2) & -(W_{sc}A_{21})^{\mathrm{T}} \\ -(P_{11}Q_1 + P_{12}Q_2)^{\mathrm{T}} & -\gamma_o I_h & (W_{sc}Q_2)^{\mathrm{T}} \\ -W_{sc}A_{21} & W_{sc}Q_2 & -\gamma_o I_q \end{bmatrix} < 0 \qquad (4.140)$$

which has the same structure as the Bounded Real Lemma (4.128). Let $\gamma_{\min}$ be the minimum value of γ that satisfies (4.128). Then, since (4.128) is a 'sub-block' of (4.127), $\gamma_{\min} \le \gamma_o$ always holds. Hence, the designer can effectively set the upper bound of γ by setting γ_o.

Remark 4.7 If the designer sets a value of γ_o to be small, it restricts the feasibility of (4.127) and hence typically $\gamma_{\min}$ will become large. However, from (4.131), a larger γ_o will increase the 'size' of G_l which may be undesirable as it will amplify noise. A matrix D_1 which is 'smaller' will improve the feasibility of (4.127) which will reduce $\gamma_{\min}$, but from (4.131) will increase the 'size' of G_l.

For a given B_d, D_d, E and H, inequality (4.136) can be viewed as resulting from an $\mathcal{H}_\infty$ filtering problem (page 462 of Zhou et al. [297]), the idea being to minimise the effect of ξ on Δ_z (see Fig. 4.14). However, in this chapter, E and H are regarded as design variables (which in particular depend on W_{sc} from (4.117)) and help determine a value for L from (4.99) which defines the optimal choice of sliding mode for fault reconstruction purposes. Once a sliding mode is established, the choice of the linear gain G_l is technically not relevant since the linear output error injection term $G_l e_y$ disappears because $e_y \equiv 0$.

Remark 4.8 In the robust fault reconstruction scheme that has been presented, the disturbance distribution matrix Q is essential to the design, otherwise there is no basis for the robustness optimisation. Therefore, in a real system, the nonlinearities (unmodelled dynamics, parametric uncertainty, external disturbances etc.) will need to be expressed as $Q\xi$. A solution to this problem is given in Chapter 5 of [51]. Examples of this approach appear in later chapters of this book.

4.5.2 Example: VTOL Aircraft Model

The robust fault reconstruction scheme described in this chapter will now be demonstrated with a 'Vertical Take-Off and Landing' (VTOL) aircraft model taken from [219, 276]. The states are horizontal velocity (kts), vertical velocity (kts), pitch rate (deg/s) and pitch angle (deg). The outputs are horizontal velocity (kts), vertical velocity (kts) and pitch angle (deg) while the inputs are collective pitch control and longitudinal cyclic pitch control.

The system is modelled as

$$\dot{x}(t) = (A + \Delta A)x(t) + (B + \Delta B)u(t) + Mf(t) \tag{4.141}$$

$$y(t) = Cx(t) \tag{4.142}$$

where

$$A = \begin{bmatrix} -9.9477 & -0.7476 & 0.2632 & 5.0337 \\ 52.1659 & 2.7452 & 5.5532 & -24.4221 \\ 26.0922 & 2.6361 & -4.1975 & -19.2774 \\ 0 & 0 & 1.0000 & 0 \end{bmatrix},$$

$$B = \begin{bmatrix} 0.4422 & 0.1761 \\ 3.5446 & -7.5922 \\ -5.5200 & 4.4900 \\ 0 & 0 \end{bmatrix}$$

$$C = \begin{bmatrix} 1 & 0 & 0 & 0 \\ 0 & 1 & 0 & 0 \\ 0 & 0 & 0 & 1 \end{bmatrix}, \qquad M = \begin{bmatrix} 0.4422 \\ 3.5446 \\ -5.5200 \\ 0 \end{bmatrix}$$

The uncertain matrices are

$$\Delta A = \begin{bmatrix} 0 & 0 & 0 & 0 \\ 0 & 0 & 0 & 0 \\ 0 & \Delta a_{32} & 0 & \Delta a_{34} \\ 0 & 0 & 0 & 0 \end{bmatrix}, \qquad \Delta B = \begin{bmatrix} 0 & 0 \\ \Delta b_{21} & 0 \\ 0 & 0 \\ 0 & 0 \end{bmatrix}$$

Hence in the notation of (4.91)–(4.92)

$$Q = \begin{bmatrix} 0 & 0 \\ 0 & 1 \\ 1 & 0 \\ 0 & 0 \end{bmatrix}, \qquad \xi = \begin{bmatrix} \begin{bmatrix} 0 \ \Delta a_{32} \ 0 \ \Delta a_{34} \end{bmatrix} x \\ \begin{bmatrix} \Delta b_{21} \ 0 \end{bmatrix} u \end{bmatrix}$$

where $\Delta a_{32} = 0.5$, $\Delta a_{34} = 2$, $\Delta b_{21} = 2$ [219]. Then, a suitable value for T_o is

$$T_o = \begin{bmatrix} -0.1913 & -1.5334 & -1 & 0 \\ 0 & 0 & 0 & 1 \\ -0.9923 & 0.1238 & 0 & 0 \\ -0.1238 & -0.9923 & 0 & 0 \end{bmatrix}$$

The sliding mode observer was designed using the method presented in Sect. 4.5.1. The design parameters were chosen as $D_1 = I_3$ and $\gamma_o = 1$. The optimisation routine gives a value of $\gamma = 8.1968 \times 10^{-4}$. The associated gains for the sliding mode observer are

$$G_l = G_n = \begin{bmatrix} 0.1914 & 2.0408 & 0.5413 \\ 2.0408 & 46.4792 & 9.2975 \\ 4.5304 & 98.5651 & 20.3670 \\ 0.5413 & 9.2975 & 2.0644 \end{bmatrix}$$

and the matrix W_{sc} from (4.117) associated with the scaling of ν_{eq}, is given by

$$W_{sc} = \begin{bmatrix} -0.5952 & -2.2440 & -0.2799 \end{bmatrix}$$

The scalar function ρ from (4.96) was chosen to be 100, and $\delta = 1 \times 10^{-5}$ was used in the smoothing of the unit vector.

Fig. 4.15 shows the sliding mode observer faithfully reconstructing the actuator fault, rejecting the effect of the uncertainty. This is to be expected, due to the small value of γ obtained.

4.6 Observer Variation

The following short section presents a variation of the observer given in (4.94)–(4.96) in Sect. 4.5. Consider an observer of the form

$$\dot{\hat{x}}(t) = A\hat{x}(t) + Bu(t) - G_l e_y(t) + G_n \nu \tag{4.143}$$

$$\hat{y}(t) = C\hat{x}(t) \tag{4.144}$$

with

$$\nu = -\rho \frac{P_o e_y}{\|P_o e_y\|} \quad \text{if } e_y(t) \neq 0 \tag{4.145}$$

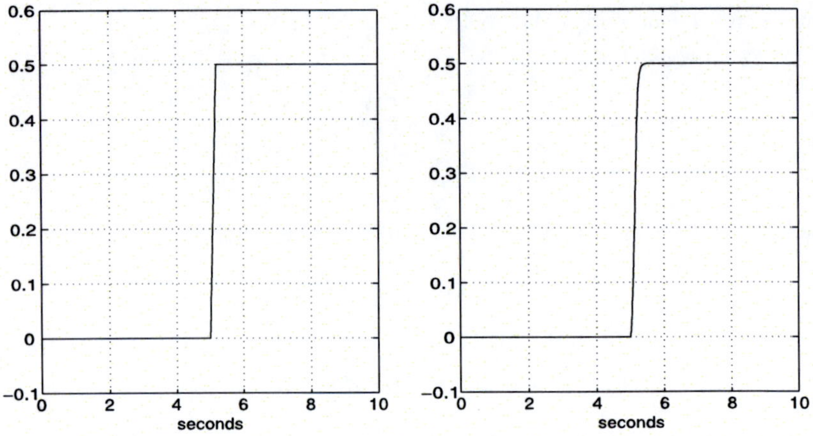

Fig. 4.15 The fault on the first actuator (*left*) and its reconstruction (*right*)

where $e_y(t) := \hat{y}(t) - y(t)$ is the output estimation error. In (4.143) the observer gain

$$G_n = \begin{bmatrix} -LT^{\mathrm{T}} \\ T^{\mathrm{T}} \end{bmatrix} \tag{4.146}$$

where $L = [L^o \ 0]$, $L^o \in \mathbb{R}^{(n-p)\times(p-q)}$ and $P_o = P_o^{\mathrm{T}}$ are design matrices and T is defined in Lemma 4.2. Now the reconstruction signal for $f(t)$ is

$$\hat{f} := W_{sc} T^{\mathrm{T}} \nu_{eq} \tag{4.147}$$

The main difference is in the definition of ν where in (4.145), the unit vector output error injection term is constructed by scaling the output error with a symmetric positive definite matrix P_o. This difference translates to a difference in the definition of G_n (4.146) and $\hat{f}$ in (4.147) compared to (4.99) and (4.118). The rest of the analysis remains unchanged. All the results from Sect. 4.3 onwards can be obtained by suitable modifications of the arguments.

4.7 Comparisons with UIOs

In this section a linear fault reconstruction scheme proposed in [219] using Unknown Input Observers (UIOs) will be described. The approach also seeks to estimate fault signals $f(t)$ in systems of the form (4.91)–(4.92). The work in [219] uses a minimal order (Luenberger) observer of the form

$$\dot{w}(t) = \hat{F}w(t) + \hat{E}y(t) + \hat{L}u(t) \tag{4.148}$$

$$\hat{x}(t) = \hat{R}w(t) + \hat{T}y(t) \tag{4.149}$$

where $\hat{F} \in \mathbb{R}^{(n-p)\times(n-p)}$ is stable. The quantity $w \in \mathbb{R}^{(n-p)}$ is the state of the observer and $\hat{x}$ is an estimate of the unknown state in (4.91). In terms of the UIO literature, the work of Saif and Guan [219] has been chosen because the problem it tackles is the closest to the one associated with the sliding mode schemes discussed earlier i.e., *fault reconstruction* rather than residual generation, and taken at face value, the assumptions required in [219] appear similar to those from Sect. 4.5.

For the uncertain system in (4.91)–(4.92), under the assumption $q + h \leq p$, the following is true:

Proposition 4.4 ([61, 219]) *A minimal order observer of the form* (4.148)–(4.149) *exists for* (4.91) *such that the estimation error* $\hat{e} := \hat{x} - x$ *is independent of* $f(\cdot)$ *and* $\xi(\cdot)$ *if and only if*

(UIO1) *the matrix* $C\bar{Q}$ *has rank* $q + h$ *(i.e., full rank) where*

$$\bar{Q} := [M \quad Q] \tag{4.150}$$

(UIO2) *no invariant zeros of* $(A, \bar{Q}, C)$ *are in* $\mathbb{C}_+$.

Furthermore $\hat{e} \to 0$ *as* $t \to \infty$ *and* $\hat{x} \to x$ *as* $t \to \infty$ *where* $\hat{x}$ *is defined in* (4.149).

Under assumption UIO1, using the results from Lemma 4.2 the system can be written as

$$A = \begin{bmatrix} A_{11} & A_{12} \\ A_{21} & A_{22} \end{bmatrix}, \qquad \bar{Q} = \begin{bmatrix} 0 \\ Q_o \end{bmatrix}, \qquad C = [0 \quad T] \tag{4.151}$$

where $A_{11} \in \mathbb{R}^{(n-p)\times(n-p)}$ and $Q_o \in \mathbb{R}^{(q+h)\times(q+h)}$ is nonsingular and $T \in \mathbb{R}^{p\times p}$ is orthogonal. The observer (4.148)–(4.149) can then be explicitly parameterised as

$$\hat{F} = A_{11} - LA_{21} \tag{4.152}$$

$$\hat{E} = \big(A_{12} - LA_{22} + (A_{11} - LA_{21})L\big)T^{\mathrm{T}} \tag{4.153}$$

$$\hat{L} = B_1 - LB_2 \tag{4.154}$$

$$\hat{R}^{\mathrm{T}} = \big[I_{(n-p)} \quad 0\big] \tag{4.155}$$

$$\hat{T}^{\mathrm{T}} = T\big[L^{\mathrm{T}} \quad I_p\big] \tag{4.156}$$

where B_1 and B_2 represent a partition of the input distribution matrix B commensurate with (4.151). The matrix L has the structure

$$L = \big[L_o \quad 0\big] \tag{4.157}$$

where $L_o \in \mathbb{R}^{(n-p)\times(p-h-q)}$. Thus the canonical form from (4.151), inherited from Lemma 4.2, is unifying from a design standpoint.

To estimate $f(t)$, Saif and Guan propose using a discrete version of the uncertain continuous time system equation (4.91) written as

$$\dot{x}(t) = Ax(t) + Bu(t) + \bar{Q}\xi_a(t, x, u) \tag{4.158}$$

where $\xi_a = \text{col}(f, \xi)$. The discrete time version

$$x(k+1) = A_d x(k) + B_d u(k) + Q_d \xi_a(k) \tag{4.159}$$

is then manipulated to yield an estimate

$$\left(\hat{f}(k), \hat{\xi}(k)\right)^{\mathrm{T}} = \left(Q_d^{\mathrm{T}} Q_d\right)^{-1} Q_d^{\mathrm{T}} \left(\hat{x}(k+1) - A_d \hat{x}(k) - B_d u(k)\right) \tag{4.160}$$

where $\hat{x}$ is the estimate of the state x from the reduced order observer.

Remark 4.9 The work in Saif and Guan [219] and Darouach [61] considers reduced order UIOs. Full order UIOs are considered in [50, 52, 62]. The formulation associated with Propositions 4.4 could equally well be posed as full order UIOs and solved using the design methodologies of [50, 52]. The two constraints developed in the work described in [50, 52] can be shown to be equivalent to UIO1 and UIO2 as follows: the rank condition in [50, 52] is identical to UIO1; the second condition in [50, 52] is that the pair (A_1, C) is observable where

$$A_1 := A - \bar{Q}\left((C\bar{Q})^{\mathrm{T}} C\bar{Q}\right)^{-1} (C\bar{Q})^{\mathrm{T}} CA \tag{4.161}$$

This is equivalent to the pair (A_{11}, A_{211}) being observable, since after some algebra, in the coordinates of (4.46), it can be shown that

$$A_1 = \begin{bmatrix} A_{11} & A_{12} \\ A_{211} & A_{221} \\ 0 & 0 \end{bmatrix}$$

where A_{221} represents the top $p - q - h$ rows of A_{22}. From the PBH rank test, the pair (A_1, C) is detectable if and only if for all $s \in \mathbb{C}_+$

$$\text{rank} \begin{bmatrix} sI - A_1 \\ C \end{bmatrix} = n \quad \Leftrightarrow \quad \text{rank} \begin{bmatrix} sI - A_{11} & -A_{12} \\ -A_{211} & A_{221} \\ 0 & T \end{bmatrix} = n$$

$$\Leftrightarrow \quad \text{rank} \begin{bmatrix} sI - A_{11} \\ -A_{211} \end{bmatrix} = n - p$$

i.e., if and only if the pair (A_{11}, A_{211}) is detectable. Thus the second condition in [50, 52] is given in terms of the detectability of (A_{11}, A_{211}) rather than a system theoretic interpretation in terms of invariant zeros, but the overall requirements are the same.

The UIO of [219] is powerful since it provides perfect reconstructions and is completely insensitive to the uncertainty. However, the structural requirements UIO1 and UIO2 on the system are stronger than the ones for the sliding mode observer (requirements A1 and A2) in Sect. 4.3. The following proposition shows that if the requirements of [219] are met, i.e., UIO1 and UIO2 are satisfied, then the approach in Sect. 4.5 yields an arbitrarily small γ and hence provides complete insensitivity to the uncertainty too.

Proposition 4.5 *Consider the design of a sliding mode observer of the form given in (4.94), (4.96) and (4.99) with a reconstruction gain (4.117) where $p > q + h$. Assume the matrix $\bar{Q} = [M \quad Q]$ obtained by aggregating the distribution matrices from (4.91) is such that conditions UIO1 and UIO2 are satisfied (i.e., the conditions for a Saif and Guan unknown input observer). Then the design freedom W_1 and L from (4.117) and (4.99) can be chosen so that $\|\hat{G}(s)\|_\infty = 0$ where $\hat{G}(s)$ is defined in (4.123) and hence the reconstruction $\hat{f}(t)$ from (4.122) is independent of the uncertainty $\xi(\cdot)$.*

Proof Since by assumption conditions UIO1 and UIO2 hold for the triple $(A, \bar{Q}, C)$, using the results from Sect. 4.3, there exists a change of coordinates such that

$$A = \begin{bmatrix} A_{11} & A_{12} \\ A_{21} & A_{22} \end{bmatrix}, \qquad \bar{Q} = \begin{bmatrix} 0 \\ Q_o \end{bmatrix}, \qquad C = [0 \quad T] \tag{4.162}$$

where $A_{11} \in \mathbb{R}^{(n-p)\times(n-p)}$, $Q_o \in \mathbb{R}^{(q+h)\times(q+h)}$ is nonsingular and $T \in \mathbb{R}^{p\times p}$ is orthogonal. Furthermore, following the arguments in Sect. 4.3 the matrix T can be chosen so that

$$Q_o = \begin{bmatrix} 0 & Q_{12} \\ Q_{21} & Q_{22} \end{bmatrix} \tag{4.163}$$

where $Q_{21} \in \mathbb{R}^{q\times q}$ and $Q_{12} \in \mathbb{R}^{h\times h}$ satisfy $\det(Q_{21}) \neq 0$ and $\det(Q_{12}) \neq 0$. One way to see this is that, as argued in Sect. 4.3, the matrix T can be obtained by 'QR reduction' of the matrix $C\bar{Q}$, which naturally produces a lower triangular structure for Q_o. Since the $\operatorname{rank}(C\bar{Q}) = h + q$ the form for Q_o in (4.163) follows immediately.

If A_{211} represents the top $p-q-h$ rows of A_{21} in (4.162), then UIO2 guarantees that the pair (A_{11}, A_{211}) is detectable. Again this follows from arguments similar to those in Sect. 4.3.

Partitioning $\bar{Q}$ to recover M and Q, in the notation of Sect. 4.5, it follows

$$M = \begin{bmatrix} 0_{(n-q)\times q} \\ Q_{21} \end{bmatrix} \quad \text{and} \quad Q = \begin{bmatrix} 0_{(n-q-h)\times h} \\ Q_{12} \\ Q_{22} \end{bmatrix}$$

From the partitions of M in (4.46) and Q in (4.123), it follows that in the notation of Sect. 4.5, $M_o = Q_{21}$, $Q_1 = 0$ and

$$Q_2 = \begin{bmatrix} 0_{(p-h-q)\times h} \\ Q_{12} \\ Q_{22} \end{bmatrix} \tag{4.164}$$

Also the gain L from (4.99) has the specific form

$$L = [L_o \quad 0] \tag{4.165}$$

where $L_o \in \mathbb{R}^{(n-p)\times(p-q-h)}$. It follows by direct evaluation that

$$Q_1 + LQ_2 = 0 \tag{4.166}$$

for all L_o. Also from UIO2 it follows L_o can always be chosen so that the matrix $A_{11} + L_o A_{211}$ is stable since (A_{11}, A_{211}) is detectable. Write the gain matrix W_{sc} from (4.117) as

$$W_{sc} = \begin{bmatrix} W_{11} & W_{12} & Q_{21}^{-1} \end{bmatrix}, \quad W_{11} \in \mathbb{R}^{q\times(p-q-h)}$$

where W_{11} and W_{12} represent design freedom and the term Q_{21}^{-1} appears since $M_o = Q_{21}$. It follows from (4.164) by direct evaluation that

$$W_{sc} Q_2 = W_{12} Q_{12} + Q_{21}^{-1} Q_{22}$$

and consequently, since $\det(Q_{12}) \neq 0$, choosing $W_{12} = -Q_{21}^{-1} Q_{22} Q_{12}^{-1}$ ensures

$$W_{sc} Q_2 = 0 \tag{4.167}$$

The gain L_o from (4.165) must be selected so that $A_{11} + L_o A_{211}$ is stable, since it can be shown that the reduced order sliding motion associated with the $n - p$ states in the $\mathcal{N}(C)$ are governed by the matrix $A_{11} + L_o A_{211}$ (see Sect. 4.5), which therefore must be stable to ensure that a sliding motion can be maintained. It follows from (4.166) and (4.167) that the transfer function matrix from (4.123) has the property that $\|\hat{G}(s)\|_\infty = 0$ and the disturbance/uncertainty $\xi(\cdot)$ is decoupled from the estimate $\hat{f}_i$ as claimed. □

Remark 4.10 Consider the example from [52] which represents a well stirred chemical reactor. The states represent the concentration of the chemical product; the temperature of the product; the temperature of the water in the heat exchanger; and the coolant temperature. The inputs are the inlet concentration of reactants; the inlet temperature; and the coolant temperature. In a time scale of hours, the system matrices from [52], in the notation of this section, are given by

$$A = \begin{bmatrix} -3.6000 & 0 & 0 & 0 \\ 0 & -3.6702 & 0 & 0.0702 \\ 0 & 0 & -36.2588 & 0.2588 \\ 0 & 0.6344 & 0.7781 & -1.4125 \end{bmatrix}, \qquad Q = \begin{bmatrix} 1.000 \\ 20.758 \\ 0 \\ 0 \end{bmatrix}$$

$$C^{\mathrm{T}} = B = \begin{bmatrix} 1 & 0 & 0 \\ 0 & 1 & 0 \\ 0 & 0 & 1 \\ 0 & 0 & 0 \end{bmatrix}$$

As in the second UIO design problem from Section 5.2.2 in [52], let the fault distribution matrix M be the third column of the input distribution matrix (which represents a fault in the coolant system). As in Proposition 4.5 let $\bar{Q} = [M \;\; Q]$. Since this example is used to demonstrate the UIO design procedure in [52], as argued in Remark 4.9, the conditions UIO1 and UIO2 must be satisfied, and the results of Proposition 4.5 are valid. In the canonical form of (4.46) the triple can be written as

$$A = \left[\begin{array}{c|ccc} -1.4125 & 0.0305 & -0.6337 & -0.7781 \\ \hline 0.0034 & -3.6002 & 0.0034 & 0 \\ -0.0701 & 0.0034 & -3.6700 & 0 \\ -0.2588 & 0 & 0 & -36.2588 \end{array}\right],$$

$$\bar{Q} = \begin{bmatrix} 0 & 0 \\ 0 & 0 \\ 0 & -20.7821 \\ -1.0000 & 0 \end{bmatrix} \tag{4.168}$$

$$C = \left[\begin{array}{c|ccc} 0 & -0.9988 & -0.0481 & 0 \\ 0 & 0.0481 & -0.9988 & 0 \\ 0 & 0 & 0 & -1.0000 \end{array}\right] \tag{4.169}$$

It follows from (4.168) that the sub-matrix Q_2 from (4.164) is

$$Q_2^{\mathrm{T}} = [0 \quad -20.7821 \quad 0] \tag{4.170}$$

and $M_o = -1.0000$. Choosing the gain matrix from (4.165) to be

$$L = [l_1 \quad 0 \quad 0] \tag{4.171}$$

where l_1 is any scalar satisfying $l_1 < 418.1578$, by direct computation, it can be shown that $Q_1 + LQ_2 = 0$ and $A_{11} + LA_{21}$ is stable. Finally if the gain matrix from (4.117) is written as

$$W_{sc} = [w_{11} \quad 0 \quad -1] \tag{4.172}$$

where w_{11} is an arbitrary scalar, then $W_{sc}Q_2 = 0$ and $\|\hat{G}(s)\|_\infty = 0$. Consequently the disturbances/uncertainty are decoupled from the actuator fault estimate $\hat{f}$ and 'perfect' reconstruction can be achieved. Using the values of L from (4.171) and the orthogonal matrix T associated with (4.169) the gain G_n from (4.99) is completely determined. Using the approach described in Sect. 4.3 the gain G_l and the Lyapunov matrix P (and hence P_o) can be easily computed without the need for further optimisation, and the observer design is complete.

Remark 4.11 In fact UIO1 and UIO2 are not necessary conditions for a value of $\gamma = 0$. In the example

$$A = \begin{bmatrix} -1 & 1 & 0 \\ 1 & 2 & 1 \\ 0 & 1 & 1 \end{bmatrix}, \quad M = \begin{bmatrix} 0 \\ 0 \\ 1 \end{bmatrix}, \quad Q = \begin{bmatrix} 1 \\ 0 \\ 0 \end{bmatrix}, \quad C = \begin{bmatrix} 0 & 1 & 0 \\ 0 & 0 & 1 \end{bmatrix}$$

it is easy to check that the conditions A1 and A2 in Sect. 4.3 are satisfied but UIO1 is not. However, it is possible to design a sliding mode observer for this system for which $\gamma = 0$. For example,

$$G_l = \begin{bmatrix} 1 & 0 \\ 4 & 1 \\ 1 & 2 \end{bmatrix}, \qquad G_n = \begin{bmatrix} 0 & 0 \\ 2 & 0 \\ 0 & 1 \end{bmatrix}, \qquad W_{sc} = [0 \quad 1]$$

ensures 'perfect' reconstruction with $\gamma = 0$.

4.7.1 Comparison Based on a Crane System

The remainder of the section will compare the two approaches described earlier using a nonlinear model of a crane system in Fig. 4.16. The equations of motion that govern the crane are

$$\left(I + m_p l^2\right)\ddot{\theta} + c\dot{\theta} + m_p g l \sin\theta + m_p l \ddot{d} \cos\theta = 0 \tag{4.173}$$

$$(m_t + m_p)\ddot{d} + b\dot{d} + m_p l \ddot{\theta} \cos\theta - m_p l \dot{\theta}^2 \sin\theta = u \tag{4.174}$$

The quantities m_p and m_t represent the mass of the pendulum and truck, respectively, and l is the length of the pendulum. The parameters c and b are friction coefficients and I is the inertia. The variable d represents the displacement of the truck and θ represents the angular displacement of the load from the (downward) vertical. The (nominal) model parameters used in the following simulations are given by $m_t = 3$ kg, $m_p = 0.5$ kg, $b = 6.2$ kg/s, $c = 0.01$ kg m^2, $I = 0.06$ kg m^2, $g = 9.81$ m/s^2 and $l = 0.35$ m. The input u associated with the model in (4.174) represents the acceleration of the cart. In reality, this would be achieved by a belt/pulley/motor arrangement. These implicit equations have been implemented in an explicit form in SIMULINK® as a basis for the comparison.

Choosing $(\theta, \dot{\theta}, d, \dot{d})$ as the state vector, and using standard small angle approximations, the nonlinear equations (4.173)–(4.174) can be written as a nominal linear system in the form of (4.91)–(4.92) where

$$A = \begin{bmatrix} 0 & 1 & 0 & 0 \\ -\frac{(m_p+m_t)m_p g l}{\Upsilon} & -\frac{c(m_p+m_t)}{\Upsilon} & 0 & \frac{m_p l b}{\Upsilon} \\ 0 & 0 & 0 & 1 \\ -\frac{(m_p l)^2 g}{\Upsilon} & \frac{c m_p l}{\Upsilon} & 0 & -\frac{b(I+m_p l^2)}{\Upsilon} \end{bmatrix}, \qquad B = \begin{bmatrix} 0 \\ -\frac{m_p l}{\Upsilon} \\ 0 \\ \frac{I+m_p l^2}{\Upsilon} \end{bmatrix} \tag{4.175}$$

and $\Upsilon = (m_p + m_t)(I + m_p l^2) - (m_p g l)^2$. In what follows, it is assumed that the quantities θ, d and $\dot{d}$ are measured. Associated with the choice of state representation, the output distribution matrix

$$C = \begin{bmatrix} 1 & 0 & 0 & 0 \\ 0 & 0 & 1 & 0 \\ 0 & 0 & 0 & 1 \end{bmatrix} \tag{4.176}$$

Fig. 4.16 Crane system

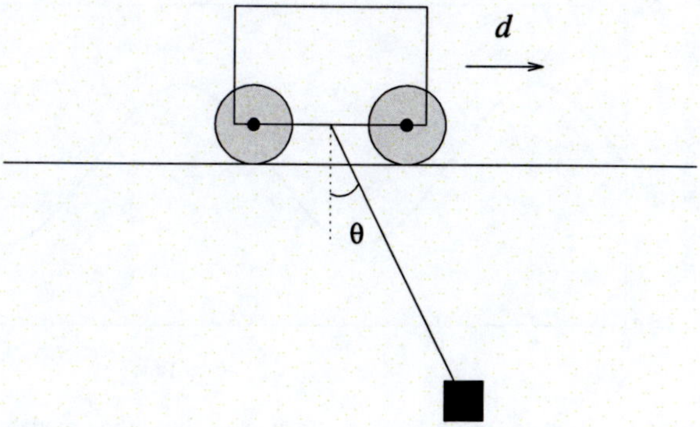

Because assumptions and simplifications have been made in the modelling process, at the FDI scheme design stage, these will be accounted for by introducing an unknown disturbance signal acting through a given (known) distribution matrix. A direct actuator fault will be considered and so in this specific case $M = B$ in (4.91).

In all the simulations a sinusoidal input u has been applied to the cart of amplitude 3 $\mathrm{m/s^2}$ and frequency 2 rad/s. This excites the system and in particular generates oscillations in the pendulum. Although typically in practice a control scheme would be in place which would aim to prevent such an oscillation, in the context of this section it provides a more stringent test of the reconstruction schemes because the system behaviour does not always satisfy small angle assumptions and the differences between the linear model and nonlinear model dynamics are exposed.

4.7.1.1 Results Using the Sliding Mode Observer

For the design of the sliding mode FDI scheme for actuator faults, the exogenous disturbance distribution matrix from (4.91) was chosen as

$$Q^{\mathrm{T}} = [0 \quad 0.9935 \quad 0 \quad -0.1137] \tag{4.177}$$

using the approach of Patton and Chen [207]. The matrix in (4.177) represents a vector of unit magnitude which has been calculated from exciting both the nonlinear and linear models with a PRBS signal and forming a series of error vectors at discrete points in time. Then using principal component analysis, a direction vector has been calculated which accounts for most of the variation. This is the approach advocated by Patton and Chen [207] for computing the distribution matrix associated with the unknown input representing the nonlinear/linear model mismatch. The structure in (4.177) follows from the fact that the modelling mismatches occur in the equations for $\dot{x}_2$ and $\dot{x}_4$ when the differential equations (4.173) and (4.174) are written out in state-space terms as $(x_1, x_2, x_3, x_4) := (\theta, \dot{\theta}, d, \dot{d})$. The entries in the first and third components of (4.177) are zero since by construction $\dot{x}_1 = x_2$ and $\dot{x}_3 = x_4$ in both the linear and nonlinear representations, and so there is no mismatch in these channels.

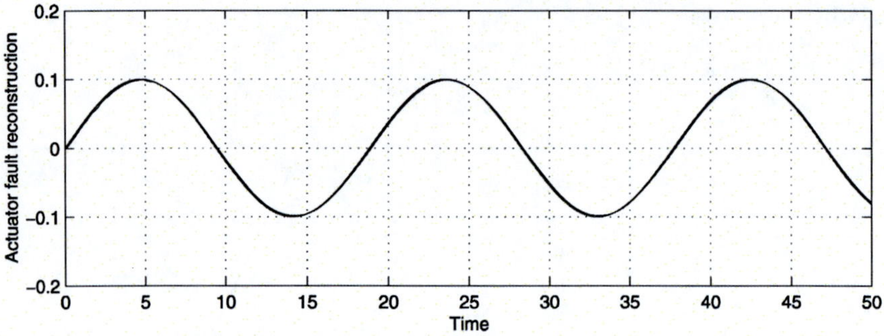

Fig. 4.17 Actuator reconstruction signal using the sliding mode scheme

The key indicator of likely quality of the reconstructions is the size of the $\mathcal{L}_2$-gain γ obtained from the optimisation (the smaller the better). For this example it was found that the choice of the design parameters D_1 and γ_o do not have a significant effect on the values of γ obtained from the optimisation. Generally speaking increasing D_1 has the effect of reducing the gain G_l. Here specifying $D_1 = I_3$ and $\gamma_o = 10$ in (4.131) the LMI synthesis procedure in Sect. 4.5 yields $\gamma = 0.3694$. The resulting observer gains are given by

$$G_l = \begin{bmatrix} 1.1742 & -0.0616 & 0.2099 \\ 0.5458 & -0.3236 & -0.1439 \\ -0.0616 & 0.3883 & -0.0176 \\ 0.2099 & -0.0176 & 0.2084 \end{bmatrix},$$

$$G_n = \begin{bmatrix} 2.3485 & -0.1233 & 0.4198 \\ 1.0916 & -0.6472 & -0.2877 \\ -0.1233 & 0.7767 & -0.0351 \\ 0.4198 & -0.0351 & 0.4167 \end{bmatrix}$$

and the symmetric positive definite matrix P_o is

$$P_o = \begin{bmatrix} 0.5220 & 0.0593 & -0.5208 \\ 0.0593 & 1.2992 & 0.0498 \\ -0.5208 & 0.0498 & 2.9284 \end{bmatrix}$$

In this case the matrix, scaling ν_{eq} in (4.117) to generate the fault reconstruction signal, is given by

$$W_{sc} = [2.4580 \quad -0.1716 \quad 1.5490]$$

In the following experiment the modulation gain associated with ν in (4.96) is $\rho = 0.3$. A sinusoidal 'actuator fault' of amplitude 0.1 and frequency 1/3 rad/s was introduced. The reconstruction signal from the sliding mode observer together with the actual fault is given in Fig. 4.17.

Almost perfect reconstruction has been obtained. In all the sliding mode simulations the unit vector term from (4.96) has been smoothed by adding a small constant to the denominator to remove the discontinuity; this facilitates the numerical simulation of the nonlinear equations: see Sect. 4.2.4. The small phase shift in the sine wave, results from the filter used to extract the equivalent output error injection signal from (4.96).

4.7.1.2 Results Using the UIO

For the crane system an ideal representation of the uncertainty distribution would be Q as given in (4.177). However, using the notation associated with equation (4.150),

$$\text{rank}(C\bar{Q}) := \text{rank}[CM \quad CQ] = 1$$

since the range space of CQ is contained within the range space of CM. Thus the choice of Q given in (4.177) is not appropriate since the rank requirement UIO1 from Sect. 4.7 is not satisfied. Another choice, which would account for parametric uncertainties in the modelling and friction effects would be

$$Q = \begin{bmatrix} 0 & q_{22} & 0 & 0 \\ 0 & 0 & 0 & q_{41} \end{bmatrix}^{\mathrm{T}} \tag{4.178}$$

where q_{22} and q_{41} are nonzero scalars. This is the most general uncertainty distribution since, as argued in Sect. 4.7.1.1, the elements in the first and third rows of Q will be zero because for both the linear and nonlinear set of equations $\dot{x}_1 = x_2$ and $\dot{x}_3 = x_4$, and hence there is no discrepancy in these channels. However, for the structure of Q in (4.178), the matrix CQ is not full column rank for any choice of q_{22} and q_{41}, and once again $\text{rank}(C\bar{Q}) = 1$. The most sensible choice is to let $Q = \phi$ (the empty matrix), which implies $Q = M$ in (4.150) and so no uncertainty will be taken into account explicitly (which is of course not entirely desirable). A similar situation would arise if the full order UIOs from [50, 52] were employed, since as argued in Remark 4.9, condition UIO1 must also be satisfied for the full order UIOs from [50, 52] to exist.

In this example the observer system in (4.148)–(4.149) will be first order. Following the approach in [219], and using the canonical form from (4.151), by choosing $\hat{F} = -2$, it follows from (4.152)–(4.156) that $\hat{L} = 0$ and

$$\hat{E} = [17.9938 \quad 0 \quad -2.8866], \qquad \hat{T} = \begin{bmatrix} 1.0000 & 0 & 0 \\ 1.9175 & 0 & -1.4433 \\ 0 & 1.0000 & 0 \\ 0 & 0 & 1.0000 \end{bmatrix},$$

$$\hat{R} = \begin{bmatrix} 0 \\ -1 \\ 0 \\ 0 \end{bmatrix}$$

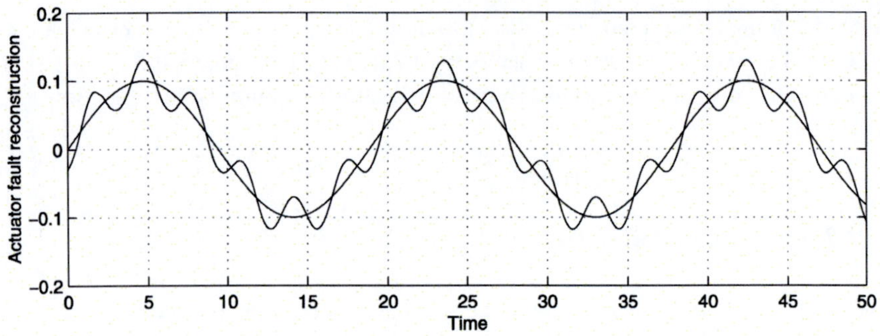

Fig. 4.18 Actuator reconstruction using the UIO scheme

The approach suggested by Saif and Guan in [219] ensures the state estimation error $e = \hat{x} - x$ is independent of the uncertainty and the fault, so that $\hat{x} \to x$ asymptotically. To estimate $f(t)$, Saif and Guan propose using a discrete version of the differential equation (4.91) as in equation (4.159), which is then manipulated to yield an estimate. In this example, using a zero-order-hold with sample period 0.01, yields the matrices

$$A_d = \begin{bmatrix} 0.9992 & 0.0100 & 0 & 0.0001 \\ -0.1524 & 0.9984 & 0 & 0.0273 \\ 0.0000 & 0.0000 & 1.0000 & 0.0099 \\ 0.0076 & 0.0001 & 0 & 0.9811 \end{bmatrix} \quad \text{and} \quad B_d = M_d = \begin{bmatrix} -0.0000 \\ -0.0044 \\ 0.0000 \\ 0.0031 \end{bmatrix}$$

An estimate for $f(t)$ can then be obtained by manipulating the discrete equation (4.159) to give $\hat{f}(k) = (M_d^{\mathrm{T}} M_d)^{-1} M_d^{\mathrm{T}}(\hat{x}(k+1) - A_d\hat{x}(k)) - u(k)$ where $\hat{x}$ are the estimates of the states x from the reduced order observer. The fault scenarios considered here are identical to those which were used in Sect. 4.7.1.1. The results are given in Fig. 4.18. A less accurate reconstruction (compared to Fig. 4.17) has been obtained. This is perhaps not surprising since the reconstruction method given in (4.160) may be viewed as a type of numerical differentiation. Of course here the quality of the robustness of the estimation is solely dependent on the choice of L.

Remark 4.12 For the approach in [219] both the uncertainty and faults are lumped together to form an aggregate unknown input signal. This means the requirements UIO1 and UIO2 are stronger than those associated with the sliding mode observer (A1 and A2 in Sect. 4.3). Also there is technically an additional constraint that $h + q < p$ for the UIO, which again is stronger than the one required for the sliding mode scheme. As argued in Remark 4.9, the conditions UIO1 and UIO2 are also necessary and sufficient for the design of the *full order* UIOs considered in [50, 52] and so formally the conditions for complete insensitivity will not be met in this case also. It could be argued that the higher order allows more design freedom which could be exploited to achieve robustness, but there are no theoretical results in this area, and the authors in [50, 52] argue the benefits of full order UIOs from an improved speed of state tracking perspective.

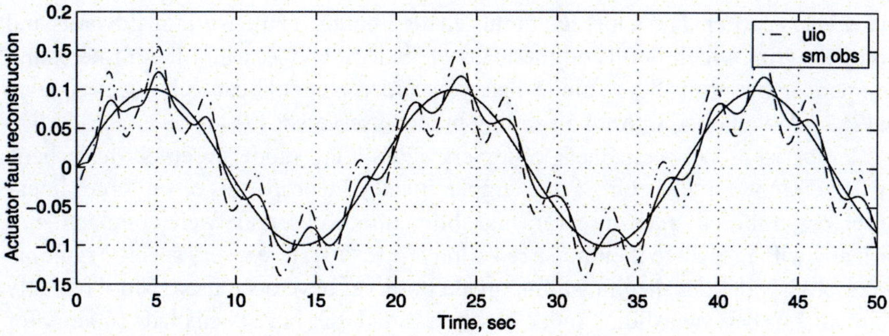

Fig. 4.19 Actuator fault reconstruction

To explore the effect of parametric uncertainty the mass of the pendulum has been changed by 5% to be $m_p = 0.525$. Figure 4.19 shows the results of the reconstructions from using the sliding mode observer and the UIO designed in Sect. 4.7.1.1 and Sect. 4.7.1.2 about the linearisation using $m_p = 0.5$. Perfect reconstruction is lost; however, the sliding mode observer performs significantly better. Again this indicates that the choice of poles for the UIO is crucial.

Remark 4.13 In the UIO method, when uncertainty cannot be included as part of the unknown inputs, the choice of L is crucial to the quality of the reconstruction. To the authors' knowledge there is no universally accepted approach for achieving this (even thinking of the UIO as a special case of a minimal order observer). Difficulties in trading-off the freedom in the gain matrix L and the effect to which noise impacts on the state reconstruction is well known for minimal order observers and is discussed for example in [6].

4.8 Summary

This chapter has introduced the concept of a sliding mode observer and the properties of the sliding motion (especially with respect to disturbance rejection). This chapter initially presented the classical Utkin observer, and then argued why the inclusion of a linear term is advantageous. The concept of fault reconstruction was presented using the Edwards–Spurgeon observer. Observer design frameworks using LMIs have also been introduced which allow robust fault reconstruction in the presence of uncertainty. Finally a comparison with a linear unknown input observer was presented.

4.9 Notes and References

The earliest work in terms of sliding mode methods applied to observer problems is attributed to Utkin [256] although these ideas had appeared in the Russian literature many years earlier. The approach in this chapter builds on the work of Edwards and Spurgeon [82] which builds on the ideas of Walcott and Zak [266]. An interesting comparison between sliding mode observers and other nonlinear techniques appears in [265]. A well cited tutorial on sliding mode observers is [74].

Comparisons between linear observers and sliding mode observers have been made before in the literature. For example, in [71] the properties of full order linear observers, minimal order observers and sliding mode observers were compared empirically with respect to their noise rejection properties. Here a very specific minimal order observer and a specific sliding mode observer have been used. More recently comparisons between sliding mode observers and UIOs have been made (mainly for the discrete case) in [246]. The canonical form in (4.46), based on assumptions A1 and A2 in Sect. 4.3, has been shown to be a useful starting point for the design of the reduced order and full order UIOs from [50, 161, 219]. As such it may be viewed as a unifying framework for UIOs and sliding mode observers in continuous time.

Although it is fair to say the majority of sliding mode observers have been based on underlying linear system representations there are notable exceptions for example Barbot and co-workers [18, 19], the early work of Slotine et al. [229] and also by Chen and Saif [277].

Recent work involving sliding mode observers for fault detection in nonlinear systems appears in Jiang et al. [141], Yan and Edwards [102, 279–282].

Chapter 5
Robust Fault Reconstruction using Observers in Cascade

In the previous chapter, sliding mode observer designs for robust fault reconstruction were presented. In this chapter, the concept of robust fault reconstruction using multiple observers in cascade will be presented. The idea is to alleviate some of the system requirements specified in the previous chapter, thus allowing the scheme which will be presented in this chapter to be applicable to a wider class of systems. Necessary and sufficient conditions, in terms of the original system matrices, will be provided. The approach will also indicate precisely the number of observers in cascade which will be required.

5.1 Introduction

In Chap. 4, particularly Sect. 4.5 which presented the use of sliding mode observers for robust fault reconstruction, one of the necessary and sufficient conditions was that the first Markov parameter (the product of the output and fault matrices) must have full column rank.

This chapter presents ideas that build on the work of [197] i.e., using multiple observers in cascade. Using similar techniques as in [197], measurable signals from an observer are used as outputs of a fictitious system; the next observer is designed for the fictitious system and the measurable signals from this observer are used as outputs of another fictitious system. The process is repeated until a fictitious system whose (first) Markov parameter is full rank is obtained. It is found that the (final) fictitious system is in the framework of the system (4.91)–(4.92) in Sect. 4.5. Hence the technique presented in Sect. 4.5 can be applied to the final system and an observer is implemented for fault reconstruction and designed such that the $\mathcal{L}_2$ gain from the disturbance to the fault reconstruction is minimised. This achieves robust fault reconstruction without reconstructing the disturbances and enables the algorithm to be applied to systems which have fewer outputs than the sum of faults and disturbance channels. Also, it is found that the design of previous observers do not affect the sliding motion of the final observer, which implies that the $\mathcal{L}_2$ gain from the disturbances to the fault reconstruction is affected only by the design of

H. Alwi et al., *Fault Detection and Fault-Tolerant Control Using Sliding Modes*, Advances in Industrial Control,
DOI 10.1007/978-0-85729-650-4_5,

the final observer. Furthermore, necessary and sufficient conditions are investigated and presented in terms of the original system matrices, so that the designer can determine at the outset whether the method is applicable or not. The results in this chapter also indicate precisely the number of observers in cascade that are required.

Throughout the chapter, a superscript will be used to represent the recursion level in the cascade; for example X^i indicates that X is a parameter for observer i. To raise a variable to a power, it will be placed in brackets first; for example $(X)^i$ means that the variable X is raised to the power of i.

5.2 The Robust Fault Reconstruction Scheme

Consider the problem considered in Sect. 4.5: specifically fault reconstruction in a system

$$\dot{x}(t) = Ax(t) + Bu(t) + Mf(t) + Q\xi(t) \tag{5.1}$$

$$y(t) = Cx(t) \tag{5.2}$$

where $A \in \mathbb{R}^{n\times n}, B \in \mathbb{R}^{n\times m}, C \in \mathbb{R}^{p\times n}$, $M \in \mathbb{R}^{n\times q}$ and $Q \in \mathbb{R}^{n\times h}$ where $n > p \geq q$. Assume that $\text{rank}(M) = q$, $\text{rank}(C) = p$ and $\text{rank}(CM) = \bar{r} \leq q$, which implies that $\bar{r} \leq \min\{p, q\}$. The function representing the faults $f \in \mathbb{R}^q$ is unknown but bounded so that

$$\|f(t)\| \leq \alpha \tag{5.3}$$

where α is a known function. The signal $\xi \in \mathbb{R}^h$ encapsulates the uncertainty present in the system such as unmodelled dynamics, parametric uncertainties and external disturbances. It is assumed to be unknown but bounded subject to $\|\xi(t)\| < \beta$ where β is known.

The signal ξ is assumed to be smooth and an upper bound on its bandwidth is assumed to be known.[1] As a result of the bandwidth assumption it is possible to write the disturbance as

$$\xi = \Omega(\mathrm{s})\xi^h \tag{5.4}$$

where $\Omega(\mathrm{s})$ represents a known filter with low-pass characteristics of appropriate bandwidth and ξ^h is a bounded unknown signal. As in other frequency domain based paradigms such as $\mathcal{H}_\infty$ and μ-synthesis, the filter $\Omega(\mathrm{s})$ can be viewed as a 'weighting function' [297]. The frequency information about the disturbance associated with $\Omega(\mathrm{s})$ will then be incorporated into the observer design. Furthermore it is assumed that ξ, together with an appropriate number of its derivatives are bounded.

[1] The assumption that a bound on the frequency content of the disturbances is known, is common in the applications literature. This sort of information has been used in the development of models of practical engineering systems where typically the disturbances are assumed to be low frequency in character. Insight from the underlying physics is usually employed to decide on the meaningful frequency range of the disturbance.

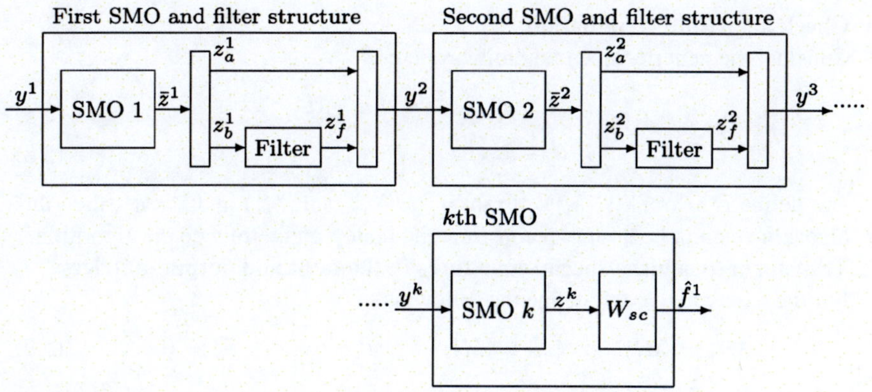

Fig. 5.1 The proposed scheme formed from a cascaded observer/filter structure

Specific details pertaining to the weighting function $\Omega(s)$ will be given in the next section. Also the first derivative of f is assumed to be bounded by a known constant. This assumption is not overly restrictive as it only implies that f cannot be an abrupt step, which is easy to detect (slow incipient faults are much more difficult to detect [51]). *The objective is to reconstruct f whilst minimising the effects of ξ on the fault reconstruction.* If $\bar{r} = q$ then the single-observer method described in Sect. 4.5 can be used directly. However, if $\bar{r} < q$, then an alternative approach is required. In this situation, this chapter proposes the cascade observer scheme shown schematically in Fig. 5.1. The next subsection describes the fault reconstruction algorithm and a systematic way of designing the components in Fig. 5.1.

5.2.1 Design Algorithm

To create the recursive observer design procedure, write (5.1) and (5.2) as

$$\dot{x}^1 = A^1 x^1 + M^1 f^1 + Q^1 \xi^1 \tag{5.5}$$

$$y^1 = C^1 x^1 \tag{5.6}$$

where without loss of generality, it has been assumed $u(t) \equiv 0$ in (5.1). Since $\text{rank}(C^1) = p$, then C^1 can be written without loss of generality in the form $C^1 = [0 \quad I_p]$. Firstly partition the matrices from (5.5) as

$$A^1 = \begin{bmatrix} A_1^1 & A_2^1 \\ A_3^1 & A_4^1 \end{bmatrix} \begin{matrix} \updownarrow n^1 - p \\ \updownarrow p \end{matrix}, \qquad M^1 = \begin{bmatrix} M_1^1 \\ M_2^1 \end{bmatrix} \begin{matrix} \updownarrow n^1 - p \\ \updownarrow p \end{matrix}, \qquad Q^1 = \begin{bmatrix} Q_1^1 \\ Q_2^1 \end{bmatrix} \begin{matrix} \updownarrow n^1 - p \\ \updownarrow p \end{matrix}$$

where A_1^1 is square. Since by assumption $C^1 = [0 \; I_p]$ and $\text{rank}(C^1 M^1) = \bar{r}^1$, then it follows that $\text{rank}(M_2^1) = \bar{r}^1$. In the representation above, Q^1 has no particular structure.

Set the index variable $i = 1$ and proceed with the following algorithm:

1 **Check algorithm termination**
Consider the generic uncertain faulty system

$$\dot{x}^i = A^i x^i + M^i f^i + Q^i \xi^i \tag{5.7}$$

$$y^i = C^i x^i \tag{5.8}$$

and define $\bar{r}^i := \text{rank}(C^i M^i)$. If $\text{rank}(C^i M^i) < \text{rank}(M^i)$ and $i = n^1$, then this algorithm cannot be used to reconstruct the faults[2] and terminate the algorithm.

2 **Transform to achieve special structures in the fault and output matrices**
For the case when $i = 1$, define the following

$$\bar{M}_{11}^0 := M_1^1, \qquad \bar{M}_{12}^0 := M_2^1, \qquad m^1 := p, \qquad \bar{r}^0 := 0 \tag{5.9}$$

Let $r^i := \text{rank}(\bar{M}_{12}^{i-1})$ and define two orthogonal matrices $D^i \in \mathbb{R}^{m^i \times m^i}$ and $T_2^i \in \mathbb{R}^{(q-\bar{r}^{i-1})\times(q-\bar{r}^{i-1})}$ such that

$$\begin{bmatrix} I_{n^i-p-(i-1)h} & 0 \\ 0 & (D^i)^{-1} \end{bmatrix} \begin{bmatrix} \bar{M}_{11}^{i-1} \\ \bar{M}_{12}^{i-1} \end{bmatrix} (T_2^i)^{-1} = \left[\begin{array}{cc} M_{11}^i & M_{12}^i \\ \hline 0 & 0 \\ 0 & M_{22}^i \end{array}\right] \begin{array}{l} \updownarrow n^i-p-(i-1)h \\ \updownarrow m^i - r^i \\ \updownarrow r^i \end{array} \tag{5.10}$$

where $M_{22}^i \in \mathbb{R}^{r^i \times r^i}$ is square and invertible. Then define $T_1^i := T_{12}^i T_{11}^i$ where

$$T_{11}^i := \left[\begin{array}{c|cc} I_{n^i-p} & 0 & 0 \\ \hline 0 & (D^i)^{-1} & 0 \\ 0 & 0 & I_{p-m^i} \end{array}\right] \tag{5.11}$$

and

$$T_{12}^i := \left[\begin{array}{cc|cccc} I_{(i-1)h} & 0 & 0 & 0 & 0 & 0 \\ 0 & I_{n^i-p-(i-1)h} & 0 & -M_{12}^i (M_{22}^i)^{-1} & 0 & 0 \\ \hline 0 & 0 & I_{m^i-r^i} & 0 & 0 & 0 \\ 0 & 0 & 0 & 0 & I_{p-\bar{r}^{i-1}-m^i} & 0 \\ 0 & 0 & 0 & I_{r^i} & 0 & 0 \\ 0 & 0 & 0 & 0 & 0 & I_{\bar{r}^{i-1}} \end{array}\right] \tag{5.12}$$

When $i = 1$, define

$$\tilde{A}_{13}^0 := A_3^1, \qquad \tilde{A}_{11}^0 := A_1^1, \qquad \bar{A}_{\Omega}^0 = \alpha^0 = \bar{M}_{22}^0 = \phi \tag{5.13}$$

where ϕ is the empty matrix. Define the partitions

$$\tilde{A}_3^i := (D^i)^{-1} \tilde{A}_{13}^{i-1} = \begin{bmatrix} \tilde{A}_{31}^i \\ \tilde{A}_{32}^i \end{bmatrix} \begin{array}{l} \updownarrow m^i - r^i \\ \updownarrow r^i \end{array} \tag{5.14}$$

[2]The justification of this will be given in Theorem 5.1 in the sequel.

and

$$\tilde{A}_1^i := \tilde{A}_{11}^{i-1} - M_{12}^i \big(M_{22}^i\big)^{-1} \tilde{A}_{32}^i \tag{5.15}$$

Define $T_f^i := \mathrm{diag}(T_2^i, I_{\bar{r}^{i-1}})$. Perform the coordinate transformations

$$x^i \mapsto T_1^i x^i, \qquad f^i \mapsto f^{i+1} := T_f^i f^i \tag{5.16}$$

then the matrix triple A^i, M^i, C^i will be transformed into

$$A^i \mapsto \begin{bmatrix} A_1^i & A_2^i \\ A_3^i & A_4^i \end{bmatrix} = \left[\begin{array}{cc|c} \bar{A}_\Omega^{i-1} & 0 & \star \\ \star & \tilde{A}_1^i & \star \\ \hline \star & \tilde{A}_{31}^i & \star \\ \star & 0 & \star \\ \star & \star & \star \end{array}\right] \begin{array}{l} \updownarrow (i-1)h \\ \updownarrow n^i - p - (i-1)h \\ \updownarrow m^i - r^i \\ \updownarrow p - m^i - \bar{r}^{i-1} \\ \updownarrow \bar{r}^i \end{array} \tag{5.17}$$

$$M^i \mapsto \begin{bmatrix} M_1^i \\ M_2^i \end{bmatrix} = \left[\begin{array}{cc} 0 & 0 \\ M_{11}^i & 0 \\ \hline 0 & 0 \\ 0 & 0 \\ 0 & \bar{M}_{22}^i \end{array}\right] \begin{array}{l} \updownarrow (i-1)h \\ \updownarrow n^i - p - (i-1)h \\ \updownarrow m^i - r^i \\ \updownarrow p - m^i - \bar{r}^{i-1} \\ \updownarrow \bar{r}^i \end{array} \tag{5.18}$$

where

$$\bar{M}_{22}^i = \begin{bmatrix} M_{22}^i & 0 \\ 0 & \alpha^{i-1} \bar{M}_{22}^{i-1} \end{bmatrix} \begin{array}{l} \updownarrow r^i \\ \updownarrow \bar{r}^{i-1} \end{array} \tag{5.19}$$

and finally

$$C^i \mapsto \begin{bmatrix} 0 & C_2^i \end{bmatrix}, \quad C_2^i = \begin{bmatrix} D^i & 0 \\ 0 & I_{p-m^i} \end{bmatrix} \begin{bmatrix} I_{m^i - r^i} & 0 & 0 & 0 \\ 0 & 0 & I_{p-\bar{r}^{i-1}-m^i} & 0 \\ 0 & I_{r^i} & 0 & 0 \\ 0 & 0 & 0 & I_{\bar{r}^{i-1}} \end{bmatrix} \tag{5.20}$$

It can be seen from the definition of $\bar{r}^i$ in Step 1, M^i and $\bar{M}_{22}^i$ in (5.18)–(5.19), and C^i in (5.20) that

$$\bar{r}^i := \bar{r}^{i-1} + r^i \tag{5.21}$$

In this coordinate system Q^i has no specific structure. If $\mathrm{rank}(C^i M^i) = \mathrm{rank}(M^i)$ then go to Step 7 and terminate the algorithm. Otherwise, proceed to the following step.

3 **Augment with the dynamics of the disturbance weight**

Assume that ξ^i is smooth and satisfies the following stable system

$$\dot{\xi}^i = A_\Omega^i \xi^i + B_\Omega^i \xi^{i+1} \tag{5.22}$$

where $\xi^{i+1} \in \mathbb{R}^h$ and A_Ω^i, B_Ω^i are matrices to be chosen by the designer. In addition, assume that ξ^{i+1} is bounded. (The motivation and implication of this assumption, and a way to choose A_Ω^i and B_Ω^i will be discussed in Remark 5.2.)

Augment (5.22) with (5.7)–(5.8) to obtain the following system of order $\bar{n}^i := n^i + h$:

$$\dot{\bar{x}}^i = \bar{A}^i \bar{x}^i + \bar{M}^i f^{i+1} + \bar{Q}^i \xi^{i+1} \tag{5.23}$$

$$y^i = \bar{C}^i \bar{x}^i \tag{5.24}$$

where $\bar{x}^i := \mathrm{col}(\xi^i, x^i)$ and

$$\bar{A}^i = \left[\begin{array}{ccc|c} A_\Omega^i & 0 & 0 & 0 \\ \star & \bar{A}_\Omega^{i-1} & 0 & \star \\ \star & \star & \tilde{A}_1^i & \star \\ \hline \star & \star & \tilde{A}_{31}^i & \star \\ \star & \star & 0 & \star \\ \star & \star & \star & \star \end{array}\right], \qquad \bar{M}^i = \left[\begin{array}{cc} 0 & 0 \\ 0 & 0 \\ M_{11}^i & 0 \\ \hline 0 & 0 \\ 0 & 0 \\ 0 & \bar{M}_{22}^i \end{array}\right],$$

$$\bar{Q}^i = \left[\begin{array}{c} B_\Omega^i \\ 0 \\ 0 \\ \hline 0 \\ 0 \\ 0 \end{array}\right] \begin{array}{l} \updownarrow h \\ \updownarrow (i-1)h \\ \updownarrow n^i - p - (i-1)h \\ \updownarrow m^i - r^i \\ \updownarrow p - m^i - \bar{r}^{i-1} \\ \updownarrow \bar{r}^i \end{array}$$

Then define $\bar{A}_\Omega^i := \left[\begin{smallmatrix} A_\Omega^i & 0 \\ \star & \bar{A}_\Omega^{i-1} \end{smallmatrix}\right]$, and note that $\bar{A}^i$ can be re-expressed as

$$\bar{A}^i = \left[\begin{array}{cc|c} \bar{A}_\Omega^i & 0 & 0 \\ \star & \tilde{A}_1^i & \star \\ \hline \bar{Q}_{21}^i & \tilde{A}_{31}^i & \star \\ \star & 0 & \star \\ \star & \star & \star \end{array}\right] \begin{array}{l} \updownarrow ih \\ \updownarrow n^i - p - (i-1)h \\ \updownarrow m^i - r^i \\ \updownarrow p - m^i - \bar{r}^{i-1} \\ \updownarrow \bar{r}^i \end{array}$$

4 **Transform to achieve a special structure in the system matrix**

Define $m^{i+1} := \mathrm{rank}(\tilde{A}_{31}^i)$. Let U_1^i and U_2^i be invertible matrices of dimensions $m^i - r^i$ and $n^i - p - (i-1)h$, respectively, such that

$$U_1^i \tilde{A}_{31}^i (U_2^i)^{-1} = \begin{bmatrix} 0 & I_{m^{i+1}} \\ 0 & 0 \end{bmatrix} \tag{5.25}$$

and

$$U_1^i \bar{Q}_{21}^i = \begin{bmatrix} \bar{Q}_{211}^i \\ \bar{Q}_{212}^i \end{bmatrix} \begin{array}{l} \updownarrow m^{i+1} \\ \updownarrow m^i - r^i - m^{i+1} \end{array} \tag{5.26}$$

where $\bar{Q}_{211}^i$, $\bar{Q}_{212}^i$ are general matrices with no particular structure. Also partition

$$U_2^i \tilde{A}_1^i (U_2^i)^{-1} = \begin{bmatrix} \tilde{A}_{11}^i & \tilde{A}_{12}^i \\ \tilde{A}_{13}^i & \tilde{A}_{14}^i \end{bmatrix} \begin{array}{l} \updownarrow n^i - p - (i-1)h - m^{i+1} \\ \updownarrow m^{i+1} \end{array} \tag{5.27}$$

Introduce the coordinate transformation $\bar{x}^i \mapsto \bar{T}^i \bar{x}^i$ where $\bar{T}^i := \bar{T}_2^i \bar{T}_1^i$ with

$$\bar{T}_1^i := \left[\begin{array}{cc|cc} I_{ih} & 0 & 0 & 0 \\ 0 & U_2^i & 0 & 0 \\ \hline 0 & 0 & U_1^i & 0 \\ 0 & 0 & 0 & I_{p+r^i-m^i} \end{array}\right], \qquad (5.28)$$

$$\bar{T}_2^i := \left[\begin{array}{ccc|c} I_{ih} & 0 & 0 & 0 \\ 0 & I_{n^i-p-(i-1)h-m^{i+1}} & 0 & 0 \\ \bar{Q}_{211}^i & 0 & I_{m^{i+1}} & 0 \\ \hline 0 & 0 & 0 & I_p \end{array}\right]$$

Then $\bar{A}^i$, $\bar{M}^i$, $\bar{Q}^i$ and $\bar{C}^i$ will be transformed to

$$\bar{A}^i \mapsto \begin{bmatrix} \bar{A}_1^i & \bar{A}_2^i \\ \bar{A}_3^i & \bar{A}_4^i \end{bmatrix} = \left[\begin{array}{ccc|c} \bar{A}_\Omega^i & 0 & 0 & \star \\ \star & \tilde{A}_{11}^i & \tilde{A}_{12}^i & \star \\ \star & \tilde{A}_{13}^i & \tilde{A}_{14}^i & \star \\ \hline 0 & 0 & I & \star \\ \star & 0 & 0 & \star \\ \star & 0 & 0 & \star \end{array}\right] \begin{array}{l} \updownarrow ih \\ \updownarrow n^i-p-m^{i+1}-(i-1)h \\ \updownarrow m^{i+1} \\ \updownarrow m^{i+1} \\ \updownarrow p-m^{i+1}-\bar{r}^i \\ \updownarrow \bar{r}^i \end{array} \qquad (5.29)$$

$$\bar{M}^i \mapsto \begin{bmatrix} \bar{M}_1^i \\ \bar{M}_2^i \end{bmatrix} = \left[\begin{array}{cc} 0 & 0 \\ \bar{M}_{11}^i & 0 \\ \bar{M}_{12}^i & 0 \\ \hline 0 & 0 \\ 0 & 0 \\ 0 & \bar{M}_{22}^i \end{array}\right] \begin{array}{l} \updownarrow ih \\ \updownarrow n^i-p-m^{i+1}-(i-1)h \\ \updownarrow m^{i+1} \\ \updownarrow m^{i+1} \\ \updownarrow p-m^{i+1}-\bar{r}^i \\ \updownarrow \bar{r}^i \end{array} \qquad (5.30)$$

$$\bar{Q}^i \mapsto \left[\begin{array}{c} \bar{Q}_1^i \\ \hline 0 \end{array}\right] \begin{array}{l} \updownarrow n^i-p+h \\ \updownarrow p \end{array} \qquad (5.31)$$

and finally

$$\bar{C}^i \mapsto \begin{bmatrix} 0 & \bar{C}_2^i \end{bmatrix}, \qquad \det(\bar{C}_2^i) \neq 0 \qquad (5.32)$$

Also partition

$$\bar{A}_3^i = \begin{bmatrix} \bar{A}_{31}^i \\ \bar{A}_{32}^i \end{bmatrix} \begin{array}{l} \updownarrow m^{i+1} \\ \updownarrow p-m^{i+1} \end{array}$$

which from (5.29) results in $\bar{A}_{31}^i = [0 \; I_{m^{i+1}}]$.

5 **Implement the ith observer for the augmented system**

A sliding mode observer building on second order supertwisting observer ideas from [166, 190] for (5.23)–(5.24) is

$$\dot{\hat{\bar{x}}}^i = \bar{A}^i \hat{\bar{x}}^i - \bar{G}_l^i \bar{e}_y^i + \bar{G}_n^i \bar{\nu}^i, \qquad \bar{e}_y^i := \bar{C}^i \hat{\bar{x}}^i - y^i \qquad (5.33)$$

where the matrices $\bar{G}_l^i, \bar{G}_n^i \in \mathbb{R}^{\bar{n}^i \times p}$ are to be designed. In particular, choose $\bar{G}_n^i$ as

$$\bar{G}_n^i = \begin{bmatrix} -\bar{L}^i \\ I_p \end{bmatrix} (\bar{C}_2^i)^{-1}, \quad \bar{L}^i = \begin{bmatrix} \bar{L}_o^i & 0 \end{bmatrix} \tag{5.34}$$

where $\bar{L}^i \in \mathbb{R}^{(\bar{n}^i - p) \times p}$ and $\bar{L}_o^i \in \mathbb{R}^{(\bar{n}^i - p) \times m^{i+1}}$ are chosen such that the matrix $\bar{A}_1^i + \bar{L}_o^i \bar{A}_{31}^i$ is stable. Partition the output estimation error componentwise as $\bar{e}_y^i = \mathrm{col}(\bar{e}_{y,1}^i, \ldots, \bar{e}_{y,p}^i)$. As in [190] the nonlinear injection term $\bar{\nu}^i := \mathrm{col}(\bar{\nu}_1^i, \ldots, \bar{\nu}_p^i)$ is defined by

$$\bar{\nu}_j^i = -\psi_j^i \operatorname{sgn}(\bar{e}_{y,j}^i) |\bar{e}_{y,j}^i|^{\frac{1}{2}} + z_j^i, \quad j = 1, \ldots, p \tag{5.35}$$

$$\dot{z}_j^i = -\beta_j^i \operatorname{sgn}(\bar{e}_{y,j}^i) - \gamma_j^i \bar{e}_{y,j}^i, \quad j = 1, \ldots, p \tag{5.36}$$

where ψ_j^i, β_j^i and γ_j^i are scalars to be selected by the designer. Define $\bar{e}^i := \hat{\bar{x}}^i - \bar{x}^i$ and combine (5.23), (5.24) and (5.33) to obtain

$$\dot{\bar{e}}^i = (\bar{A}^i - \bar{G}_l^i \bar{C}^i)\bar{e}^i + \bar{G}_n^i \bar{\nu}^i - \bar{M}^i f^{i+1} - \bar{Q}^i \xi^{i+1} \tag{5.37}$$

Apply a change of coordinates associated with T_L^i to the triple (5.29)–(5.32) and $\bar{G}_n^i$ in (5.34) where

$$T_L^i := \begin{bmatrix} I_{\bar{n}^i - p} & \bar{L}^i \\ 0 & \bar{C}_2^i \end{bmatrix}$$

then $\bar{A}^i, \bar{M}^i, \bar{C}^i$ from (5.29)–(5.32) and $\bar{G}_n^i$ from (5.34) are transformed to be

$$\begin{aligned} &\bar{A}^i \to \begin{bmatrix} \bar{A}_1^i + \bar{L}_o^i \bar{A}_{31}^i & \mathcal{A}_{12}^i \\ \bar{C}_2^i \bar{A}_3^i & \mathcal{A}_{22}^i \end{bmatrix}, \qquad \bar{M}^i \to \begin{bmatrix} \bar{M}_1^i \\ \bar{C}_2^i \bar{M}_2^i \end{bmatrix}, \\ &\bar{C}^i \to \begin{bmatrix} 0 & I_p \end{bmatrix}, \qquad \bar{G}_n^i \to \begin{bmatrix} 0 \\ I_p \end{bmatrix} \end{aligned} \tag{5.38}$$

whilst $\bar{Q}^i$ retains the structure in (5.31) after the transformation. Define

$$T_L^i \bar{e}^i =: \begin{bmatrix} \bar{e}_1^i \\ \bar{e}_y^i \end{bmatrix}, \qquad T_L^i \bar{G}_l^i =: \begin{bmatrix} \mathcal{G}_1^i \\ \mathcal{G}_2^i \end{bmatrix} \begin{matrix} \updownarrow \bar{n}^i - p \\ \updownarrow p \end{matrix} \tag{5.39}$$

and choose $\bar{G}_l^i$ so that $\mathcal{G}_1^i = \mathcal{A}_{12}^i$, $\mathcal{G}_2^i = \mathcal{A}_{22}^i + \mathcal{A}_s^i$ where $\mathcal{A}_s^i := \mathrm{diag}(\lambda_1^i, \ldots, \lambda_p^i)$ and the scalars $\lambda_j^i > 0$, $j = 1, \ldots, p$. Partitioning (5.37) according to (5.38)–(5.39) results in

$$\dot{\bar{e}}_1^i = (\bar{A}_1^i + \bar{L}_o^i \bar{A}_{31}^i)\bar{e}_1^i + \bar{M}_1^i f^{i+1} + \bar{Q}_1^i \xi^{i+1} \tag{5.40}$$

$$\dot{\bar{e}}_y^i = \bar{C}_2^i \bar{A}_3^i \bar{e}_1^i + \bar{C}_2^i \bar{M}_2^i f^{i+1} - \mathcal{A}_s^i \bar{e}_y^i + \bar{\nu}^i \tag{5.41}$$

where $\bar{M}_1^i$, $\bar{M}_2^i$ and $\bar{Q}_1^i$ are defined in (5.30)–(5.31). Equation (5.41) can be written as

$$\dot{\bar{e}}_y^i = \zeta^i - \mathcal{A}_s^i \bar{e}_y^i + \bar{\nu}^i \tag{5.42}$$

where

$$\zeta^i = \hat{G}(s)\begin{bmatrix} f^{i+1} \\ \xi^{i+1} \end{bmatrix}$$

and the transfer function matrix

$$\hat{G}(s) := -\begin{bmatrix} \bar{C}_2^i \bar{M}_2^i & 0 \end{bmatrix} - \bar{C}_2^i \bar{A}_3^i \big(sI - (\bar{A}_1^i + \bar{L}_o^i \bar{A}_{31}^i)\big)^{-1} \begin{bmatrix} \bar{M}_1^i & \bar{Q}_1^i \end{bmatrix}$$

It is obvious ζ^i and $\dot{\zeta}^i$ are bounded since $\bar{A}_1^i + \bar{L}_o^i \bar{A}_{31}^i$ is stable and f^{i+1}, $\dot{f}^{i+1}$ and ξ^{i+1} are bounded by assumption.

Let $\zeta^i = \mathrm{col}(\zeta_1^i, \ldots, \zeta_p^i)$ and define $\hat{z}_j^i := z_j^i + \zeta_j^i$. Substitute (5.35) into (5.42) and combine with (5.36) to obtain

$$\dot{\bar{e}}_{y,j}^i = -\psi_j^i \,\mathrm{sgn}(\bar{e}_{y,j}^i)|\bar{e}_{y,j}^i|^{\frac{1}{2}} - \lambda_j^i \bar{e}_{y,j}^i + \hat{z}_j^i \tag{5.43}$$

$$\dot{\hat{z}}_j^i = -\beta_j^i \,\mathrm{sgn}(\bar{e}_{y,j}^i) - \gamma_j^i \bar{e}_{y,j}^i + \dot{\zeta}_j^i \tag{5.44}$$

where $j = 1, \ldots, p$. Define constants $d_j^i > |\dot{\zeta}_j^i|$ and choose the gains from (5.35) and (5.36) as

$$\psi_j^i > 2\sqrt{d_j^i} \tag{5.45}$$

$$\lambda_j^i > 0 \tag{5.46}$$

$$\beta_j^i > d_j^i \tag{5.47}$$

$$\gamma_j^i > \frac{(\lambda_j^i)^2((\psi_j^i)^3 + \frac{5}{4}(\psi_j^i)^2 + \frac{5}{2}(\beta_j^i - d_j^i))}{\psi_j^i(\beta_j^i - d_j^i)} \tag{5.48}$$

Then, it can be proved from Theorem 5 in [190] that if the inequalities (5.45)–(5.48) are satisfied, a sliding motion will take place and $\bar{e}_{y,j}^i = \dot{\bar{e}}_{y,j}^i = 0$ in finite time.

6 **Create the output for observer $i+1$**

Assume that a sliding motion has taken place, then (5.35) and (5.42) yields $z^i = -\zeta^i$ where $z^i := \mathrm{col}(z_1^i, \ldots, z_p^i)$. Note that z^i is an available continuous signal since it is generated from $\bar{e}_{y,j}^i$ according to (5.36). Define $w^i := -e_1^i$ and partition (5.37) using (5.38)–(5.39) to obtain

$$\dot{w}^i = (\bar{A}_1^i + \bar{L}_o^i \bar{A}_{31}^i) w^i + \bar{M}_1^i f^{i+1} + \bar{Q}_1^i \xi^{i+1} \tag{5.49}$$

$$z^i = \bar{C}_2^i \bar{A}_3^i w^i + \bar{C}_2^i \bar{M}_2^i f^{i+1} \tag{5.50}$$

Define and partition

$$\bar{z}^i := (\bar{C}_2^i)^{-1} z^i := \begin{bmatrix} z_a^i \\ z_b^i \end{bmatrix} \begin{matrix} \updownarrow m^{i+1} \\ \updownarrow p - m^{i+1} \end{matrix}$$

Substituting for $\bar{A}_3^i$ from Step 4 and $\bar{M}_2^i$ from (5.30) into (5.50) results in

$$z_a^i = [0 \quad I_{m^{i+1}}]w^i \tag{5.51}$$

$$z_b^i = \bar{A}_{32}^i w^i + \begin{bmatrix} 0 & 0 \\ 0 & \bar{M}_{22}^i \end{bmatrix} f^{i+1} \tag{5.52}$$

Create a signal z_f^i in real-time as the output from a stable filter driven by z_b^i from (5.52) according to

$$\dot{z}_f^i := -\alpha^i z_f^i + \alpha^i z_b^i = -\alpha^i z_f^i + \alpha^i \bar{A}_{32}^i w^i + \begin{bmatrix} 0 & 0 \\ 0 & \alpha^i \bar{M}_{22}^i \end{bmatrix} f^{i+1},$$
$$\alpha^i \in \mathbb{R}_+ \tag{5.53}$$

Combine (5.49), (5.53) and (5.51) to obtain the system

$$\dot{x}^{i+1} = A^{i+1}x^{i+1} + M^{i+1}f^{i+1} + Q^{i+1}\xi^{i+1} \tag{5.54}$$

$$y^{i+1} = C^{i+1}x^{i+1} \tag{5.55}$$

where the state vector $x^{i+1} \in \mathbb{R}^{n^{i+1}}$ with $n^{i+1} := \bar{n}^i - m^{i+1}$ and

$$x^{i+1} := \begin{bmatrix} w^i \\ z_f^i \end{bmatrix}, \qquad y^{i+1} := \begin{bmatrix} z_a^i \\ z_f^i \end{bmatrix}, \qquad C^{i+1} := [0 \quad I_p] \tag{5.56}$$

By substituting (5.29) and (5.30) into (5.49) and (5.53), A^{i+1} and M^{i+1} can be expanded to be

$$A^{i+1} = \left[\begin{array}{cc|ccc} \bar{A}_\Omega^i & 0 & \star & 0 & 0 \\ \star & \tilde{A}_{11}^i & \star & 0 & 0 \\ \hline \star & \tilde{A}_{13}^i & \star & 0 & 0 \\ \star & 0 & \star & -\alpha^i I & 0 \\ \star & \star & \star & 0 & -\alpha^i I \end{array}\right],$$

$$M^{i+1} = \left[\begin{array}{cc} 0 & 0 \\ \bar{M}_{11}^i & 0 \\ \hline \bar{M}_{12}^i & 0 \\ 0 & 0 \\ 0 & \alpha^i \bar{M}_{22}^i \end{array}\right] \begin{array}{l} \updownarrow ih \\ \updownarrow n^i - p - (i-1)h - m^{i+1} \\ \updownarrow m^{i+1} \\ \updownarrow p - m^{i+1} - \bar{r}^i \\ \updownarrow \bar{r}^i \end{array} \tag{5.57}$$

while Q^{i+1} has no specific structure. The structure of C^{i+1} in (5.56) is due to the special structure of $\bar{A}_3^i$ obtained in (5.29). Now increment the counter i by 1 and return to Step 1.

7 **Reconstruct the fault if the Markov parameter is full rank**

Set $k = i$. Since $\text{rank}(C^k M^k) = \text{rank}(M^k)$, M_{11}^k in (5.10) and (5.18)–(5.19) does not exist since $\bar{r}^k = q$. As a result, choose $T_2^k = I_{q-\bar{r}^{k-1}} \Rightarrow f^{k+1} = f^k$ (see (5.16)). Set $\bar{A}^k = A^k$, $\bar{M}^k = M^k$, $\bar{C}^k = C^k$, $\bar{Q}^k = Q^k$, $m^{k+1} = p - q$. Also define Q_1^k, Q_2^k to be, respectively, the top $n^k - p$ and bottom p rows of Q^k. Implement and design the observer in Step 5 on the system (5.7)–(5.8). However, now design $\bar{L}_o^k$ as follows:

minimise γ with respect to the variables $\bar{P}_{11}, \bar{P}_{12}, \bar{W}_1, \gamma$ subject to:

$$\begin{bmatrix} \bar{P}_{11}A_1^k + \bar{P}_{12}A_3^k + (\star) & \bar{P}_{11}Q_1^k + \bar{P}_{12}Q_2^k & \bar{W}A_3^k \\ \star & -\gamma I_h & 0 \\ \star & 0 & -\gamma I_q \end{bmatrix} < 0, \quad \bar{P}_{11} = \bar{P}_{11}^{\mathrm{T}} > 0 \tag{5.58}$$

where $\star$ are terms that make the inequality (5.58) symmetric.

The matrices $\bar{W} := [\,\bar{W}_1\ (\bar{M}_{22}^k)^{-1}\,]$, $\bar{P}_{12} := [\,\bar{P}_{121}\ 0\,]$, $\bar{P}_{121} \in \mathbb{R}^{(n^k-p)\times(p-q)}$, are similar to the structure of W_{sc} and P_{12} in Sect. 4.5.1. Then calculate $\bar{L}_o^k = (\bar{P}_{11})^{-1}\bar{P}_{121}$. When a sliding motion has occurred, reconstruct the fault using $\hat{f}^k := \bar{W}(C_2^k)^{-1}z^k$. From the results in Sect. 4.5, $\hat{f}^k$ will reconstruct f^k together with some corruption due to ξ^k; the design of $\bar{L}_o^k$ and $\bar{W}_1$ in this step minimises the $\mathcal{L}_2$ gain from ξ^k to $\hat{f}^k$. The reconstruction of f^1 can be obtained from

$$\hat{f}^1 := (T_f^{k-1})^{-1} \ldots (T_f^2)^{-1}(T_f^1)^{-1}\hat{f}^k \tag{5.59}$$

where the T_f^i are defined in (5.16).

Remark 5.1 The design method for the final observer in Steps 5 and 7 is similar to the design method in Sect. 4.5.1. However, the linear gain $\bar{G}_l^k$ is designed somewhat differently due to the supertwisting structure employed in the observer. This is of no consequence though, as it is only the reduced order motion that governs the quality of the fault reconstruction.

Remark 5.2 The purpose of the assumption that the (unknown) signal ξ^i is obtained as the output of the low-pass filter in (5.22), is to achieve the recursive formulation in (5.54)–(5.55), where the disturbances appear in the 'state' equation. This is a generalisation of the formulation of (4.91)–(4.92) in Sect. 4.5. It should be noted that there is no 'physical' filtration of the disturbances: the filter in (5.22) only implies that ξ^i is smooth and can be considered to be the output of a low-pass filter $G^i(\mathrm{s}) := (\mathrm{s}I - A_\Omega^i)^{-1}B_\Omega^i$ driven by an unknown signal ξ^{i+1}. The choice of A_Ω^1 and B_Ω^1 is not unique. In this chapter, first order linear filter realisations have been chosen, although higher order linear filters could equally well have been used. The crucial decision is the choice of the filter bandwidth and not the particular choice of the filter itself. The relationship between the filter pairs (A^i_Ω, B^i_Ω) and the original weighting function is $\Omega(\mathrm{s}) = C_\Omega(\mathrm{s}I - A_\Omega)^{-1}B_\Omega$ where $C_\Omega = [\,I_h \quad 0_{h\times(k-2)h}\,]$ and

$$\begin{bmatrix} \dot{\xi}^1 \\ \dot{\xi}^2 \\ \vdots \\ \dot{\xi}^{k-2} \\ \dot{\xi}^{k-1} \end{bmatrix} = \underbrace{\begin{bmatrix} A^1_\Omega & B^1_\Omega & 0 & \dots & 0 \\ 0 & A^2_\Omega & B^2_\Omega & \dots & 0 \\ \vdots & \ddots & \ddots & \ddots & \vdots \\ 0 & \dots & 0 & A^{k-2}_\Omega & B^{k-2}_\Omega \\ 0 & \dots & 0 & 0 & A^{k-1}_\Omega \end{bmatrix}}_{A_\Omega} \begin{bmatrix} \xi^1 \\ \xi^2 \\ \vdots \\ \xi^{k-2} \\ \xi^{k-1} \end{bmatrix} + \underbrace{\begin{bmatrix} 0 \\ 0 \\ \vdots \\ 0 \\ B^{k-1}_\Omega \end{bmatrix}}_{B_\Omega} \xi^k$$

Modelling the characteristics of the exogenous disturbances using filters is the basis of all the $\mathcal{H}_\infty$ and μ-synthesis paradigms which are based on frequency domain assumptions on the uncertainty. There are also some parallels with the work of [227] in the sense that the uncertainty belongs to a restricted class of signals. In terms of fault estimation, it is the low frequency components that are important; for example slow incipient faults are the most difficult to identify [51]. To decouple these low frequency faults from low frequency disturbances is very important (and non-trivial).

To choose reasonable values of A^i_Ω and B^i_Ω, let the assumed bandwidth of ξ^i be ω^i_c, and choose $A^i_\Omega = -\kappa I_h$, $B^i_\Omega = \kappa I_h$ where $\kappa \in \mathbb{R}_+$. If κ is chosen to be much larger than ω^i_c, then $\xi^i \approx \xi^{i+1}$ and ultimately $\xi^k \approx \xi^1$. At the final step of the algorithm, the effect of ξ^1 on $\hat{f}^k$ is formally minimised in Step 7.

Remark 5.3 This approach is akin to the so-called 'step-by-step' methods [19, 22, 123, 277]. As the number of cascade operations increases, in practice, the accuracy of the estimation which is achieved degrades [100]. However, as argued in [22], the use of the supertwisting structure gives optimal performance at each step at least, and obviates the need to approximate the equivalent injection signals via sigmoidal approximations or low-pass filtering of discontinuous injection signals.

Remark 5.4 Since $\bar{n}^i = n^i + h$ (Step 3) and $n^{i+1} = \bar{n}^i - m^{i+1}$ (Step 6), it can be shown that

$$n^{i+1} = n^i + h - m^{i+1} \Rightarrow n^i = (i-1)h - \sum_{j=2}^{i} m^j + n^1 \tag{5.60}$$

Theorem 5.1 *If* $\mathrm{rank}(C^{n^1} M^{n^1}) < \mathrm{rank}(M^{n^1})$ *then the fault can never be fully reconstructed.*

Proof From (5.17), it can be seen that $\tilde{A}^i_1$ has $n^i - (i-1)h - p$ rows and therefore $n^i - (i-1)h - p \geq 0$. Substituting for n^i from (5.60) results in

$$n^1 - \sum_{j=2}^{i} m^j - p \geq 0 \tag{5.61}$$

Since $m^{i+1} = \mathrm{rank}(\tilde{A}^i_{31})$ and knowing that $\tilde{A}^i_{31}$ has $m^i - r^i$ rows (see Step 4), it is obvious that $m^{i+1} \leq m^i$, resulting in $0 \leq m^i \leq m^{i-1} \leq \cdots \leq m^2 \leq m^1 = p$. It follows from (5.61) that $m^i = 0$ when $i > n^1$. From (5.10), it is clear that $r^i \leq m^i$ and therefore $r^i = 0$ when $i > n^1$. Then from (5.21), $\bar{r}^i = \bar{r}^{n^1}$ when $i > n^1$ which results in $\mathrm{rank}(C^i M^i) = \mathrm{rank}(C^{n^1} M^{n^1})$ when $i > n^1$. This means that if observer n^1 is unable to reconstruct the fault, then subsequent observers will not be able to either, and the cascaded observer scheme in this chapter is not feasible. □

Remark 5.5 From (5.25)–(5.26), since $\tilde{A}^i_{31}$ has $n^i - p - (i-1)h$ columns, it follows that $n^i - p - (i-1)h \geq m^{i+1}$. Then substituting for n^i from (5.60) yields $n^{i+1} - ih > p \Rightarrow n^{i+1} > p$. This implies that for every observer, there are more states than outputs.

Remark 5.6 Notice from the structure of A^{i+1} in (5.57), the matrix $\bar{L}^i_o$ appears only in the last p columns of A^i. From the structure of C^{i+1} in (5.56), it is clear that $\bar{L}^i_o$ affects only the last p output states of x^{i+1}, and hence $\bar{L}^i_o$ will not affect the sliding motion of observer $i+1$ and also all subsequent observers. Also, it is obvious that $\bar{G}^i_l$ does not affect subsequent observers as it vanishes during sliding motion ($\bar{e}^i_y = 0$). As the fault reconstruction in Step 7 is performed during the sliding motion of observer k, it can therefore be concluded that the gains of previous observers ($\bar{L}^i_o$, $\mathcal{A}^i_s$ and subsequently $\bar{G}^i_l$, $\bar{G}^i_n$) can be arbitrarily designed as they will not affect the quality of the fault reconstruction, and only observer k needs to be designed as described in Step 7.

5.3 Existence Conditions

The method proposed in Sect. 5.2 is feasible if and only if the following are satisfied

A1. $\mathrm{rank}(C^k M^k) = \mathrm{rank}(M^k)$, i.e., $\bar{r}^k = q$ for some $1 \leq k \leq n^1$
A2. All observers have a stable sliding motion.

It is of interest to find existence conditions for the method proposed in this chapter in terms of the original matrices A^1, M^1, C^1, so that it can be easily ascertained from the beginning whether the method proposed in this chapter is applicable or not. To conveniently analyse the existence conditions, A^1, M^1, C^1 will be transformed into a special structure.

5.3.1 Overall Coordinate Transformation

In the following analysis, i is an integer $1 \leq i \leq k$ unless otherwise specified. To achieve a convenient representation of A^1, M^1, C^1, parts of the transformations T^i_1, T^i_2 and $\bar{T}^i$ (from Steps 2 and 4 in the algorithm in Sect. 5.2.1) will be used.

However, some modifications need to be made to $T_1^i, T_2^i, \bar{T}^i$ as the structure that will be aimed for will be of different order from the original system. Notice that for each observer, the system undergoes two transformations; the first one involves T_1^i and T_2^i which transforms the state and fault, respectively, so that the structures of M^i and C^i in (5.18)–(5.20) are achieved; the second transformation involves $\bar{T}^i$, applied to the augmented system to obtain the structure of $\bar{A}^i$ in (5.29). It can be seen from the process described in Sect. 5.2.1 that to get to the system for the next observer design, there is an augmentation of h states (Step 3), followed by the removal of the bottom m^1 (or p) states due to the sliding motion, and finally the addition of $m^1 - m^i$ states to the bottom of the state vector to obtain the next intermediate system (Step 6). To obtain the system for the ith observer, this process is repeated $i-1$ times on the original system (of order n^1). In order to obtain the transformation matrices for the original system, the process needs to be reversed and applied $i-1$ times to T_1^i, T_2^i and $\bar{T}^i$.

From T_1^i remove (from T_{11}^i and T_{12}^i from (5.11)–(5.12)) the sub-blocks associated with the last $m^1 - m^i$ states (i.e., the last $m^1 - m^i$ columns together with the relevant rows to make T_{11}^i and T_{12} square and invertible). Then add m^1 states to the bottom of the state space, by augmenting the truncated T_{11}^i, T_{12}^i with I_{m^1}, and then remove the first h rows and columns. Repeat this process $i-1$ times. Define the first transformation to be applied to the state of the original system as

$$T_a^i = \begin{array}{c} \\ \\ \end{array}\begin{matrix} \overset{n^1-\sum_{j=1}^{i} m^j}{\leftrightarrow} & \overset{m^i-r^i}{\leftrightarrow} & \overset{r^i}{\leftrightarrow} & \overset{\sum_{j=1}^{i-1} m^j}{\leftrightarrow} \end{matrix} \quad \begin{matrix} \overset{n^1-\sum_{j=1}^{i} m^j}{\leftrightarrow} & \overset{m^i}{\leftrightarrow} & \overset{\sum_{j=1}^{i-1} m^j}{\leftrightarrow} \end{matrix}$$

$$T_a^i = \begin{bmatrix} I & 0 & -M_{12}^i(M_{22}^i)^{-1} & 0 \\ 0 & I & 0 & 0 \\ 0 & 0 & I & 0 \\ 0 & 0 & 0 & I \end{bmatrix} \begin{bmatrix} I & 0 & 0 \\ 0 & (D^i)^{-1} & 0 \\ 0 & 0 & I \end{bmatrix}$$

Notice that for systems 1 to i, the number of potential faults remain as q. Therefore, the transformation for the fault applied to the original system is identical to T_f^i defined in (5.16).

From $\bar{T}^i$ in (5.28), remove the first h rows and columns (because it is applied to the augmented system) and repeat the process that was applied to T_1^i. The second transformation of the state to be applied to the original system is

$$T_b^i = \text{diag}\big(U_2^i, U_1^i, I_{\sum_{j=1}^{i-1} m^j + r^i}\big)$$

As the algorithm is exited at Step 2 of the kth iteration, it is clear that the coordinate transformation in Step 2 is performed k times, whereas the transformation in Step 4 is performed only $k-1$ times. For convenience of analysis in this section, the transformations T_b and T_a (Steps 2 and 4 of the algorithm) are also performed on the kth system. Define the following matrices:

$$T_x := \big(T_b^k T_a^k\big)\big(T_b^{k-1} T_a^{k-1}\big)\big(T_b^{k-2} T_a^{k-2}\big)\cdots\big(T_b^3 T_a^3\big)\big(T_b^2 T_a^2\big)\big(T_b^1 T_a^1\big) \tag{5.62}$$

$$T_f := T_f^k T_f^{k-1} \cdots T_f^3 T_f^2 T_f^1 \tag{5.63}$$

Then perform the change of coordinates $x^1 \mapsto T_x x^1$, $f^1 \mapsto f^k := T_f f^1$. By using the relationship in (5.10), (5.14) and (5.15) when applying the transformation T_a^i, and (5.25)–(5.26) and (5.27) when applying the transformation T_b^i, the following structure for A^1 is obtained:

$$A^1 \mapsto T_x A^1 T_x^{-1} = \left[\begin{array}{cc|c|c|c|c}
U_2^k \tilde{A}_1^k (U_2^k)^{-1} & \star & \dots & \star & \star & \star \\
U_1^k \tilde{A}_{31}^k (U_2^k)^{-1} & \star & \dots & \star & \star & \star \\
\tilde{A}_{32}^k & \star & \dots & \star & \star & \star \\
\hline
0 & J^k & \dots & \star & \star & \star \\
0 & 0 & \dots & \star & \star & \star \\
\star & \star & \dots & \star & \star & \star \\
\hline
\vdots & \vdots & \ddots & \vdots & \vdots & \vdots \\
\hline
0 & 0 & \dots & J^3 & \star & \star \\
0 & 0 & \dots & 0 & \star & \star \\
\star & \star & \dots & \star & \star & \star \\
\hline
0 & 0 & \dots & 0 & J^2 & \star \\
0 & 0 & \dots & 0 & 0 & \star \\
\star & \star & \dots & \star & \star & \star
\end{array}\right]
\begin{array}{l}
\updownarrow n^1 - \sum_{j=1}^k m^j \\
\updownarrow m^k - r^k \\
\updownarrow r^k \\
\updownarrow m^k \\
\updownarrow m^{k-1} - m^k - r^{k-1} \\
\updownarrow r^{k-1} \\
\vdots \\
\updownarrow m^3 \\
\updownarrow m^2 - m^3 - r^2 \\
\updownarrow r^2 \\
\updownarrow m^2 \\
\updownarrow m^1 - m^2 - r^1 \\
\updownarrow r^1
\end{array} \tag{5.64}$$

Then using (5.10), M^1 is transformed to have the structure

$$M^1 \mapsto T_x M^1 T_f^{-1} = \left[\begin{array}{c|c|c|c|c|c}
U_2^k M_{11}^k & 0 & \dots & 0 & 0 & 0 \\
0 & 0 & \dots & 0 & 0 & 0 \\
0 & M_{22}^k & \dots & 0 & 0 & 0 \\
\hline
\vdots & \vdots & \ddots & \vdots & \vdots & \vdots \\
\hline
0 & 0 & \dots & 0 & 0 & 0 \\
0 & 0 & \dots & M_{22}^3 & 0 & 0 \\
\hline
0 & 0 & \dots & 0 & 0 & 0 \\
0 & 0 & \dots & 0 & M_{22}^2 & 0 \\
\hline
0 & 0 & \dots & 0 & 0 & 0 \\
0 & 0 & \dots & 0 & 0 & M_{22}^1
\end{array}\right]
\begin{array}{l}
\updownarrow n^1 - \sum_{j=1}^k m^j \\
\updownarrow m^k - r^k \\
\updownarrow r^k \\
\vdots \\
\updownarrow m^3 - r^3 \\
\updownarrow r^3 \\
\updownarrow m^2 - r^2 \\
\updownarrow r^2 \\
\updownarrow m^1 - r^1 \\
\updownarrow r^1
\end{array} \tag{5.65}$$

$$C^1 \mapsto C^1 T_x^{-1} = \begin{bmatrix} 0 & D^1 \end{bmatrix}, \qquad \det(D^1) \neq 0 \tag{5.66}$$

where $\operatorname{rank}(M_{11}^k) = q - \sum_{j=1}^k r^j$ and $J^i := D^i \operatorname{diag}((U_1^i)^{-1}, I_{r^i})$. Note that J^i, M_{22}^i and $\tilde{A}_1^k$ are square matrices (which determine the column widths in (5.64) and (5.65)), and that $\tilde{A}_1^k$, $\tilde{A}_{31}^k$, $\tilde{A}_{32}^k$ have no particular structure.

For ease of analysis, it is convenient to first perform a change of coordinates using the following:

Proposition 5.1 *There exists a change of coordinates such that A^1 in (5.64) can be written as*

$$
A^1 = \left[\begin{array}{c|c|c|c|c|c}
U_2^k \tilde{A}_1^k (U_2^k)^{-1} & \star & \cdots & \star & \star & \star \\
U_1^k \tilde{A}_{31}^k (U_2^k)^{-1} & \star & \cdots & \star & \star & \star \\
\tilde{A}_{32}^k & \star & \cdots & \star & \star & \star \\
\hline
0 & J^k & \cdots & 0 & 0 & \star \\
0 & 0 & \cdots & 0 & 0 & \star \\
\star & \star & \cdots & \star & \star & \star \\
\hline
\vdots & \vdots & \cdots & \vdots & \vdots & \vdots \\
\hline
0 & 0 & \cdots & J^3 & 0 & \star \\
0 & 0 & \cdots & 0 & 0 & \star \\
\star & \star & \cdots & \star & \star & \star \\
\hline
0 & 0 & \cdots & 0 & J^2 & \star \\
0 & 0 & \cdots & 0 & 0 & \star \\
\star & \star & \cdots & \star & \star & \star
\end{array}\right]
\begin{array}{l}
\updownarrow n^1 - \sum_{j=1}^{k} m^j \\
\updownarrow m^k - r^k \\
\updownarrow r^k \\
\updownarrow m^k \\
\updownarrow m^{k-1} - m^k - r^{k-1} \\
\updownarrow r^{k-1} \\
\vdots \\
\updownarrow m^3 \\
\updownarrow m^2 - m^3 - r^2 \\
\updownarrow r^2 \\
\updownarrow m^2 \\
\updownarrow m^1 - m^2 - r^1 \\
\updownarrow r^1
\end{array}
\qquad (5.67)
$$

In this coordinate system, the structures of M^1 in (5.65) and C^1 from (5.66) remain unchanged.

Proof Define a transformation matrix H^i $(0 \leq i < k)$ with the structure

$$
\left[\begin{array}{c|c|c|c|c|c}
I_{n^1 - \sum_{j=1}^{k-1} m^j} & 0 & \cdots & 0 & 0 & 0 \\
\hline
0 & I_{m^{k-1}} & \cdots & 0 & \begin{array}{cc} -E_{1(i-1)}^1 (J^{k-i+1})^{-1} & 0 \\ -E_{1(i-1)}^2 (J^{k-i+1})^{-1} & 0 \\ 0 & 0 \end{array} & 0 \\
\hline
\vdots & \vdots & \ddots & \vdots & \vdots & \vdots \\
\hline
0 & 0 & \cdots & I_{m^{k-i+1}} & \begin{array}{cc} -E_{(i-1)(i-1)}^1 (J^{k-i+1})^{-1} & 0 \\ -E_{(i-1)(i-1)}^2 (J^{k-i+1})^{-1} & 0 \\ 0 & 0 \end{array} & 0 \\
\hline
0 & 0 & \cdots & 0 & I_{m^{k-i}} & 0 \\
\hline
0 & 0 & \cdots & 0 & 0 & I_{\sum_{j=1}^{k-i-1} m^j}
\end{array}\right]
\qquad (5.68)
$$

where the elements E will be formally defined below. Define

$$
\bar{H}^i := H^i H^{i-1} \cdots H^2 H^1
$$

where $H^1 = I_{n^1}$. Then define

$$\breve{A}^i := \left[\begin{array}{cc|c|c|c|c|c|c}
U_2^k \tilde{A}_1^k (U_2^k)^{-1} & \star & \dots & \star & \star & \dots & \star & \star \\
U_1^k \tilde{A}_{31}^k (U_2^k)^{-1} & \star & \dots & \star & \star & \dots & \star & \star \\
\hline
0 & J^k & \dots & 0 & E_{1i}^1 & \dots & \star & \star \\
0 & 0 & \dots & 0 & E_{1i}^2 & \dots & \star & \star \\
\star & \star & \dots & \star & \star & \dots & \star & \star \\
\hline
\vdots & \vdots & \ddots & \vdots & \vdots & \ddots & \vdots & \vdots \\
\hline
0 & 0 & \dots & J^{k-i+1} & E_{ii}^1 & \dots & \star & \star \\
0 & 0 & \dots & 0 & E_{ii}^2 & \dots & \star & \star \\
\star & \star & \dots & \star & \star & \dots & \star & \star \\
\hline
0 & 0 & \dots & 0 & J^{k-i} & \dots & \star & \star \\
0 & 0 & \dots & 0 & 0 & \dots & \star & \star \\
\star & \star & \dots & \star & \star & \dots & \star & \star \\
\hline
\vdots & \vdots & \ddots & \vdots & \vdots & \ddots & \vdots & \vdots \\
\hline
0 & 0 & \dots & 0 & 0 & \dots & J^2 & \star \\
0 & 0 & \dots & 0 & 0 & \dots & 0 & \star \\
\star & \star & \dots & \star & \star & \dots & \star & \star
\end{array}\right] \tag{5.69}$$

Firstly, perform the coordinate transformation $\bar{H}^1 A^1 (\bar{H}^1)^{-1}$ to obtain $\breve{A}^1 = A^1$ in (5.64) (because $H^1 = I_{n^1}$), from which E_{11}^1, E_{11}^2 can be obtained. Then, H^2 (and $\bar{H}^2$) can be calculated, and it can be shown that $\bar{H}^2 A^1 (\bar{H}^2)^{-1} = \breve{A}^2$. The matrices $E_{12}^1, E_{12}^2, E_{22}^1, E_{22}^2$ can then be obtained from $\breve{A}^2$, and H^3 (and $\bar{H}^3$) can be calculated to get $\bar{H}^3 A^1 (\bar{H}^3)^{-1} = \breve{A}^3$. Repeat the process until $\breve{A}^{k-1} := \bar{H}^{k-1} A^1 (\bar{H}^{k-1})^{-1}$ is obtained. It can be shown that $\breve{A}^{k-1}$ is identical to A^1 in (5.67). □

From this canonical form, the objective is to recast Conditions A1 and A2 in terms of the original system matrices A^1, M^1, C^1. The main results in this chapter are summarised in the following theorems.

Theorem 5.2 *Condition A1 is satisfied if and only if the following is satisfied*

$$\operatorname{rank}(\Xi^k) - \operatorname{rank}(\Xi^{k-1}) = \operatorname{rank}(M^1) \tag{5.70}$$

where $\Xi^i \in \mathbb{R}^{ip \times iq}$ $(0 \le i \le k)$ *is defined by*

$$\Xi^i = \begin{bmatrix}
C^1 M^1 & 0 & 0 & \dots & 0 \\
C^1 A^1 M^1 & C^1 M^1 & 0 & \dots & 0 \\
C^1 (A^1)^2 M^1 & C^1 A^1 M^1 & C^1 M^1 & \dots & 0 \\
\vdots & \vdots & \vdots & \ddots & \vdots \\
C^1 (A^1)^{i-1} M^1 & C^1 (A^1)^{i-2} M^1 & C^1 (A^1)^{i-3} M^1 & \dots & C^1 M^1
\end{bmatrix} \tag{5.71}$$

Theorem 5.3 *Condition A2 is satisfied if and only if* (A^1, M^1, C^1) *is minimum phase.*

The following subsections present constructive proofs of Theorems 5.3 and 5.2.

5.3.2 *Proof of Theorem 5.2*

Condition A1 is satisfied if and only if $\bar{r}^k = q$ which implies that $M_{11}^k = \phi$ (the empty matrix). Let K^1 be the last m^1 columns of A^1 in (5.67) and define $A_o := A^1 - K^1(C_2^1)^{-1}C^1$. Therefore A_o is identical to A^1 in (5.67) except that the last m^1 columns of A_o are zero. It can then be shown that

$$C^1 A_o^{i-1} = F^i \begin{array}{c} \begin{array}{cccccc} \overset{n^1-\sum_{j=1}^i m^j}{\leftrightarrow} & \overset{m^i}{\leftrightarrow} & \overset{m^{i-1}}{\leftrightarrow} & & \overset{m^2}{\leftrightarrow} & \overset{m^1}{\leftrightarrow} \end{array} \\ \begin{bmatrix} 0 & 0 & 0 & \dots & 0 & 0 \\ 0 & I_{m^i} & 0 & \dots & 0 & 0 \\ \star & \star & \star & \dots & \star & 0 \end{bmatrix} \end{array} \begin{array}{l} \updownarrow p-m^i-\bar{r}^{i-1} \\ \updownarrow m^i \\ \updownarrow \bar{r}^{i-1} \end{array} \tag{5.72}$$

where F^i is invertible and is defined as $F^i := \bar{D}^1\bar{D}^2 \cdots \bar{D}^{i-1}\bar{D}^i$ with

$$\bar{D}^j := \text{diag}\big(I_{p-m^j-\bar{r}^{j-1}}, J^j, I_{\bar{r}^{j-1}}\big)$$

By multiplying $C^1 A_o^{i-1}$ with M^1 in (5.65) it can be shown that $C^1 A_o^{i-1} M^1 = F^i N^i$ where $N^i \in \mathbb{R}^{p\times q}$ is defined by

$$N^i = \begin{array}{c} \begin{array}{cccccc} \overset{q-\sum_{j=1}^i r^j}{\leftrightarrow} & \overset{r^i}{\leftrightarrow} & \overset{r^{i-1}}{\leftrightarrow} & & \overset{r^2}{\leftrightarrow} & \overset{r^1}{\leftrightarrow} \end{array} \\ \begin{bmatrix} 0 & 0 & 0 & \dots & 0 & 0 \\ 0 & M_{22}^i & 0 & \dots & 0 & 0 \\ \star & \star & \star & \dots & \star & 0 \end{bmatrix} \end{array} \begin{array}{l} \updownarrow p-\bar{r}^i \\ \updownarrow r^i \\ \updownarrow \bar{r}^{i-1} \end{array} \tag{5.73}$$

Proposition 5.2 *For all positive integers* $v > i$ *the following matrix identity holds:* $F^i N^i = F^v N^i$

Proof It can be shown that $F^v N^i = \overbrace{\bar{D}^1\bar{D}^2 \cdots \bar{D}^{i-1}\bar{D}^i}^{F^i}\, \bar{D}^{i+1} \cdots \bar{D}^v N^i = F^i \bar{D}^{i+1} \cdots \bar{D}^v N^i$. From the definition of $\bar{D}^i$, it can be seen that pre-multiplying any matrix with $\bar{D}^i$ affects only the top $p - \bar{r}^{i-1}$ rows of the matrix. In addition, by knowing that $\bar{r}^{i+1} \geq \bar{r}^i$ (since $\bar{r}^{i+1} =: \bar{r}^i + r^{i+1}$) and that the top $p - \bar{r}^i$ rows of N^i are zero (see (5.73)), it can be concluded that $\bar{D}^{i+1} \cdots \bar{D}^v N^i = N^i$. Hence the proof is complete. □

Define

$$\Psi^i := \begin{bmatrix} C^1M^1 & 0 & 0 & \dots & 0 \\ C^1A_oM^1 & C^1M^1 & 0 & \dots & 0 \\ C^1(A_o)^2M^1 & C^1A_oM^1 & C^1M^1 & \dots & 0 \\ \vdots & \vdots & \vdots & \vdots & \vdots \\ C^1(A_o)^{i-1}M^1 & C^1(A_o)^{i-2}M^1 & C^1(A_o)^{i-3}M^1 & \dots & C^1M^1 \end{bmatrix} \tag{5.74}$$

then the following result can be established:

Proposition 5.3 *The matrix Ψ^i has rank $\sum_{j=1}^{i}(i+1-j)r^j$*

Proof It can be easily shown that

$$\Psi^i = \begin{bmatrix} F^1N^1 & 0 & \dots & 0 & 0 \\ F^2N^2 & F^1N^1 & \dots & 0 & 0 \\ \vdots & \vdots & \ddots & \vdots & \vdots \\ F^{i-1}N^{i-1} & F^{i-2}N^{i-2} & \dots & F^1N^1 & 0 \\ F^iN^i & F^{i-1}N^{i-1} & \dots & F^2N^2 & F^1N^1 \end{bmatrix} \tag{5.75}$$

By using Proposition 5.2, Ψ^i in (5.75) is equivalent to

$$\Psi^i = \begin{bmatrix} F^i & 0 & 0 & \dots & 0 & 0 \\ 0 & F^i & 0 & \dots & 0 & 0 \\ \vdots & \vdots & \vdots & \ddots & \vdots & \vdots \\ 0 & 0 & 0 & \dots & F^i & 0 \\ 0 & 0 & 0 & \dots & 0 & F^i \end{bmatrix} \underbrace{\begin{bmatrix} N^1 & 0 & \dots & 0 & 0 \\ N^2 & N^1 & \dots & 0 & 0 \\ \vdots & \vdots & \ddots & \vdots & \vdots \\ N^{i-1} & N^{i-2} & \dots & N^1 & 0 \\ N^i & N^{i-1} & \dots & N^2 & N^1 \end{bmatrix}}_{N}$$

By expanding N from (5.73), it can be shown that $\operatorname{rank}(N) = r^i + 2r^{i-1} + 3r^{i-2} + \dots + (i-1)r^2 + ir^1$. Since F^i is square and invertible, the proof is complete. □

Proposition 5.4 *For any positive integer i the following identity holds*

$$C^1(A^1)^i = C^1A_o^i - \sum_{h=1}^{i} C^1(A)^{h-1}R^1C^1A_o^{i-h}, \quad R^1 := -K^1\left(C_2^1\right)^{-1} \tag{5.76}$$

Proof This will be proven inductively. Assume that the following holds for any positive integer j:

$$C^1\left(A^1\right)^j = C^1\left(A^1 + R^1C^1\right)^j - \sum_{h=1}^{j} C^1\left(A^1\right)^{h-1}R^1C^1\left(A^1 + R^1C^1\right)^{j-h} \tag{5.77}$$

It is then desired to prove that if (5.77) holds, then the following holds:

$$C^1\left(A^1\right)^{j+1} = C^1\left(A^1 + R^1C^1\right)^{j+1} - \sum_{h=1}^{j+1} C^1\left(A^1\right)^{h-1}R^1C^1\left(A^1 + R^1C^1\right)^{j+1-h} \tag{5.78}$$

The right-hand-side (RHS) of (5.78) can be expanded to become

$$\begin{aligned}
&C^1(A^1+R^1C^1)^j(A^1+R^1C^1) - C^1(A^1)^jR^1C^1 \\
&\quad - \sum_{h=1}^{j} C^1(A^1)^{h-1}R^1C^1(A^1+R^1C^1)^{j-h}(A^1+R^1C^1) \\
&= \left(C^1(A^1+R^1C^1)^j - \sum_{h=1}^{j} C^1(A^1)^{h-1}R^1C^1(A^1+R^1C^1)^{j-h}\right)(A^1+R^1C^1) \\
&\quad - C^1(A^1)^jR^1C^1
\end{aligned} \tag{5.79}$$

If (5.77) holds, then (5.79) becomes

$$C^1(A^1)^j(A^1+R^1C^1) - C^1(A^1)^jR^1C^1 = C^1(A^1)^{j+1} \tag{5.80}$$

which is the LHS of (5.78). Therefore it is clear that if (5.77) holds, then (5.78) holds.

It can be easily shown that (5.77) holds when $j=1$. Hence, by induction, it can be concluded that (5.77) holds for any positive integer of j. Consequently (5.76) holds and the proof is complete. □

Corollary 5.1 *The matrices Ξ^i from (5.71) and Ψ^i from (5.74) have equal rank.*

Proof Define the following matrix which by construction is square and invertible:

$$\Phi^i := \begin{bmatrix} I_p & 0 & 0 & \dots & 0 \\ -C^1K^1(C_2^1)^{-1} & I_p & 0 & \dots & 0 \\ -C^1A^1K^1(C_2^1)^{-1} & -C^1K^1(C_2^1)^{-1} & I_p & \dots & 0 \\ \vdots & \vdots & \vdots & \vdots & \vdots \\ -C^1(A^1)^{i-2}K^1(C_2^1)^{-1} & -C^1(A^1)^{i-3}K^1(C_2^1)^{-1} & -C^1(A^1)^{i-4}K^1(C_2^1)^{-1} & \dots & I_p \end{bmatrix}$$

From Proposition 5.4, it is clear that $\Phi^i\Psi^i = \Xi^i$ and hence $\text{rank}(\Psi^i) = \text{rank}(\Xi^i)$ since Φ^i is square and invertible. □

From Corollary 5.1 and Proposition 5.3, $\text{rank}(\Xi^i) = \sum_{j=1}^{i}(i+1-j)r^j$. Then it follows that

$$\begin{aligned}
\text{rank}(\Xi^k) - \text{rank}(\Xi^{k-1}) &= \sum_{j=1}^{k}(k+1-j)r^j - \sum_{j=1}^{k-1}(k-j)r^j \\
&= r^k + \sum_{j=1}^{k-1}(k+1-j)r^j - \sum_{j=1}^{k-1}(k-j)r^j \\
&= \sum_{j=1}^{k} r^j = \bar{r}^k
\end{aligned} \tag{5.81}$$

Notice that the LHS of (5.81) is given in terms of the original system matrices A^1, M^1, C^1. Hence, Condition A1 can be recast in terms of the original system matrices as

$$\operatorname{rank}(\Xi^k) - \operatorname{rank}(\Xi^{k-1}) = \operatorname{rank}(M^1) \tag{5.82}$$

From the algorithm in Sect. 5.2.1, note that for each iteration, one observer is needed. Furthermore, the algorithm is exited at the kth iteration, which therefore implies that k observers are necessary and sufficient to reconstruct the fault. Hence, the results in this section also indicate precisely the number of observers that are required. Using the results of Theorem 5.1, the scheme in this chapter can never reconstruct the faults when $\operatorname{rank}(\Xi^{n^1}) - \operatorname{rank}(\Xi^{n^1-1}) < \operatorname{rank}(M^1)$ which results in $k \leq n^1$. Hence Theorem 5.2 is proven. □

The results of this section now enable the designer to systematically investigate the likelihood of success of this scheme. The designer can construct Ξ^i and increment i systematically from 1 until $\operatorname{rank}(\Xi^i) - \operatorname{rank}(\Xi^{i-1}) = \operatorname{rank}(M^1)$ is satisfied, and that value of i is set to be k. In addition, the user can also determine the number of observers required, as well as when the scheme in this chapter will fail.

5.3.3 *Stability of the Reduced Order Sliding Modes*

Assume that Condition A1 is already satisfied, i.e., $M_{11}^k = \phi$ (the empty matrix). Then from Sect. 4.5, observer k will have a stable sliding motion if and only if (A^k, M^k, C^k) is minimum phase.

Proposition 5.5 *(A^k, M^k, C^k) is minimum phase if and only if (A^1, M^1, C^1) is minimum phase.*

Proof The invariant zeros of (A^k, M^k, C^k) are given by the values of s that make the following matrix pencil lose rank

$$\bar{P}_{11}(\mathrm{s}) := \begin{bmatrix} \mathrm{s}I - A^k & M^k \\ C^k & 0 \end{bmatrix}$$

Using the expressions in (5.17)–(5.20) and taking $M_{11}^k = \phi$, the matrix pencil $\bar{P}_{11}(\mathrm{s})$ becomes

$$\bar{P}_{11}(\mathrm{s}) = \begin{bmatrix} \mathrm{s}I - \bar{A}_{\Omega}^{k-1} & 0 & \star & 0 \\ \star & \mathrm{s}I - \tilde{A}_1^k & \star & 0 \\ \star & -\tilde{A}_{31}^k & \star & 0 \\ \star & 0 & \star & 0 \\ \star & \star & \star & \bar{M}_{22}^k \\ 0 & 0 & C_2^k & 0 \end{bmatrix}$$

Since C_2^k and $\bar{M}_{22}^k$ are square and invertible, then $\bar{P}_{11}(\mathrm{s})$ loses rank if and only if $\bar{P}_{12}(\mathrm{s})$ loses rank, where

$$\bar{P}_{12}(\mathrm{s}) := \begin{bmatrix} \mathrm{s}I - \bar{A}_{\Omega}^{k-1} & 0 \\ \star & \mathrm{s}I - \tilde{A}_1^k \\ \star & -\tilde{A}_{31}^k \\ \star & 0 \end{bmatrix}$$

However, the matrix $\bar{A}_{\Omega}^{k-1}$ is stable, and hence the only possible unstable zeros of (A^k, M^k, C^k) are the unobservable modes of $(\tilde{A}_1^k, \tilde{A}_{31}^k)$.

The invariant zeros of (A^1, M^1, C^1) are the values of s that make the following matrix lose rank:

$$P_{21}(\mathrm{s}) = \begin{bmatrix} \mathrm{s}I - A^1 & -M^1 \\ C^1 & 0 \end{bmatrix}$$

Substitute for A^1, M^1, C^1 from (5.64)–(5.66) into $P_{21}(\mathrm{s})$. Because J^i and M_{22}^i are nonsingular, and assuming that Condition A1 is already satisfied (M_{11}^k does not exist), then it can be shown that $P_{21}(\mathrm{s})$ loses rank if and only if the following matrix pencil loses rank:

$$P_{22}(\mathrm{s}) = \begin{bmatrix} \mathrm{s}I - U_2^k\tilde{A}_1^k(U_2^k)^{-1} \\ -U_1^k\tilde{A}_{31}^k(U_2^k)^{-1} \end{bmatrix} = \begin{bmatrix} U_2^k & 0 \\ 0 & U_1^k \end{bmatrix} \begin{bmatrix} \mathrm{s}I - \tilde{A}_1^k \\ -\tilde{A}_{31}^k \end{bmatrix} (U_2^k)^{-1}$$

Since U_1^k, U_2^k are invertible, the invariant zeros of (A^1, M^1, C^1) are the unobservable modes of $(\tilde{A}_1^k, \tilde{A}_{31}^k)$. It follows that (A^k, M^k, C^k) and (A^1, M^1, C^1) have the same unstable zeros. □

From (5.49), the reduced order sliding motion matrix for the ith observer ($i < k$) is $\bar{A}_1^i + L_o^i\bar{A}_{31}^i$. In order for the sliding motion matrix to be stable, it requires that $(\bar{A}_1^i, \bar{A}_{31}^i)$ be detectable.

Proposition 5.6 *The undetectable modes (if any) of observer i are given by the undetectable modes of* $(\tilde{A}_1^i, \tilde{A}_{31}^i)$.

Proof The unobservable modes of observer i are the unobservable modes of the pair $(\bar{A}_1^i, \bar{A}_{31}^i)$, which are given by the values of s that make the following matrix pencil lose rank:

$$P_{31}^i(\mathrm{s}) = \begin{bmatrix} \mathrm{s}I - \bar{A}_1^i \\ -\bar{A}_{31}^i \end{bmatrix}$$

Substituting from (5.29) results in

$$P_{31}^i(\mathrm{s}) = \left[\begin{array}{ccc} \mathrm{s}I - \bar{A}_{\Omega}^i & 0 & 0 \\ \star & \mathrm{s}I - \tilde{A}_{11}^i & -\tilde{A}_{12}^i \\ \star & -\tilde{A}_{13}^i & \mathrm{s}I - \tilde{A}_{14}^i \\ \hline 0 & 0 & I_{m^{i+1}} \end{array}\right]$$

It is clear that $P_{31}^i(\mathrm{s})$ loses rank if and only if $P_{32}^i(\mathrm{s})$ loses rank, where

$$P_{32}^i(\mathrm{s}) := \begin{bmatrix} \mathrm{s}I - \bar{A}_\Omega^i & 0 \\ \star & \mathrm{s}I - \tilde{A}_{11}^i \\ \star & -\tilde{A}_{13}^i \end{bmatrix}$$

However, $\bar{A}_\Omega^i$ is stable, hence the values of $\mathrm{s} \in \mathbb{C}_+$ at which $P_{31}^i(\mathrm{s})$ loses rank are the undetectable modes of $(\tilde{A}_{11}^i, \tilde{A}_{13}^i)$.

From (5.25), (5.26) and (5.27), it is clear that

$$\begin{bmatrix} U_2^i & 0 \\ 0 & U_1^i \end{bmatrix} \begin{bmatrix} \mathrm{s}I - \tilde{A}_1^i \\ -\tilde{A}_{31}^i \end{bmatrix} (U_2^i)^{-1} = \begin{bmatrix} \mathrm{s}I - U_2^i \tilde{A}_1^i (U_2^i)^{-1} \\ -U_1^i \tilde{A}_{31}^i (U_2^i)^{-1} \end{bmatrix}$$

$$= \begin{bmatrix} \mathrm{s}I - \tilde{A}_{11}^i & -\tilde{A}_{12}^i \\ -\tilde{A}_{13}^i & \mathrm{s}I - \tilde{A}_{14}^i \\ 0 & I_{m^{i+1}} \\ 0 & 0 \end{bmatrix}$$

which implies that $(\tilde{A}_{11}^i, \tilde{A}_{13}^i)$ and $(\tilde{A}_1^i, \tilde{A}_{31}^i)$ have the same unobservable modes. Therefore, the undetectable modes of observer i are the undetectable modes of $(\tilde{A}_1^i, \tilde{A}_{31}^i)$. □

Proposition 5.7 *For $i < k$, the unobservable modes of $(\tilde{A}_1^i, \tilde{A}_{31}^i)$ are a subset of the unobservable modes of $(\tilde{A}_1^{i+1}, \tilde{A}_{31}^{i+1})$.*

Proof From the proof of Proposition 5.6, $(\tilde{A}_1^i, \tilde{A}_{31}^i)$ and $(\tilde{A}_{11}^i, \tilde{A}_{13}^i)$ have the same unobservable modes. Define D_x^{i+1} to be the bottom r^{i+1} rows of $(D^{i+1})^{-1}$. From (5.14) and (5.15), it can be shown that

$$\begin{bmatrix} I & -M_{12}^{i+1}(M_{22}^{i+1})^{-1} D_x^{i+1} \\ 0 & D^{i+1} \end{bmatrix} \begin{bmatrix} \mathrm{s}I - \tilde{A}_{11}^i \\ -\tilde{A}_{13}^i \end{bmatrix} = \begin{bmatrix} \mathrm{s}I - \tilde{A}_1^{i+1} \\ -\tilde{A}_3^{i+1} \end{bmatrix} = \begin{bmatrix} \mathrm{s}I - \tilde{A}_1^{i+1} \\ -\tilde{A}_{31}^{i+1} \\ -\tilde{A}_{32}^{i+1} \end{bmatrix} \tag{5.83}$$

Since D^{i+1} is nonsingular, any unobservable modes of $(\tilde{A}_{11}^i, \tilde{A}_{13}^i)$ (or equivalently, the unobservable modes of $(\tilde{A}_1^i, \tilde{A}_{31}^i)$) will be a subset of the unobservable modes of $(\tilde{A}_1^{i+1}, \tilde{A}_{31}^{i+1})$. □

If (A^1, M^1, C^1) is not minimum phase, then a stable sliding motion for observer k does not exist. However, if (A^1, M^1, C^1) is minimum phase, then a stable sliding motion exists for observer k, and $(\tilde{A}_1^k, \tilde{A}_{31}^k)$ is detectable. Then from Proposition 5.7, $(\tilde{A}_1^i, \tilde{A}_{31}^i)$ is also detectable for $i < k$, which implies that stable sliding motions exist for all the previous observers (Proposition 5.6). Therefore, A2 is satisfied if and only if (A^1, M^1, C^1) is minimum phase and Theorem 5.3 is proven. □

5.4 Design Example

The method proposed in this chapter will now be demonstrated using a model of an aircraft with state space matrices given as follows:

$$A^1 = \begin{bmatrix} -0.5137 & -0.5831 & -0.6228 & 0.0004 & 0 \\ 1.0064 & -0.6284 & -0.0352 & -0.0021 & 0 \\ 0 & 0 & -37.0000 & 0 & 0 \\ 0 & 1.7171 & 0 & -0.0166 & -9.8046 \\ 1.0000 & 0 & 0 & 0 & 0 \end{bmatrix},$$

$$M^1 = \begin{bmatrix} 0 \\ 0 \\ 37 \\ 0 \\ 0 \end{bmatrix}$$

where the states are q, α, elevator position, V_{tas}, θ and the input is the elevator command. It is assumed that the first and second rows of the matrix A^1 contain uncertainties, hence a suitable choice of Q^1 will be $Q^1 = \mathrm{col}(I_2, 0)$. Assuming that V_{tas} and θ are measurable, the matrix C^1 is already in the form $C^1 = [0 \quad I_2]$. Therefore the state equation of the system becomes

$$\dot{x}^1 = (A^1 + \Delta A^1)x + M^1 f^1 \tag{5.84}$$

where ΔA^1 is the discrepancy between the known matrix A^1 and its actual value. Equation (5.84) can be placed in the framework of (5.5) by writing

$$\Delta A^1 x^1 = Q^1 \underbrace{E x^1}_{\xi^1} \tag{5.85}$$

where $E \in \mathbb{R}^{2\times 5}$ consists of the first and second rows of ΔA^1.

From (5.85), the disturbance ξ^1 will be generated by the states x^1, which is in turn generated by the fault f^1. It can be verified that

$$C^1 M^1 = C^1 A^1 M^1 = \begin{bmatrix} 0 \\ 0 \end{bmatrix}, \qquad C^1 (A^1)^2 M^1 = \begin{bmatrix} -2.2364 \\ -23.0436 \end{bmatrix}$$

and resulting in $\mathrm{rank}(\Xi^3) - \mathrm{rank}(\Xi^2) = \mathrm{rank}(M^1)$ (hence $k = 3$), hence the fault can be reconstructed using the method in this chapter: specifically 3 observers in cascade. It can be established that $n^1 = 5, p = 2, q = 1, h = 2, \bar{r}^1 = 0$.

5.4.1 Design of Observers

It can be seen that A^1, M^1, C^1 are already in the form of (5.17)–(5.20), and hence no transformation is required resulting in $T_1^1 = I_5, T_2^1 = 1$.

The filter matrices that appropriately describe the characteristics of ξ^1 are chosen as $A_\Omega^1 = -10I_2, B_\Omega^1 = 10I_2$. Note the choice of A_Ω^1 and B_Ω^1 is not unique. In this example, first order filter linear realisations have been chosen although higher order linear filters could equally well have been chosen resulting in a different (A_Ω^1, B_Ω^1) pair. The crucial decision is the choice of the filter bandwidth and not the particular choice of filter itself. Here choosing first order filter representations minimise the order of $\bar{n}^1$. With this choice of A_Ω^1 and B_Ω^1 an augmented system of dimension $\bar{n}^1 = n^1 + h = 7$ is produced (as in (5.23)–(5.24)). It can be shown by direct calculation that $m^2 = 2$. Then, to obtain the structures in (5.29)–(5.32), a suitable transformation $\bar{T}^1$ is

$$\bar{T}^1 = \begin{bmatrix} 1 & 0 & 0 & 0 & 0 & 0 & 0 \\ 0 & 1 & 0 & 0 & 0 & 0 & 0 \\ 0 & 0 & 0 & 0 & 1 & 0 & 0 \\ 0 & 0 & 1 & 0 & 0 & 0 & 0 \\ 0 & 0 & 0 & 1.7171 & 0 & 0 & 0 \\ 0 & 0 & 0 & 0 & 0 & 1 & 0 \\ 0 & 0 & 0 & 0 & 0 & 0 & 1 \end{bmatrix}$$

For the first observer, $\bar{L}_o^1$ was calculated such that $\bar{A}_1^1 + \bar{L}_o^1\bar{A}_{31}^1$ has eigenvalues at $\{-1, -2, -3, -4, -5\}$. Next $\mathcal{A}_s^1 = \mathrm{diag}(6, 7)$ was chosen. This results in the observer gains

$$\bar{G}_l^1 = \begin{bmatrix} 38.27 & 32.22 \\ -10.97 & 9.87 \\ 11933.35 & 68749.50 \\ 199.02 & 1170.80 \\ 32.16 & 57.55 \\ 6.00 & -30.96 \\ 0.80 & -11.33 \end{bmatrix}, \qquad \bar{G}_n^1 = \begin{bmatrix} -9.57 & -10.74 \\ 27.43 & -3.29 \\ -384.95 & -2291.65 \\ -6.00 & -7.96 \\ -5.18 & -1.53 \\ 0 & 1.00 \\ 1.00 & 0 \end{bmatrix}$$

Since $p - m^2 = 0$, α^1 does not exist. It follows that the parameters for the system associated with the second observer (with order $n^2 = \bar{n}^1 - m^2 = 5$ and the number of outputs $p = 2$) are

$$A^2 = \begin{bmatrix} -10.00 & 0 & 0 & 10.74 & 9.57 \\ 0 & -10.00 & 0 & 3.29 & -27.43 \\ 0 & 0 & -37.00 & 2291.65 & 384.95 \\ 10.00 & 0 & -0.62 & 37.45 & 5.66 \\ 0 & 1.72 & -0.06 & 3.25 & 0 \end{bmatrix},$$

$$M^2 = \begin{bmatrix} 0 \\ 0 \\ 37 \\ 0 \\ 0 \end{bmatrix}, \qquad Q^2 = \begin{bmatrix} 10I_2 \\ 0_{3\times 2} \end{bmatrix}$$

It is clear that $C^2M^2 = 0$, and hence $\bar{r}^2 = 0$ which results in $r^2 = 0$. Since A^2, M^2, C^2 are already in the form of (5.17)–(5.20), then $T_1^2 = I_5, T_2^2 = 1$. Here the matrices A_Ω^2, B_Ω^2 that describe ξ^2 are chosen as $A_\Omega^2 = -10I_2$, $B_\Omega^2 = 10I_2$. The augmented system (5.23)–(5.24) can then be formed. It can be shown that $m^3 = 1$. To obtain the structure (5.29)–(5.32) as in Step 4, a suitable transformation matrix is

$$\bar{T}^2 = \begin{bmatrix} 1.00 & 0 & 0 & 0 & 0 & 0 & 0 \\ 0 & 1.00 & 0 & 0 & 0 & 0 & 0 \\ 0 & 0 & 1.00 & 0 & 0 & 0 & 0 \\ 0 & 0 & 0 & 1.00 & 0 & 0 & 0 \\ 0 & 0 & -1.59 & -0.26 & 1.00 & 0 & 0 \\ 0 & 0 & 0 & 0 & 0 & -1.59 & -0.15 \\ 0 & 0 & 0 & 0 & 0 & -0.15 & 1.59 \end{bmatrix}$$

The gain $\bar{L}_o^2$ was chosen such that $\bar{A}_1^2 + \bar{L}_o^2 \bar{A}_{31}$ has eigenvalues at $\{-1, -2, -3, -4, -5\}$. Then choosing $\mathcal{A}_s^2 = \text{diag}(3, 4)$ results in

$$\bar{G}_l^2 = \begin{bmatrix} 7.26 & 0.53 \\ 1.21 & 0.09 \\ -37.64 & 5.61 \\ -4.77 & -28.09 \\ -544.97 & 111.32 \\ 11.51 & -2.91 \\ 0.74 & 17.72 \end{bmatrix}, \qquad \bar{G}_n^2 = \begin{bmatrix} -1.82 & -0.18 \\ -0.30 & -0.03 \\ 7.56 & 0.73 \\ 1.26 & 0.12 \\ 81.12 & 7.87 \\ -1.59 & -0.15 \\ -0.15 & 1.59 \end{bmatrix}$$

The filter scalar α^2 was chosen as 1. It follows that the system for observer 3 will be of order $n^3 = \bar{n}^2 - m^3 = 6$ and the number of outputs is $p = 2$. The system matrices are

$$A^3 = \begin{bmatrix} -10.00 & 0 & 0 & 0 & -1.14 & 0 \\ 0 & -10.00 & 0 & 0 & -0.19 & 0 \\ 10.00 & 0 & -10.00 & -0 & 4.75 & 0 \\ 0 & 10.00 & 0 & -10.00 & 0.79 & 0 \\ -15.91 & -2.65 & -42.95 & -7.16 & 14.00 & 0 \\ 0 & 0 & -0.15 & 2.73 & 0 & -1.00 \end{bmatrix},$$

$$M^3 = \begin{bmatrix} 0 \\ 0 \\ 0 \\ 0 \\ 37 \\ 0 \end{bmatrix}, \qquad Q^3 = \begin{bmatrix} 10I_2 \\ 0_{4\times 2} \end{bmatrix}$$

It is obvious that $\text{rank}(C^3 M^3) = \text{rank}(M^3)$, which confirms the initial check that three observers are necessary and sufficient to reconstruct the fault f^1. Finally, a sliding mode observer can be designed based on A^3, M^3, C^3, Q^3 using Step 7 of the algorithm. It is clear that a choice of

$$D^3 = \begin{bmatrix} 0 & 1 \\ 1 & 0 \end{bmatrix}$$

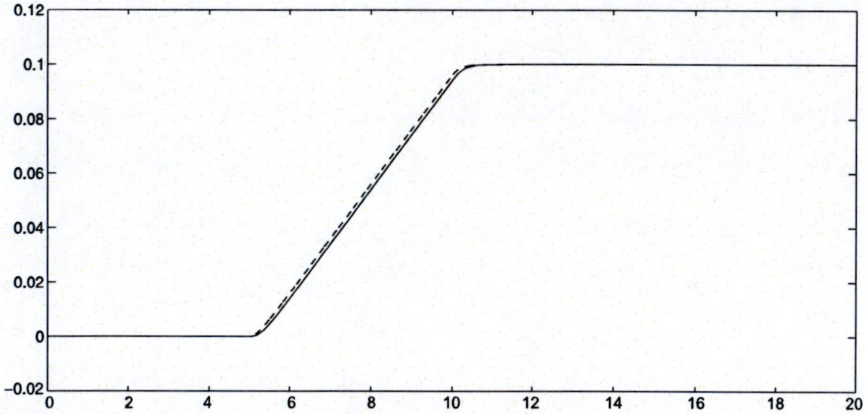

Fig. 5.2 The fault applied to the actuator (*dashed*) and its reconstruction (*solid*) for the case when $\Delta A^1 = 0$

places A^3, M^3 in the structure of (5.17)–(5.19). In minimising γ subject to (5.58), two additional inequalities were imposed so that the sliding motion of the third observer has eigenvalues that lie the following regions:

- within a conic sector centred at the origin, symmetric about the real axis with an incline of $\pm 60°$, to ensure a damping ratio of no less than 0.5;
- to the right of a vertical strip at -20, so that the eigenvalues and gains do not become excessively large.

Then by choosing $\mathcal{A}_s^3 = \{6, 7\}$ and minimising γ, yields $\gamma = 1.6050$ and

$$\bar{G}_l^3 = \begin{bmatrix} -1.14 & 0.29 \\ -0.19 & -3.37 \\ 4.75 & -1.06 \\ 0.79 & 12.67 \\ 20.00 & 0.72 \\ 0 & 4.69 \end{bmatrix}, \quad \bar{G}_n^3 = \begin{bmatrix} 0 & -0.10 \\ 0 & 1.12 \\ 0 & 0.03 \\ 0 & -0.48 \\ 1.00 & 0 \\ 0 & 1.00 \end{bmatrix}, \quad W_1 = 0.0669$$

5.4.2 *Simulation Results*

For observer 1, the gains were chosen as $\psi_1^1 = \psi_2^1 = 12.65$, $\beta_1^1 = \beta_2^1 = 20$, $\gamma_1^1 = 35.56 = \gamma_2^1 = 35.56$. For observer 2 and 3, the same gains were chosen. Firstly, the nominal uncertainty-free situation will be considered, where $\Delta A^1 = 0 \Rightarrow \xi^1 = 0$. Figure 5.2 shows the applied fault, and its reconstruction. It is clear that the reconstruction is a visually perfect replica of the fault, which shows that any degradation

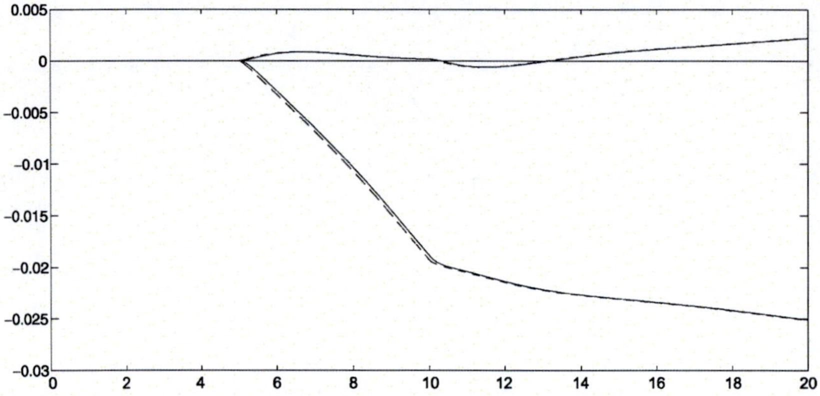

Fig. 5.3 The components of ξ^1 (*solid*) and the fictitious signal ξ^3 (*dashed*)

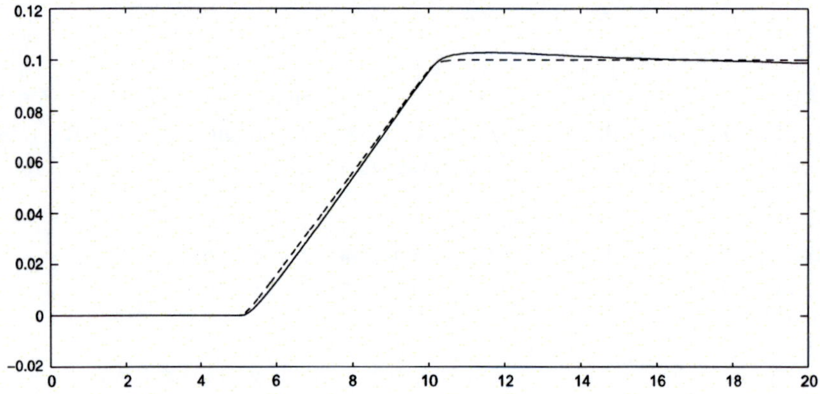

Fig. 5.4 The fault (*dashed*) and its reconstruction (*solid*) for when ΔA^1 as in (5.86)

in accuracy due to the cascade observer scheme is not significant. Next ΔA^1 was set to be

$$\Delta A^1 = \begin{bmatrix} 0.10 & 0 & 0.05 & 0 & 0.07 \\ 0 & 0 & 0 & 0 & 0 \\ -0.10 & 0 & 0.05 & 0 & -0.20 \\ 0 & 0 & 0 & 0 & 0 \\ 0 & 0 & 0 & 0 & 0 \end{bmatrix} \tag{5.86}$$

and the same fault was applied. Figure 5.3 shows the disturbances ξ^1 that arise from the applied fault, and ξ^3, which is a fictitious signal obtained from ξ^1 by performing the operation $\xi^2 = \frac{1}{10}\dot{\xi}^1 + \xi^1, \xi^3 = \frac{1}{10}\dot{\xi}^2 + \xi^2$ (which is the reverse of the fictitious filtering of ξ^3 to obtain ξ^1 using $A^1_\Omega = A^2_\Omega = -10I_2, B^1_\Omega = B^2_\Omega = 10I_2$). It can be seen in Fig. 5.3 that ξ^3 is almost identical to ξ^1 which implies the weighting function for the disturbance using the values of $A^1_\Omega = A^2_\Omega = -10I_2, B^1_\Omega = B^2_\Omega = 10I_2$ is valid for this example. Figure 5.4 shows the fault and its reconstruction. Although there is a slight degradation due to ΔA^1, the reconstruction is not severely affected by ξ^1 (which is significant—being more than 10% of the magnitude of the fault) because the fault reconstruction scheme has been designed to minimise the upper bound of the $\mathcal{L}_2$ gain from ξ^3 to $\hat{f}^1$ (where $\xi^3 \approx \xi^1$).

5.5 Summary

This chapter has presented a scheme for robust fault reconstruction using multiple observers in cascade. Signals from one observer are used as outputs of a fictitious system, and the next observer is designed based on the fictitious system. The important consequence of this setup is that it can reconstruct faults for a wider class of systems, compared to previous methods. In addition, the scheme is formulated in a framework which enables the minimisation of disturbances on the fault reconstruction. This is particularly useful in cases when the number of outputs is less than the number of disturbances and faults combined, a scenario that will render many other multiple observer methods inapplicable. Necessary and sufficient conditions, in terms of the original system matrices, have been investigated. This enables the designer to immediately determine if the scheme is applicable. In addition, the results in this chapter indicate precisely the number of observers in cascade that are required. A simulation example verifies the effectiveness of the scheme.

5.6 Notes and References

Recently, there have been other developments in fault reconstruction for systems whose first Markov parameter is not full rank. Floquet and Barbot [93, 94] transformed the system into an 'output information' form such that existing techniques can be implemented to reconstruct the faults. Higher order sliding mode schemes have also been suggested [23, 54, 100]. The work in [100] uses the concept of 'strong observability' together with higher order sliding mode observers. Strong observability has also been exploited in [23] using a hierarchy of observers. Chen and Saif [54] use a bank of high-order sliding mode differentiators to differentiate the outputs and estimate the faults from the output derivatives [54]. Floquet et al. [95, 96] suggests the use of exact differentiators to generate derivatives of the measurements to 'create' additional outputs to circumvent relative degree assumptions. However all the work in [23, 54, 93, 94, 96, 100] does not consider disturbances or uncertainty—unless the faults and disturbances are augmented and treated as 'unknown inputs' in which case the number of disturbances plus faults must not exceed the number of outputs. This results in stronger constraints which must be satisfied, and hence a smaller class of systems for which the results are applicable. Ng et al. [197] extended the work of Tan and Edwards [248] in Sect. 4.5 to relax the requirement of a full rank first Markov parameter, by exploiting two sliding mode observers in cascade.

Chapter 6
Reconstruction of Sensor Faults

Sensor fault reconstruction will be discussed in detail in this chapter. The main idea is to transform the sensor fault reconstruction scenario into an actuator fault reconstruction problem, and therefore the ideas presented in earlier chapters can directly be used. Furthermore, a design synthesis for sensor fault reconstruction in non-minimum phase systems will be considered and its significance will be discussed at length. Simulation examples will be given for each approach, to provide some insight into the design process.

6.1 Introduction

The previous chapters demonstrated the use of sliding mode observers for actuator fault reconstruction. Sensor faults are faults associated with the sensors that measure the system states and do not affect the process directly, but will do so when the measured outputs are used for feedback control. The source of these faults could be wear and tear of the sensor, prolonged use without calibration or even the total failure of a sensor. This chapter presents techniques to reconstruct sensor faults. The approaches are all based on the actuator fault reconstruction technique described in the previous chapter. Existence conditions for each technique are also discussed.

6.2 Sensor Fault Reconstruction Schemes

In this section a number of different strategies for detection of sensor faults will be discussed. The same structure of observer will be considered with suitable modifications. Indeed one approach involves transforming the sensor fault problem into a fictitious actuator fault framework so that the schemes discussed in the previous chapter can be used directly. In this chapter a version of the sliding mode observer from Sect. 4.6 will be used whereby the output estimation error is scaled by a symmetric positive definite matrix prior to use in the nonlinear injection signal.

H. Alwi et al., *Fault Detection and Fault-Tolerant Control Using Sliding Modes*,
Advances in Industrial Control,
DOI 10.1007/978-0-85729-650-4_6,

6.2.1 Preliminaries

Consider initially a nominal system subject to sensor faults as follows:

$$\dot{x}(t) = Ax(t) + Bu(t) \tag{6.1}$$

$$y(t) = Cx(t) + f_o(t) \tag{6.2}$$

where $f_o \in \mathbb{R}^p$ is the sensor fault vector. In fault-free conditions $f_o \equiv 0$. Assume that the pair (A, C) is detectable. A suitable sliding mode observer for the system in (6.1)–(6.2) is

$$\dot{\hat{x}}(t) = A\hat{x}(t) + Bu(t) - G_l e_y(t) + G_n \nu \tag{6.3}$$

$$\hat{y}(t) = C\hat{x}(t) \tag{6.4}$$

where $e_y(t) := \hat{y}(t) - y(t)$ is the output estimation error and both $G_l \in \mathbb{R}^{n\times p}$ and $G_n \in \mathbb{R}^{n\times p}$ are design matrices to be determined. The discontinuous vector ν is defined by

$$\nu = -\rho \frac{P_o e_y}{\|P_o e_y\|} \quad \text{if } e_y \neq 0 \tag{6.5}$$

where ρ is a positive scalar and $P_o = P_o^{\mathrm{T}} \in \mathbb{R}^{p\times p}$. This represents the observer formulation from Sect. 4.6

Assume without loss of generality that the system matrices A and C are in the following form:

$$A = \begin{bmatrix} A_{11} & A_{12} \\ A_{21} & A_{22} \end{bmatrix}, \qquad C = [0 \quad T] \tag{6.6}$$

where $A_{11} \in \mathbb{R}^{(n-p)\times(n-p)}$ and $T \in \mathbb{R}^{p\times p}$ is orthogonal. In the coordinates of (6.6), the matrix G_n has the structure

$$G_n = \begin{bmatrix} -LT^{\mathrm{T}} \\ T^{\mathrm{T}} \end{bmatrix} \tag{6.7}$$

where $L \in \mathbb{R}^{(n-p)\times p}$ is a design matrix.

Note that the matrix L in this section is not constrained to have the structure in (4.54).

Define $e(t) := \hat{x}(t) - x(t)$, and subtract (6.1) and (6.2) from (6.3) and (6.4), respectively, to obtain the following pair of equations:

$$\dot{e}(t) = Ae(t) - G_l e_y(t) + G_n \nu \tag{6.8}$$

$$e_y(t) = Ce(t) - f_o(t) \tag{6.9}$$

Introduce the change of coordinates $x(t) \mapsto T_L x(t)$ where

$$T_L = \begin{bmatrix} I_{n-p} & L \\ 0 & T \end{bmatrix} \tag{6.10}$$

such that A and C from (6.6) are transformed to be

$$\mathcal{A} = \begin{bmatrix} \mathcal{A}_{11} & \mathcal{A}_{12} \\ \mathcal{A}_{21} & \mathcal{A}_{22} \end{bmatrix}, \qquad \mathcal{C} = [0 \quad I_p] \tag{6.11}$$

where $\mathcal{A}_{11} = A_{11} + LA_{21}$, $\mathcal{A}_{21} = TA_{21}$. Arguing as in Sect. 4.2.1, from the PBH rank test on (A, C) in (6.6), it can be shown that the detectability of (A, C) implies that (A_{11}, A_{21}) is also detectable. Therefore, a value of L exists such that $\mathcal{A}_{11}$ is stable. Then if $\mathrm{col}(e_1, e_2) = T_L e$ where $e_1 \in \mathbb{R}^{n-p}$, (6.8) can be partitioned according to (6.11) into the following:

$$\dot{e}_1(t) = \mathcal{A}_{11}e_1(t) + \mathcal{A}_{12}\big(e_y(t) + f_o(t)\big) - \mathcal{G}_{l,1}e_y(t) \tag{6.12}$$

$$\dot{e}_y(t) + \dot{f}_o(t) = \mathcal{A}_{21}e_1(t) + \mathcal{A}_{22}\big(e_y(t) + f_o(t)\big) - \mathcal{G}_{l,2}e_y(t) + \nu \tag{6.13}$$

Assuming that $\|\dot{f}_o\|$ is bounded, for a large enough ρ a sliding motion can be attained ($e_y(t) = \dot{e}_y(t) = 0$), and the error system (6.12)–(6.13) satisfies

$$\dot{e}_1(t) = \mathcal{A}_{11}e_1(t) + \mathcal{A}_{12}f_o(t) \tag{6.14}$$

$$\dot{f}_o(t) = \mathcal{A}_{21}e_1(t) + \mathcal{A}_{22}f_o(t) + \nu_{eq} \tag{6.15}$$

where ν_{eq} is the equivalent output error injection that is required to maintain sliding motion.

6.2.2 Reconstruction in the Steady-State

Assuming that $\mathcal{A}$ is full rank, define a reconstruction for $f_o(t)$ as

$$\hat{f}_o(t) := \big(\mathcal{A}_{21}\mathcal{A}_{11}^{-1}\mathcal{A}_{12} - \mathcal{A}_{22}\big)^{-1}\nu_{eq} \tag{6.16}$$

Note that since $\mathcal{A}$ is full rank, using the Schur complement it can be seen that $(\mathcal{A}_{21}\mathcal{A}_{11}^{-1}\mathcal{A}_{12} - \mathcal{A}_{22})^{-1}$ exists. At pseudo-steady-state $\dot{e}_1(t) \approx 0$, and assuming the sensor fault is a slowly varying drift[1] such that $\dot{f}_o(t) \approx 0$, it follows from (6.14)–(6.16) that

$$\hat{f}_o(t) \to f_o(t) \tag{6.17}$$

To see the effect of neglecting $\dot{f}_o$ and $\dot{e}_1$ (i.e. analyzing only at steady-state), eliminating $e_1(t)$ from (6.14)–(6.15), it can be seen that

$$\nu_{eq} = \dot{f}_o(t) - \mathcal{A}_{21}\mathcal{A}_{11}^{-1}\dot{e}_1(t) + \big(\mathcal{A}_{21}\mathcal{A}_{11}^{-1}\mathcal{A}_{12} - \mathcal{A}_{22}\big)f_o(t) \tag{6.18}$$

[1]This is not a restrictive assumption because step sensor failures are relatively easy to detect using sliding mode observers as they usually break the sliding motion which is readily apparent from monitoring $e_y(t)$ [86].

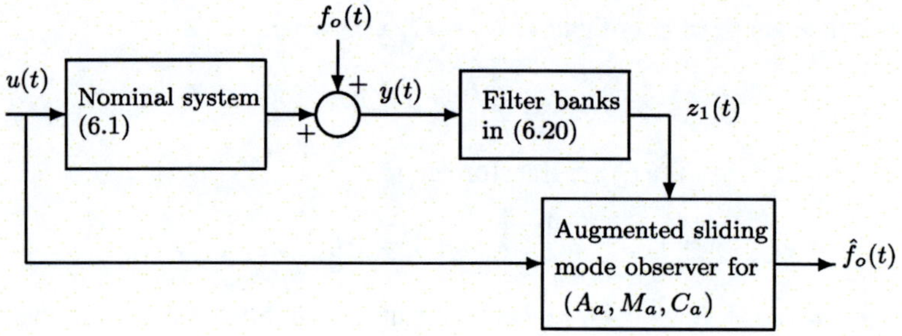

Fig. 6.1 Schematic of the sensor fault reconstruction scheme using an augmented observer method

Substituting (6.18) into (6.16), it is straightforward to see that

$$\hat{f}_o(t) = f_o(t) + \left(\mathcal{A}_{21}\mathcal{A}_{11}^{-1}\mathcal{A}_{12} - \mathcal{A}_{22}\right)^{-1}\left(\dot{f}_o(t) - \mathcal{A}_{21}\mathcal{A}_{11}^{-1}\dot{e}_1(t)\right) \quad (6.19)$$

which demonstrates how the sensor fault reconstruction in (6.16) is corrupted by $\dot{f}_o(t)$ and $\dot{e}_1(t)$. The following subsections of this chapter provide improvements to this method of sensor fault reconstruction.

6.2.3 Dynamic Sensor Fault Reconstruction

Consider a new state $z_1 \in \mathbb{R}^p$ that is a filtered version of $y(t)$

$$\dot{z}_1(t) = -A_{f,1}z_1(t) + A_{f,1}y(t) \quad (6.20)$$

where $-A_{f,1} \in \mathbb{R}^{p\times p}$ is a stable matrix. Substituting for $y(t)$ from (6.2) yields

$$\dot{z}_1(t) = -A_{f,1}z_1(t) + A_{f,1}Cx(t) + A_{f,1}f_o(t) \quad (6.21)$$

Equations (6.1) and (6.21) can be combined to form an augmented state-space system of order $n + p$:

$$\begin{bmatrix} \dot{x}(t) \\ \dot{z}_1(t) \end{bmatrix} = \underbrace{\begin{bmatrix} A & 0 \\ A_{f,1}C & -A_{f,1} \end{bmatrix}}_{A_a} \begin{bmatrix} x(t) \\ z_1(t) \end{bmatrix} + \underbrace{\begin{bmatrix} B \\ 0 \end{bmatrix}}_{B_a} u(t) + \underbrace{\begin{bmatrix} 0 \\ A_{f,1} \end{bmatrix}}_{M_a} f_o(t) \quad (6.22)$$

$$z_1(t) = \underbrace{[0 \quad I_p]}_{C_a} \begin{bmatrix} x(t) \\ z_1(t) \end{bmatrix} \quad (6.23)$$

Notice that $z_1(t)$ is a (measurable) output of a 'fictitious' system (6.22)–(6.23) which now treats the sensor fault as an 'actuator fault'. Hence (6.22)–(6.23) are in the same form as (4.43)–(4.44) and an observer of order $n + p$ can be designed for system (6.22)–(6.23) to reconstruct the fault $f_o(t)$ using the fault reconstruction technique in Sect. 4.3. A schematic diagram of this arrangement is shown in Fig. 6.1.

Compared with conditions A1 and A2 in Sect. 4.3, the sensor fault can be reconstructed accurately if and only if the following conditions are satisfied:

B1. $\text{rank}(C_a M_a) = \text{rank}(M_a)$
B2. The invariant zeros (if any) of (A_a, M_a, C_a) are stable.

By construction, from (6.22)–(6.23) it can be seen that $\text{rank}(C_a M_a) = \text{rank}(M_a)$ and hence condition B1 is satisfied. By constructing the Rosenbrock matrix for (A_a, M_a, C_a), it can be shown that the invariant zeros of (A_a, M_a, C_a) are the eigenvalues of the open-loop matrix A.

The invariant zeros of (A_a, M_a, C_a) are given by values of s for which

$$\text{rank}\begin{bmatrix} \mathrm{s}I - A & 0 & 0 \\ -A_{f,1}C & \mathrm{s}I + A_{f,1} & -A_{f,1} \\ 0 & I_p & 0 \end{bmatrix} < n + 2p$$

It is clear that

$$\text{rank}\begin{bmatrix} \mathrm{s}I - A & 0 & 0 \\ -A_{f,1}C & \mathrm{s}I + A_{f,1} & -A_{f,1} \\ 0 & I_p & 0 \end{bmatrix} = \text{rank}[\mathrm{s}I - A] + 2p$$

Therefore the invariant zeros of (A_a, M_a, C_a) are the eigenvalues of A.

Hence, in order to satisfy condition B2 for the observer formulation in this section, it is necessary and sufficient that the open-loop system be stable.

Remark 6.1 Since system (6.22)–(6.23) is square i.e. the number of faults is equal to the number of outputs, using the argument in Remark 4.4, there are no degrees of freedom remaining to ensure a robust fault reconstruction in the presence of uncertainties.

6.2.3.1 Example: A Stable Aircraft

The aircraft example from [124] is described by the following system matrices:

$$A = \begin{bmatrix} 0 & 0 & 1.0000 & 0 & 0 & 0 & 0 \\ 0 & -0.1540 & -0.0042 & 1.5400 & 0 & -0.7440 & -0.0320 \\ 0 & 0.2490 & -1.0000 & -5.2000 & 0 & 0.3370 & -1.1200 \\ 0.0386 & -0.9960 & -0.0003 & -2.1170 & 0 & 0.0200 & 0 \\ 0 & 0.5000 & 0 & 0 & -4.0000 & 0 & 0 \\ 0 & 0 & 0 & 0 & 0 & -20.0000 & 0 \\ 0 & 0 & 0 & 0 & 0 & 0 & -25.0000 \end{bmatrix}$$

$$C = \begin{bmatrix} 0 & -0.1540 & -0.0042 & 1.5400 & 0 & -0.7440 & -0.0320 \\ 0 & 0.2490 & -1.0000 & -5.2000 & 0 & 0.3370 & -1.1200 \\ 1.0000 & 0 & 0 & 0 & 0 & 0 & 0 \\ 0 & 0 & 0 & 0 & 1.0000 & 0 & 0 \end{bmatrix}$$

The states are $[\phi, r, p, \delta, x_7, \delta_r, \delta_a]$ which represent bank angle (rad), yaw rate (rad/s), roll rate (rad/s), sideslip angle (rad), washout filter state, rudder deflection

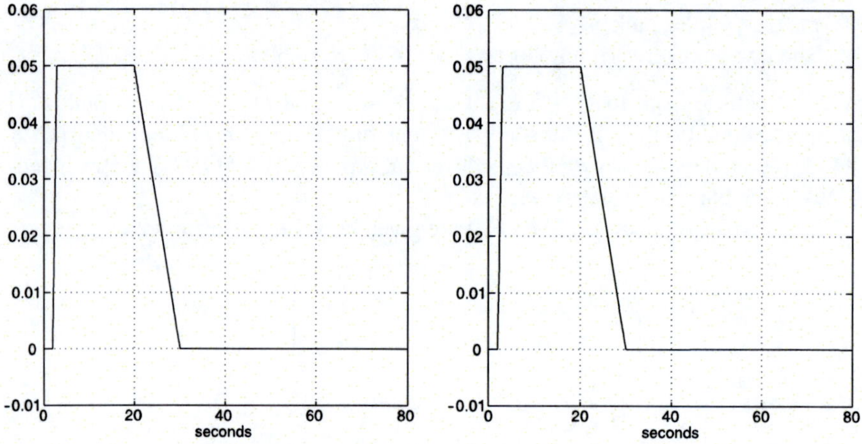

Fig. 6.2 The fault on the first sensor (*left*) and its reconstruction using the single augmented observer method in Sect. 6.2.3 (*right*)

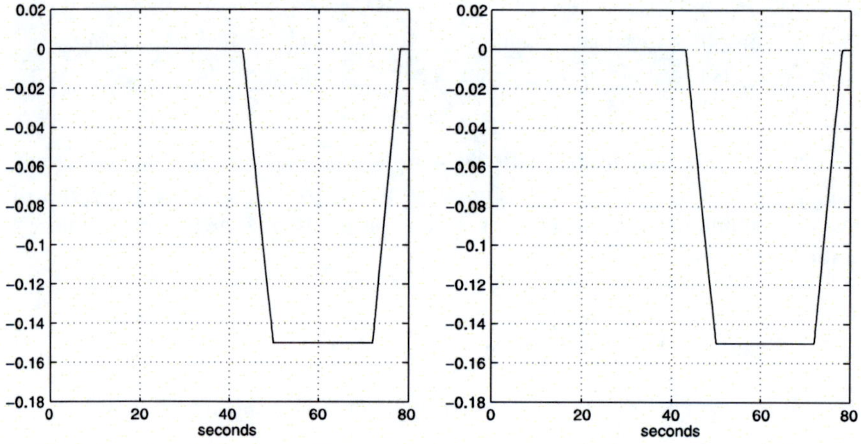

Fig. 6.3 The fault on the second sensor (*left*) and its reconstruction using the single augmented observer method in Sect. 6.2.3 (*right*)

(rad) and aileron deflection (rad). The outputs are $[r_a, \; p_a, \; \phi, x_7]$ which represent roll acceleration (rad/s^2), yaw acceleration (rad/s^2), bank angle (rad) and washout filter state. This system is open-loop stable and so the methods described above are applicable. The augmented observer has been designed using the method in Sect. 4.4. The weights $\mathbf{W}_a = 0.01 I_{11}$ and $\mathbf{V}_a = I_4$ from (4.77) and the filtering matrix $A_{f,1} = 10 I_4$ from (6.20) were chosen in designing the augmented observer. The parameters in (4.76) were chosen as $\rho_a = 50$ and $\delta_a = 1 \times 10^{-4}$.

Figures 6.2 and 6.3 show the augmented sliding mode observer faithfully reconstructing both the sensor faults.

6.2.4 Reconstructions for Unstable Systems

The only condition imposed in this section is that the system matrix A is full rank, implying that the system should not possess inherently any integral action. In the analysis in this section, the *derivative of the sensor fault is neglected* i.e. $\dot{f}_o(t) \approx 0$ which is acceptable for a slow incipient drift.

Assume a primary observer has been designed for the system (6.1)–(6.2) in Sect. 6.2.1 and that a sliding motion has been attained. The error system is then governed by (6.14)–(6.15). Consider a new state $z_2 \in \mathbb{R}^p$ which is a filtered version of ν_{eq} from (6.15)

$$\dot{z}_2(t) = -A_{f,2} z_2(t) + A_{f,2} \nu_{eq} \tag{6.24}$$

where $-A_{f,2} \in \mathbb{R}^{p \times p}$ is a stable filter matrix. Using the expression for ν_{eq} in (6.15), and setting $\dot{f}_o(t) = 0$, (6.24) becomes

$$\dot{z}_2(t) = -A_{f,2} z_2(t) - A_{f,2} \mathcal{A}_{21} e_1(t) - A_{f,2} \mathcal{A}_{22} f_o(t) \tag{6.25}$$

Combining (6.14) and (6.25), the following state-space representation is obtained:

$$\begin{bmatrix} \dot{e}_1(t) \\ \dot{z}_2(t) \end{bmatrix} = \underbrace{\begin{bmatrix} \mathcal{A}_{11} & 0 \\ -A_{f,2}\mathcal{A}_{21} & -A_{f,2} \end{bmatrix}}_{A_b} \begin{bmatrix} e_1(t) \\ z_2(t) \end{bmatrix} + \underbrace{\begin{bmatrix} \mathcal{A}_{12} \\ -A_{f,2}\mathcal{A}_{22} \end{bmatrix}}_{M_b} f_o(t) \tag{6.26}$$

$$z_2(t) = \underbrace{[0 \quad I_p]}_{C_b} \begin{bmatrix} e_1(t) \\ z_2(t) \end{bmatrix} \tag{6.27}$$

Equations (6.26)–(6.27) are now in a form similar to (4.43) and (4.44). Hence the sliding mode observer in Sect. 4.3 can be used to reconstruct the sensor fault $f_o(t)$ using the concepts described in Sect. 4.3.2. A schematic diagram of this arrangement is shown in Fig. 6.4.

From conditions A1–A2 in Sect. 4.3, the scheme described above is feasible if and only if

C1. $\text{rank}(C_b M_b) = \text{rank}(M_b)$
C2. The invariant zeros (if any) of (A_b, M_b, C_b) are stable.

From (6.26)–(6.27), $C_b M_b = -A_{f,2}\mathcal{A}_{22}$, and hence the necessary and sufficient condition for condition C1 to be satisfied, is that $\mathcal{A}_{22}$ is invertible. Applying the change of coordinates associated with the matrix

$$T_c = \begin{bmatrix} I_{n-p} & \mathcal{A}_{12}\mathcal{A}_{22}^{-1} A_{f,2}^{-1} \\ 0 & I_p \end{bmatrix} \tag{6.28}$$

to the triple (A_b, M_b, C_b) yields

$$A_b \to \begin{bmatrix} \mathcal{A}_{11} - \mathcal{A}_{12}\mathcal{A}_{22}^{-1}\mathcal{A}_{21} & \star \\ -A_{f,2}\mathcal{A}_{21} & \star \end{bmatrix}, \qquad M_b \to \begin{bmatrix} 0 \\ -A_{f,2}\mathcal{A}_{22} \end{bmatrix}, \qquad C_b \to [0 \quad I_p] \tag{6.29}$$

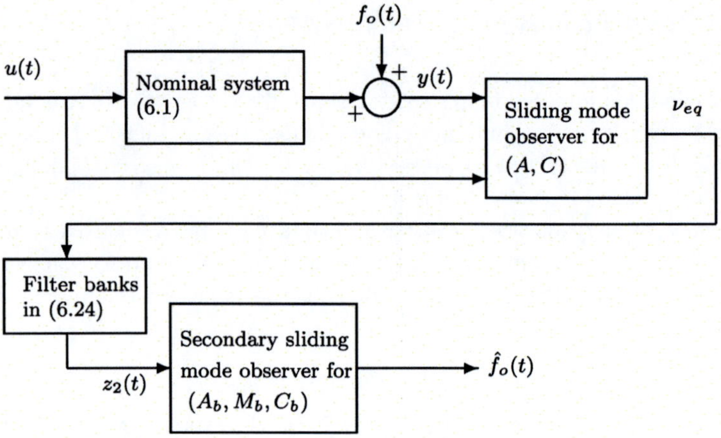

Fig. 6.4 Schematic of the sensor fault reconstruction scheme using the secondary observer method

where the entries $\star$ play no further part in the analysis. By constructing the Rosenbrock matrix for (A_b, M_b, C_b), it is clear that the invariant zeros of (A_b, M_b, C_b) are the eigenvalues of $(\mathcal{A}_{11} - \mathcal{A}_{12}\mathcal{A}_{22}^{-1}\mathcal{A}_{21})$. Therefore the existence conditions for the secondary observer are $\mathcal{A}_{22}$ must be full rank and $(\mathcal{A}_{11} - \mathcal{A}_{12}\mathcal{A}_{22}^{-1}\mathcal{A}_{21})$ is stable.

Define $\nu_{eq,b}$ as the equivalent output error injection associated with the secondary observer. Since (A_b, M_b, C_b) is already in the observer canonical form in (4.63), then from (6.29), the reconstruction for $f_o(t)$ is defined as

$$\hat{f}_o(t) := -\mathcal{A}_{22}^{-1} A_{f,2}^{-1} \nu_{eq,b} \tag{6.30}$$

which gives asymptotic reconstruction of $f_o(t)$.

It is clear that L is the only design parameter that influences the partitions $\mathcal{A}_{11}, \mathcal{A}_{12}, \mathcal{A}_{21}, \mathcal{A}_{22}$. The next section will describe a method to synthesise the design matrix so that C1 and C2 are satisfied.

6.2.4.1 Design of a Primary Observer

The following section demonstrates a systematic method for designing L in the primary observer (6.3)–(6.7) such that

- $\mathcal{A}_{22}$ is invertible (and hence C1 is satisfied);
- $\mathcal{A}_{11} - \mathcal{A}_{12}\mathcal{A}_{22}^{-1}\mathcal{A}_{21}$ is stable (and hence C2 is satisfied);
- $\mathcal{A}_{11}$ is stable so that the primary observer has a stable sliding motion.

The problem now is to make $\mathcal{A}_{22}$ invertible and $\mathcal{A}_{11} - \mathcal{A}_{12}\mathcal{A}_{22}^{-1}\mathcal{A}_{21}$ stable by choice of L, whilst retaining the property that $\mathcal{A}_{11}$ is stable. All these requirements will be incorporated with the design of the primary observer which in general can be summarised as an LMI problem as follows: Find a value of P and G_l that satisfies the following inequalities:

$$P(A - G_l C) + (A - G_l C)^{\mathrm{T}} P < 0 \tag{6.31}$$

$$P = \begin{bmatrix} P_{11} & P_{12} \\ P_{12} & P_{22} \end{bmatrix} > 0 \tag{6.32}$$

with $P_{11} \in \mathbb{R}^{(n-p)\times(n-p)}$. The gain G_n from (6.7) is then parameterised by $L = P_{11}^{-1} P_{12}$.

It is assumed throughout this section that $\det(A) \neq 0$.

The choice of L will be achieved by a two stage process. Suppose L in (6.7) is written as

$$L = L_1 + L_2 \tag{6.33}$$

where L_1 and L_2 have no particular structure. The canonical transformation in (6.10) can be decomposed as

$$T_L = T_{L,2} T_{L,1} = \begin{bmatrix} I_{n-p} & L_2 \\ 0 & T \end{bmatrix} \begin{bmatrix} I_{n-p} & L_1 \\ 0 & I_p \end{bmatrix} \tag{6.34}$$

Applying the first change of coordinates $T_{L,1}$ to the system matrix A will yield

$$\tilde{A} = \begin{bmatrix} \tilde{A}_{11} & \tilde{A}_{12} \\ \tilde{A}_{21} & \tilde{A}_{22} \end{bmatrix} := \begin{bmatrix} A_{11} + L_1 A_{21} & (-A_{11} L_1 - L_1 A_{21} L_1 + A_{12} + L_1 A_{22}) \\ A_{21} & A_{22} - A_{21} L_1 \end{bmatrix} \tag{6.35}$$

Applying the second change of coordinates $T_{L,2}$ to $\tilde{A}$, $\mathcal{A}$ can be written as

$$\mathcal{A} = \begin{bmatrix} \mathcal{A}_{11} & \mathcal{A}_{12} \\ \mathcal{A}_{21} & \mathcal{A}_{22} \end{bmatrix} = \begin{bmatrix} \tilde{A}_{11} + L_2 \tilde{A}_{21} & (-\tilde{A}_{11} L_2 - L_2 \tilde{A}_{21} L_2 + \tilde{A}_{12} + L_2 \tilde{A}_{22}) T^{\mathrm{T}} \\ T \tilde{A}_{21} & T(\tilde{A}_{22} - \tilde{A}_{21} L_2) T^{\mathrm{T}} \end{bmatrix} \tag{6.36}$$

Based on these definitions the following lemma holds.

Lemma 6.1 *The matrix $(\mathcal{A}_{11} - \mathcal{A}_{12} \mathcal{A}_{22}^{-1} \mathcal{A}_{21})$ can be expressed as $\mathcal{N} \mathcal{M}^{-1}$ where*

$$\mathcal{N} = \tilde{A}_{11} - \tilde{A}_{12} \tilde{A}_{22}^{-1} \tilde{A}_{21} \tag{6.37}$$

$$\mathcal{M} = I_{n-p} - L_2 \tilde{A}_{22}^{-1} A_{21} \tag{6.38}$$

Proof From the definitions in (6.36),

$$\begin{aligned} &\mathcal{A}_{11} - \mathcal{A}_{12} \mathcal{A}_{22}^{-1} \mathcal{A}_{21} \\ &\quad = \tilde{A}_{11} + L_2 \tilde{A}_{21} - (-\tilde{A}_{11} L_2 - L_2 \tilde{A}_{21} L_2 + \tilde{A}_{12} + L_2 \tilde{A}_{22}) \\ &\qquad \times (\tilde{A}_{22} - \tilde{A}_{21} L_2)^{-1} \tilde{A}_{21} \\ &\quad = \tilde{A}_{11} + (\tilde{A}_{11} L_2 - \tilde{A}_{12})(\tilde{A}_{22} - \tilde{A}_{21} L_2)^{-1} \tilde{A}_{21} \end{aligned} \tag{6.39}$$

From the Matrix Inversion Lemma [228],

$$(\tilde{A}_{22} - \tilde{A}_{21} L_2)^{-1} = \tilde{A}_{22}^{-1} + \tilde{A}_{22}^{-1} \tilde{A}_{21} \mathcal{M}^{-1} L_2 \tilde{A}_{22}^{-1}$$

where $\mathcal{M}$ is defined in (6.38) and hence the expression for $\mathcal{A}_{11} - \mathcal{A}_{12}\mathcal{A}_{22}^{-1}\mathcal{A}_{21}$ in (6.39) becomes

$$\tilde{A}_{11} + (\tilde{A}_{11}L_2 - \tilde{A}_{12})\tilde{A}_{22}^{-1}\tilde{A}_{21} + (\tilde{A}_{11}L_2 - \tilde{A}_{12})\tilde{A}_{22}^{-1}\tilde{A}_{21}\mathcal{M}^{-1}L_2\tilde{A}_{22}^{-1}\tilde{A}_{21} \tag{6.40}$$

By pre-multiplying (6.38) with $\mathcal{M}^{-1}$, it follows that

$$\mathcal{M}^{-1}L_2\tilde{A}_{22}^{-1}\tilde{A}_{21} = \mathcal{M}^{-1} - I_{n-p}$$

Then substituting for $\mathcal{M}^{-1}L_2\tilde{A}_{22}^{-1}\tilde{A}_{21}$ in (6.40) and expanding yields

$$\mathcal{A}_{11} - \mathcal{A}_{12}\mathcal{A}_{22}^{-1}\mathcal{A}_{21} = \tilde{A}_{11} + (\tilde{A}_{11}L_2 - \tilde{A}_{12})\tilde{A}_{22}^{-1}\tilde{A}_{21}\mathcal{M}^{-1} \tag{6.41}$$

Post-multiplying (6.38) with $\mathcal{M}^{-1}$ results in $L_2\tilde{A}_{22}^{-1}\tilde{A}_{21}\mathcal{M}^{-1} = \mathcal{M}^{-1} - I_{n-p}$. Then finally expanding (6.41) and substituting for $L_2\tilde{A}_{22}^{-1}\tilde{A}_{21}\mathcal{M}^{-1}$ results in

$$\begin{aligned}\mathcal{A}_{11} - \mathcal{A}_{12}\mathcal{A}_{22}^{-1}\mathcal{A}_{21} &= \tilde{A}_{11} - \tilde{A}_{12}\tilde{A}_{22}^{-1}\tilde{A}_{21}\mathcal{M}^{-1} + \tilde{A}_{11}\big(\mathcal{M}^{-1} - I_{n-p}\big)\\ &= \big(\tilde{A}_{11} - \tilde{A}_{12}\tilde{A}_{22}^{-1}\tilde{A}_{21}\big)\mathcal{M}^{-1}\\ &= \mathcal{N}\mathcal{M}^{-1}\end{aligned}$$

as claimed. □

Lemma 6.2 *If* $\det(A) \neq 0$ *there exists an* L_1 *such that* $\tilde{A}_{22} = (A_{22} - A_{21}L_1)$ *is full rank.*

Proof Under the assumption that $\det(A) \neq 0$ it follows

$$\operatorname{rank}[\,A_{21} \quad A_{22}\,] = p$$

Thus the matrix pencil $[\,\mathrm{s}I_p - A_{22}\; A_{21}\,]$ associated with the PBH controllability test for the fictitious pair (A_{22}, A_{21}) has full rank at $\mathrm{s} = 0$. This implies that $\mathrm{s} = 0$ is not an uncontrollable mode of (A_{22}, A_{21}). Consequently, L_1 can always be chosen so that

$$\tilde{A}_{22} = A_{22} - A_{21}L_1$$

has nonzero determinant.

□

Since by assumption $\det(A) \neq 0$ it will be assumed for the rest of this section that L_1 has been selected so that $\det(\tilde{A}_{22}) \neq 0$.

Lemma 6.3 *The matrix* $\mathcal{M}$ *defined in* (6.38) *is invertible if and only if* $\mathcal{A}_{22}$ *defined in* (6.36) *is invertible.*

Proof The matrix $\mathcal{M}$ as defined in (6.38) is a Schur complement of

$$\mathcal{M}_s := \begin{bmatrix} I_{n-p} & L_2 \\ \tilde{A}_{21} & \tilde{A}_{22} \end{bmatrix} \tag{6.42}$$

since $\tilde{A}_{22}$ is nonsingular. Therefore $\mathcal{M}$ is invertible if and only if $\mathcal{M}_s$ is invertible. However, $\mathcal{M}_s$ is invertible if and only if $\det(\tilde{A}_{22} - \tilde{A}_{21}L_2) \neq 0$ (which also represents a Schur complement). From the definitions of L, $\tilde{A}_{21}$ and $\tilde{A}_{22}$,

$$\det(\tilde{A}_{22} - \tilde{A}_{21}L_2) = \det\left(T^{\mathrm{T}}\mathcal{A}_{22}T\right) = \det(\mathcal{A}_{22})$$

since T is orthogonal. This proves that $\det(\mathcal{M}) \neq 0 \Leftrightarrow \det \mathcal{A}_{22} \neq 0$. □

Lemma 6.4 *The matrix $\mathcal{N}$ defined in (6.37) is invertible if A is invertible*

Proof If A is invertible, $\tilde{A}$ defined in (6.35) is invertible for any choice of L_1. It can be seen from (6.37) that $\mathcal{N}$ is a Schur complement for $\tilde{A}$. By design, L_1 is chosen to make $\tilde{A}_{22}$ invertible, and hence $\mathcal{N}$ is invertible. □

The problem is now to force $\mathrm{Re}(\lambda(\mathcal{N}\mathcal{M}^{-1})) < 0$ by choice of L_2. However, this can also be achieved by forcing the requirement $\mathrm{Re}(\lambda(\mathcal{M}\mathcal{N}^{-1})) < 0$. (By implication, this would mean that $\mathcal{M}\mathcal{N}^{-1}$ would have no eigenvalues at the origin, and hence will be invertible. This in turn implies that $\mathcal{M}$ will be invertible, and from Lemma 6.3, $\mathcal{A}_{22}$ will be invertible.)

The problem of selecting a L_2 so that $\mathcal{M}\mathcal{N}^{-1}$ is stable is equivalent to finding a symmetric positive definite matrix $\tilde{P} \in \mathbb{R}^{(n-p)\times(n-p)}$ and L_2 satisfying

$$\tilde{P}\mathcal{M}\mathcal{N}^{-1} + \left(\tilde{P}\mathcal{M}\mathcal{N}^{-1}\right)^{\mathrm{T}} < 0 \tag{6.43}$$

From the definition of $\mathcal{M}$ in (6.38), inequality (6.43) becomes

$$\tilde{P}\mathcal{N}^{-1} + \left(\tilde{P}\mathcal{N}^{-1}\right)^{\mathrm{T}} - \tilde{P}L_2\tilde{A}_{22}^{-1}A_{21}\mathcal{N}^{-1} - \left(\tilde{P}L_2\tilde{A}_{22}^{-1}A_{21}\mathcal{N}^{-1}\right)^{\mathrm{T}} < 0 \tag{6.44}$$

Choosing $\tilde{P} = P_{11}$ from the Lyapunov matrix in (6.32) and recalling that $L = L_1 + L_2$, substituting for L_2, inequality (6.44) is equivalent to

$$P_{11}\left(\mathcal{N}^{-1} + L_1\tilde{A}_{22}^{-1}A_{21}\mathcal{N}^{-1}\right) - P_{11}L\tilde{A}_{22}^{-1}A_{21}\mathcal{N}^{-1} + (\star) < 0 \tag{6.45}$$

where $\star$ represents the transpose of the first two terms in (6.45) so that it is symmetric. From (6.32) and $P_{11}L = P_{12}$ and hence (6.45) becomes

$$P_{11}\left(\mathcal{N}^{-1} + L_1\tilde{A}_{22}^{-1}A_{21}\mathcal{N}^{-1}\right) - P_{12}\tilde{A}_{22}^{-1}A_{21}\mathcal{N}^{-1} + (\star) < 0 \tag{6.46}$$

Therefore if inequalities (6.46) and $P_{11} > 0$ have a feasible solution, then the eigenvalues of $\mathcal{A}_{11} - \mathcal{A}_{12}\mathcal{A}_{22}^{-1}\mathcal{A}_{21}$ have negative real parts by choice of L. Inequality (6.46) can be added to inequalities (6.31) and (6.32) when designing the primary observer. (The requirement $P_{11} > 0$ is satisfied by ensuring inequality (6.32) holds.) In forcing $\mathcal{A}_{11} - \mathcal{A}_{12}\mathcal{A}_{22}^{-1}\mathcal{A}_{21}$ to be stable, the LMI variables involved are P_{11} and P_{12}.

The design problem for the primary observer to make $\mathcal{A}_{11}$ and $\mathcal{A}_{11} - \mathcal{A}_{12}\mathcal{A}_{22}^{-1}\mathcal{A}_{21}$ stable (and to incorporate the design methods from Sect. 4.4). Choose L_1 to make $\tilde{A}_{22}$ invertible. Then the design can be summarised as an LMI problem as follows:

Minimise trace(X_P) with respect to the variables $P_{11}, P_{12}, P_{22}, X_P$ subject to inequalities (6.46), (6.32) and

$$\begin{bmatrix} PA + A^{\mathrm{T}}P - C^{\mathrm{T}}\mathbf{V}^{-1}C & P \\ P & -\mathbf{W}^{-1} \end{bmatrix} < 0 \tag{6.47}$$

$$\begin{bmatrix} -P & I_n \\ I_n & -X_P \end{bmatrix} < 0 \tag{6.48}$$

where $X_P > P^{-1}$ while $\mathbf{W} \in \mathbb{R}^{n\times n}$ and $\mathbf{V} \in \mathbb{R}^{p\times p}$ are symmetric positive definite user-defined matrices.

Rather than lying in the open left-half-plane (LHP), the eigenvalues of the matrix $(\mathcal{A}_{11} - \mathcal{A}_{12}\mathcal{A}_{22}^{-1}\mathcal{A}_{21})$ can be forced to lie in a circle centred on the real axis at $(q_c, 0)$ with a radius r_c. If $q_c < 0$ and $|q_c| > r_c$, then the entire circle lies in the LHP.

Let λ_c represent an eigenvalue of $\mathcal{M}\mathcal{N}^{-1}$, and λ_c^* is its complex conjugate. The eigenvalues of $\mathcal{N}\mathcal{M}^{-1}$ will lie in the circle centred at $(q_c, 0)$ with a radius r_c if the following inequality is satisfied:

$$\begin{bmatrix} -r_c & \frac{1}{\lambda_c} - q_c \\ \frac{1}{\lambda_c^*} - q_c & -r_c \end{bmatrix} < 0 \tag{6.49}$$

From the Schur complement, inequality (6.49) implies

$$-r_c + \frac{1}{r_c}\left(\frac{1}{\lambda_c} - q_c\right)\left(\frac{1}{\lambda_c^*} - q_c\right) < 0$$

$$\Rightarrow -r_c^2 + \frac{1}{\lambda_c\lambda_c^*} - \frac{q_c}{\lambda_c} - \frac{q_c}{\lambda_c^*} + q_c^2 < 0 \tag{6.50}$$

It is clear that $\lambda_c\lambda_c^* > 0$ and consequently multiplying (6.50) by $\lambda_c\lambda_c^*$ yields

$$(q_c^2 - r_c^2)\lambda_c\lambda_c^* - q_c\lambda_c - q_c\lambda_c^* + 1 < 0 \tag{6.51}$$

If the entire circle lies in the LHP, $q_c^2 - r_c^2 > 0$ and dividing (6.51) by $q_c^2 - r_c^2$ yields

$$(\lambda_c - \tilde{q}_c)(\lambda_c^* - \tilde{q}_c) - \tilde{r}_c^2 < 0 \tag{6.52}$$

where $\tilde{r}_c = \frac{r_c}{q_c^2 - r_c^2}$ and $\tilde{q}_c = \frac{q_c}{q_c^2 - r_c^2}$. Using the Schur complement, inequality (6.52) is equivalent to the inequality

$$\begin{bmatrix} -\tilde{r}_c & \lambda_c - \tilde{q}_c \\ \lambda_c^* - \tilde{q}_c & -\tilde{r}_c \end{bmatrix} < 0 \tag{6.53}$$

From [55], there is a one-to-one mapping

$$(1, \lambda_c, \lambda_c^*) \rightarrow (\tilde{P}, \tilde{P}\mathcal{M}\mathcal{N}^{-1}, (\tilde{P}\mathcal{M}\mathcal{N}^{-1})^{\mathrm{T}})$$

and hence the eigenvalues of $\mathcal{N}\mathcal{M}^{-1}$, i.e. the eigenvalues of $\mathcal{A}_{11} - \mathcal{A}_{12}\mathcal{A}_{22}^{-1}\mathcal{A}_{21}$, can be forced to lie in the original circle if the inequality

$$\begin{bmatrix} -\tilde{r}_c\tilde{P} & \tilde{P}\mathcal{M}\mathcal{N}^{-1} - \tilde{q}_c\tilde{P} \\ (\tilde{P}\mathcal{M}\mathcal{N}^{-1})^{\mathrm{T}} - \tilde{q}_c\tilde{P} & -\tilde{r}_c\tilde{P} \end{bmatrix} < 0 \tag{6.54}$$

is satisfied. As in (6.44)–(6.45), $\tilde{P}$ is chosen to be P_{11} and hence

$$\tilde{P}\mathcal{M}\mathcal{N}^{-1} = P_{11}\mathcal{N}^{-1} - P_{12}\tilde{A}_{22}^{-1}A_{21}\mathcal{N}^{-1}$$

which is affine in the LMI variables P_{11} and P_{12}.

Lemma 6.5 *For the sensor fault reconstruction method in this section, the reconstruction of the sensor fault satisfies*

$$\hat{f}_o(t) = f_o(t) + G(\mathrm{s})\dot{f}_o(t) \tag{6.55}$$

where $G(\mathrm{s}) = \mathcal{A}_{22}^{-1}\mathcal{A}_{21}(\mathrm{s}I_{n-p} - (\mathcal{A}_{11} - \mathcal{A}_{12}\mathcal{A}_{22}^{-1}\mathcal{A}_{21}))^{-1}\mathcal{A}_{12}\mathcal{A}_{22}^{-1} - \mathcal{A}_{22}^{-1}$

Proof If $\dot{f}_o(t)$ is not neglected in the analysis, then (6.26) becomes

$$\begin{bmatrix} \dot{e}_1(t) \\ \dot{z}_2(t) \end{bmatrix} = \begin{bmatrix} \mathcal{A}_{11} & 0 \\ -A_{f,2}\mathcal{A}_{21} & -A_{f,2} \end{bmatrix} \begin{bmatrix} e_1(t) \\ z_2(t) \end{bmatrix} + \begin{bmatrix} \mathcal{A}_{12} \\ -A_{f,2}\mathcal{A}_{22} \end{bmatrix} f_o(t) + \begin{bmatrix} 0 \\ A_{f,2} \end{bmatrix} \dot{f}_o(t) \tag{6.56}$$

Applying the change of coordinates (6.28) to the system in (6.56), and assuming that the secondary sliding mode observer has been designed and a sliding motion has been achieved, then the error system associated with the secondary observer (in the observer canonical coordinates of (6.11)) will be

$$\dot{e}_{1,b}(t) = \big(\mathcal{A}_{11} - \mathcal{A}_{12}\mathcal{A}_{22}^{-1}\mathcal{A}_{21}\big)e_{1,b}(t) - \mathcal{A}_{12}\mathcal{A}_{22}^{-1}\dot{f}_o(t) \tag{6.57}$$

$$0 = -A_{f,2}\mathcal{A}_{21}e_{1,c} + A_{f,2}\mathcal{A}_{22}f_o(t) - A_{f,2}\dot{f}_o(t) + \nu_{eq,b} \tag{6.58}$$

where $e_{1,b}(t)$ is the vector that governs the sliding motion of the secondary observer. From the definition of $\hat{f}_o(t)$ in (6.30), it is easy to show from (6.58) that

$$\hat{f}_o(t) = f_o(t) - \mathcal{A}_{22}^{-1}\dot{f}_o(t) - \mathcal{A}_{22}^{-1}\mathcal{A}_{21}e_{1,b} \tag{6.59}$$

and hence combining (6.57) and (6.59) results in (6.55) as claimed. □

Lemma 6.5 shows how the sensor fault reconstruction signal in this section is corrupted by the fault derivative (which has been neglected in the analysis). From the Matrix Inversion Lemma [228],

$$\big(\mathcal{A}_{21}\mathcal{A}_{11}^{-1}\mathcal{A}_{12} - \mathcal{A}_{22}\big)^{-1} = -\mathcal{A}_{22}^{-1} - \mathcal{A}_{22}^{-1}\mathcal{A}_{21}\big(\mathcal{A}_{11} - \mathcal{A}_{12}\mathcal{A}_{22}^{-1}\mathcal{A}_{21}\big)^{-1}\mathcal{A}_{12}\mathcal{A}_{22}^{-1}$$

and hence the fault reconstruction (6.19) can be re-expressed as

$$\hat{f}_o(t) = f_o(t) + G_1\dot{f}_o(t) + G_2\dot{e}_1(t) \tag{6.60}$$

where

$$G_1 = -\mathcal{A}_{22}^{-1}\mathcal{A}_{21}\big(\mathcal{A}_{11} - \mathcal{A}_{12}\mathcal{A}_{22}^{-1}\mathcal{A}_{21}\big)^{-1}\mathcal{A}_{12}\mathcal{A}_{22}^{-1} - \mathcal{A}_{22}^{-1} \tag{6.61}$$

$$G_2 = \mathcal{A}_{22}^{-1}\mathcal{A}_{21}\big(\mathcal{A}_{11} - \mathcal{A}_{12}\mathcal{A}_{22}^{-1}\mathcal{A}_{21}\big)^{-1} \tag{6.62}$$

This can now be directly compared to the result in Lemma 6.5. In the method presented in this section, the reconstruction $\hat{f}_o(t)$ in (6.55) is not affected by $e_1(t)$. The other difference is the way the reconstructions are corrupted by $\dot{f}_o(t)$.

6.2.4.2 Example: Unstable Helicopter (All Sensors Faulty)

Consider an eighth order model of a helicopter with six outputs and four inputs, taken from [84]. The states are $[\theta,\ \phi,\ p,\ q,\ r,\ u,\ v,\ w]$ which represent pitch attitude (rad), roll attitude (rad), body roll rate (rad/s), body pitch rate (rad/s), body yaw rate (rad/s), forward velocity (ft/s), lateral velocity (ft/s), normal velocity (ft/s). The measured outputs are $[\dot{h},\ \theta,\ \phi,\ \dot{\psi},\ q,\ p]$ which represent heave velocity (ft/s), pitch attitude (rad), roll attitude (rad), heading rate (ft/s), body pitch rate (rad/s), body roll rate (rad/s). The matrices A and C that define the model are given by

$$A = \begin{bmatrix} 0 & 0 & 0 & 0.9986 & 0.0534 & 0 & 0 & 0 \\ 0 & 0 & 1.0000 & -0.0032 & 0.0595 & 0 & 0 & 0 \\ 0 & 0 & -11.5705 & -2.5446 & -0.0636 & 0.1068 & -0.0949 & 0.0071 \\ 0 & 0 & 0.4394 & -1.9982 & 0 & 0.0167 & 0.0185 & -0.0012 \\ 0 & 0 & -2.0409 & -0.4590 & -0.7350 & 0.0193 & -0.0046 & 0.0021 \\ -32.1036 & 0 & -0.5034 & 2.2979 & 0 & -0.0212 & -0.0212 & 0.0158 \\ 0.1022 & 32.0578 & -2.3472 & -0.5036 & 0.8349 & 0.0212 & -0.0379 & 0.0004 \\ -1.9110 & 1.7138 & -0.0040 & -0.0574 & 0 & 0.0140 & -0.0009 & -0.2905 \end{bmatrix}$$

$$C = \begin{bmatrix} 0 & 0 & 0 & 0 & 0 & 0.0595 & 0.0533 & -0.9968 \\ 1.0000 & 0 & 0 & 0 & 0 & 0 & 0 & 0 \\ 0 & 1.0000 & 0 & 0 & 0 & 0 & 0 & 0 \\ 0 & 0 & 0 & -0.0535 & 1.0000 & 0 & 0 & 0 \\ 0 & 0 & 1.0000 & 0 & 0 & 0 & 0 & 0 \\ 0 & 0 & 0 & 1.0000 & 0 & 0 & 0 & 0 \end{bmatrix}$$

The open-loop poles of the system are

$$\{-11.4968, -2.3036, 0.2342 \pm 0.5513i, -0.1593 \pm 0.5990i, -0.7104, -0.2923\}$$

In the following example, it is assumed that all sensors are potentially faulty. Note that det $(A) \neq 0$ and so the method described in this section is appropriate. By transforming the pair (A, C) to the coordinates of (6.6), the following sub-matrices can be extracted:

$$A_{22} = \begin{bmatrix} -0.0000 & -0.0000 & 0.0792 & -0.0127 & 1.0001 & 0.0010 \\ 0.0000 & 0.0000 & 0.1336 & 0.9919 & -0.0006 & -0.0298 \\ -0.0000 & -0.0000 & -0.7244 & 0.0760 & -0.0199 & -0.0279 \\ 0.0000 & 0.0000 & -1.8671 & -10.9450 & 4.3139 & 0.2854 \\ 0.0000 & 0.0000 & 0.1526 & 1.0816 & -2.6711 & -0.0206 \\ 0.0001 & 0.0000 & 0.1074 & 0.5664 & -0.0174 & -0.3060 \end{bmatrix}$$

$$A_{21} = \begin{bmatrix} 0.0000 & 0.0000 \\ -0.0000 & -0.0000 \\ -0.0002 & -0.0114 \\ -0.1044 & -0.0960 \\ 0.0345 & -0.0019 \\ 0.0070 & -0.0102 \end{bmatrix}$$

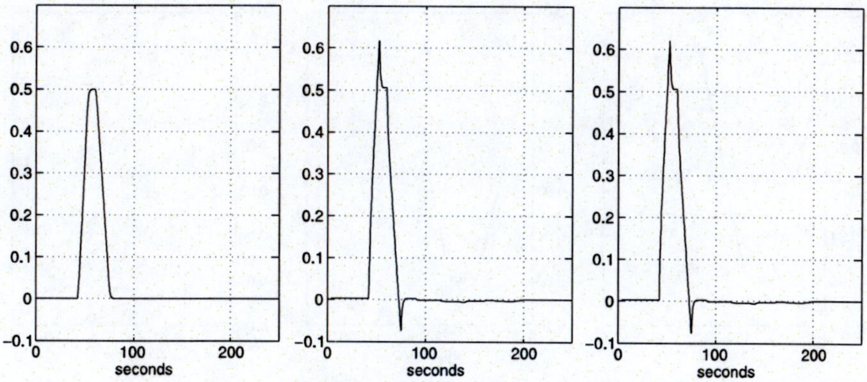

Fig. 6.5 The fault on the first sensor (*left*), its reconstruction using the secondary observer using the method in Sect. 6.2.4 (*middle*) and its reconstruction using the method in Sect. 6.2.2 (*right*)

which shows clearly that A_{22} is rank deficient. The first step is therefore to choose

$$L_1 = \begin{bmatrix} 1 & 1 & 1 & 1 & 1 & 1 \\ 1 & 1 & 1 & 1 & 1 & 1 \end{bmatrix}$$

so that $\tilde{A}_{22} = A_{22} - A_{21}L_1$ from (6.35) is full rank.

The primary observer was designed using the method in Sect. 6.2.4. The weighting matrices $\mathbf{W} = 0.1I_8$ and $\mathbf{V} = 0.1I_6$ were used. The spectrum of $\mathcal{A}_{11} - \mathcal{A}_{12}\mathcal{A}_{22}^{-1}\mathcal{A}_{21}$ is required to lie in a circle centred on $(0, -5)$ with a radius of 4.5. Implementing the synthesis procedures in Sect. 6.2.4 yields the eigenvalues of the sliding motion of the secondary observer as

$$\lambda(\mathcal{A}_{11} - \mathcal{A}_{12}\mathcal{A}_{22}^{-1}\mathcal{A}_{21}) = -8.1326 \pm 3.2306i$$

which are stable, and lie in the required circle.

The secondary observer was designed using the LQG-like procedure in Sect. 4.4 where the weighting matrices $\mathbf{W}_b = I_8$ and $\mathbf{V}_b = I_6$ and the filter $A_{f,2} = 10I_6$. In the simulations that follow, $\rho = \rho_b = 50$, $\delta = 1 \times 10^{-3}$ and $\delta_b = 1 \times 10^{-4}$. For comparison purposes, the fault reconstruction method in Sect. 6.2.2 will also be shown.

Figures 6.5–6.10 show the faults acting on the sensors and their reconstruction signals. For the fault reconstructions using the method in this section, clearly the quality of the reconstruction is not perfect. Some coupling exists between the reconstruction channels, in particular in Fig. 6.9, where the reconstruction of a fault in the fifth sensor is affected by the derivative of a fault in the third sensor, and in Fig. 6.10 where the reconstruction of a fault in the sixth sensor is affected by a fault in the second sensor. However, the faults are still reasonably well reconstructed. Importantly, all sensors could be potentially faulty. The fault reconstructions using the single observer method in Sect. 6.2.2, by comparison, are worse, in particular in the cases of the second and third sensors (Figs. 6.6–6.7).

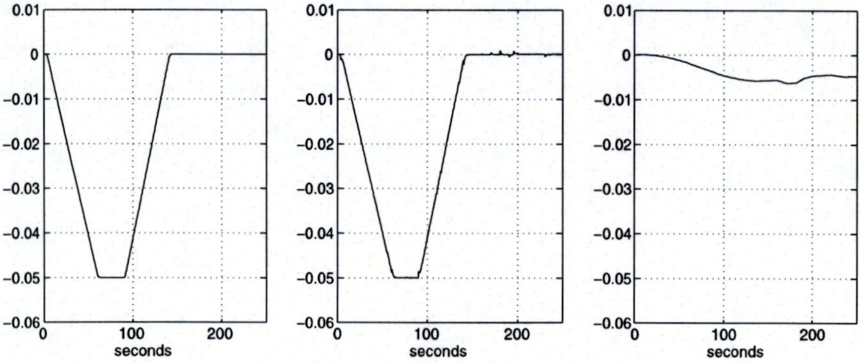

Fig. 6.6 The fault on the second sensor (*left*), its reconstruction using the secondary observer using the method in Sect. 6.2.4 (*middle*) and its reconstruction using the method in Sect. 6.2.2 (*right*)

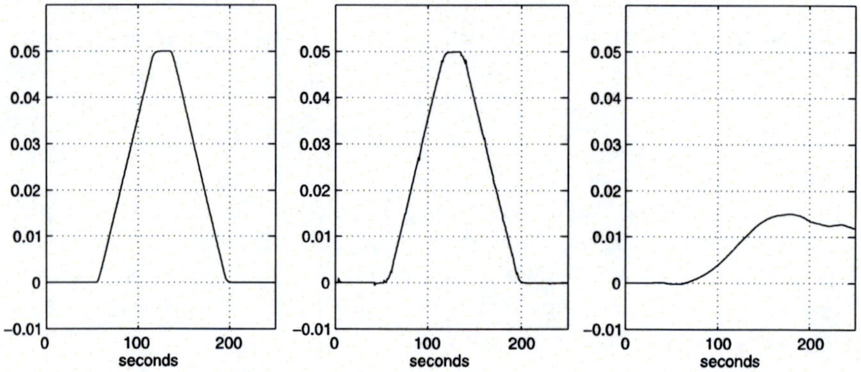

Fig. 6.7 The fault on the third sensor (*left*), its reconstruction using the secondary observer using the method in Sect. 6.2.4 (*middle*) and its reconstruction using the method in Sect. 6.2.2 (*right*)

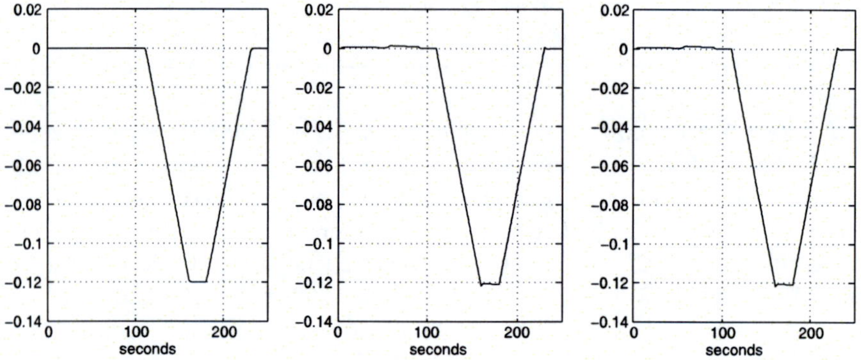

Fig. 6.8 The fault on the fourth sensor (*left*), its reconstruction using the secondary observer using the method in Sect. 6.2.4 (*middle*) and its reconstruction using the method in Sect. 6.2.2 (*right*)

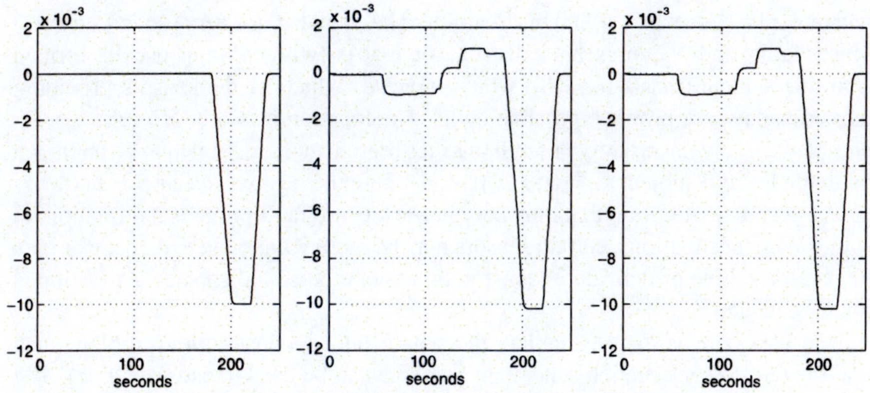

Fig. 6.9 The fault on the fifth sensor (*left*), its reconstruction using the secondary observer using the method in Sect. 6.2.4 (*middle*) and its reconstruction using the method in Sect. 6.2.2 (*right*)

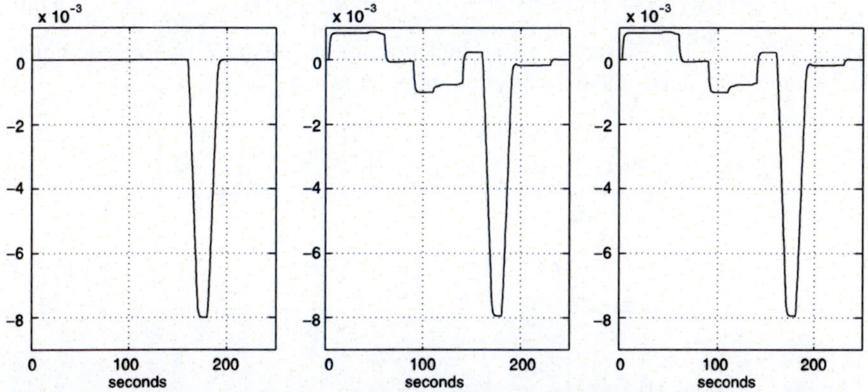

Fig. 6.10 The fault on the sixth sensor (*left*), its reconstruction using the secondary observer using the method in Sect. 6.2.4 (*middle*) and its reconstruction using the method in Sect. 6.2.2 (*right*)

6.3 Robust Sensor Fault Reconstruction

This section considers the situation in which the model of the plant is uncertain. To this end consider an uncertain system affected by sensor faults described by

$$\dot{x}(t) = Ax(t) + Bu(t) + Q\xi(t, x, u) \tag{6.63}$$

$$y(t) = Cx(t) + Nf_o(t) \tag{6.64}$$

where $A \in \mathbb{R}^{n\times n}$, $B \in \mathbb{R}^{n\times m}$, $C \in \mathbb{R}^{p\times n}$, $N \in \mathbb{R}^{p\times r}$ and $Q \in \mathbb{R}^{n\times h}$ with $n \geq p \geq r$. Assume that the matrices C and N are full row and column rank, respectively, and the function $f_o : \mathbb{R}_+ \to \mathbb{R}^r$ is unknown but bounded so that

$$\| f_o(t) \| \leq \alpha(t) \tag{6.65}$$

where $\alpha : \mathbb{R}_+ \to \mathbb{R}_+$ is a known function. The signal $f_o(t)$ represents (additive) sensor faults and N represents a distribution matrix (with columns usually formed from the standard basis for $\mathbb{R}^p$), which indicates which of the sensors providing measurements, are prone to possible faults. The function $\xi : \mathbb{R}_+ \times \mathbb{R}^p \times \mathbb{R}^m \to \mathbb{R}^h$ encapsulates any uncertainty or nonlinearities present and is assumed to be unknown but bounded and subject to $\|\xi(t,x,u)\| < \beta$ where the positive scalar β is known.

Whilst the assumption that only certain sensors are fault prone is a limitation, in some practical situations, certain sensors may be more vulnerable to damage or may be more sensitive or delicate in construction than others, and so such a situation is not unrealistic.

The idea here is to re-formulate the sensor fault reconstruction problem such that the fault reconstruction scheme in Sect. 4.5 can be used to estimate $f_o(t)$. The objective is to reconstruct the faults $f_o(t)$ as accurately as possible despite the uncertainty. As shown in Sect. 6.2.3 an effective way to do this is to first introduce a filter. Consider a new state $z_f \in \mathbb{R}^p$ that is a filtered version of y, satisfying

$$\dot{z}_f(t) = -A_f z_f(t) + A_f C x(t) + A_f N f_o(t) \tag{6.66}$$

where $-A_f \in \mathbb{R}^{p\times p}$ is a stable matrix. Equations (6.63) and (6.66) can be combined to form an augmented state-space system of order $n+p$ given by

$$\begin{bmatrix} \dot{x}(t) \\ \dot{z}_f(t) \end{bmatrix} = \underbrace{\begin{bmatrix} A & 0 \\ A_f C & -A_f \end{bmatrix}}_{A_a} \underbrace{\begin{bmatrix} x(t) \\ z_f(t) \end{bmatrix}}_{x_a} + \underbrace{\begin{bmatrix} B \\ 0 \end{bmatrix}}_{B_a} u(t) + \underbrace{\begin{bmatrix} 0 \\ A_f N \end{bmatrix}}_{M_a} f_o(t)$$

$$+ \underbrace{\begin{bmatrix} Q \\ 0 \end{bmatrix}}_{Q_a} \xi(\cdot) \tag{6.67}$$

$$z_f(t) = \underbrace{[0 \quad I_p]}_{C_a} \begin{bmatrix} x(t) \\ z_f(t) \end{bmatrix} \tag{6.68}$$

Define $x_a \in \mathbb{R}^{(n+p)}$ to be the augmented state and note that equations (6.67)–(6.68) treat the 'sensor faults' f_o as 'actuator faults'. By construction $\operatorname{rank}(C_a M_a) = r$ and it will be shown (in Proposition 6.1) that the invariant zeros of $(A_a, M_a, C_a) \subseteq \lambda(A)$ and so if the open-loop plant is stable, the system (A_a, M_a, C_a) is minimum phase.[2]

Proposition 6.1 *The invariant zeros of* (A_a, M_a, C_a) *are given by values of* s *for which*

$$\operatorname{rank}\begin{bmatrix} \mathrm{s}I_n - A & 0 \\ C & N \end{bmatrix} < n + r \tag{6.69}$$

Furthermore, the invariant zeros of $(A_a, M_a, C_a) \subseteq \lambda(A)$.

---

[2]In fact, in the set-up considered here, if $p > r$, the system (A_a, M_a, C_a) has strictly more outputs than inputs and it would be expected that the system (A_a, M_a, C_a) would have no invariant zeros.

Proof From (6.67)–(6.69), the invariant zeros of (A_a, M_a, C_a) are given by the values of s for which

$$P_a(\mathrm{s}) = \begin{bmatrix} \mathrm{s}I_n - A & 0 & 0 \\ -A_f C & \mathrm{s}I_p + A_f & -A_f N \\ 0 & I_p & 0 \end{bmatrix}$$

loses normal rank. It is straightforward to see that $P_a(\mathrm{s})$ loses normal rank if and only if

$$\operatorname{rank}\begin{bmatrix} \mathrm{s}I_n - A & 0 \\ -A_f C & -A_f N \end{bmatrix} < n + r$$

Pre-multiplying by the invertible matrix $\operatorname{diag}(I_n, -A_f^{-1})$ it is easy to see that the invariant zeros of (A_a, M_a, C_a) are given by the values of s when

$$\operatorname{rank}\begin{bmatrix} \mathrm{s}I_n - A & 0 \\ C & N \end{bmatrix} < n + r \tag{6.70}$$

Notice that if s is not an eigenvalue of A, then $\det(\mathrm{s}I_n - A) \neq 0$ and

$$\operatorname{rank}\begin{bmatrix} \mathrm{s}I_n - A & 0 \\ C & N \end{bmatrix} = n + r$$

Hence the invariant zeros of $(A_a, M_a, C_a) \subseteq \lambda(A)$ as claimed. □

Remark 6.2 If the original system matrix A is stable, then the fact that the invariant zeros of $(A_a, M_a, C_a) \subseteq \lambda(A)$ causes no difficulty. The only implication is that certain modes of the sliding motion are fixed.

For the uncertain system in (6.67)–(6.68) a sliding mode observer of the form

$$\dot{\hat{x}}(t) = A_a \hat{x}(t) + B_a u(t) - G_l e_y(t) + G_n \nu \tag{6.71}$$

will be considered. In (6.71) the discontinuous output error injection term

$$\nu = -\rho(t, y, u) \frac{P_o e_y}{\|P_o e_y\|} \quad \text{if } e_y \neq 0 \tag{6.72}$$

where $e_y(t) := C\hat{x}(t) - y(t)$ is the output estimation error and P_o is a symmetric positive definite matrix.

Equations (6.67)–(6.68) are of the same form as in Sect. 4.5 and therefore the observer design schemes from Sect. 4.5 apply.

6.3.1 An Example

The robust FDI scheme in this paper will now be demonstrated with an example, which is a VTOL aircraft model taken from [219, 276] described in Sect. 4.5.2. In the notation of (6.63)–(6.64)

$$C = \begin{bmatrix} 1 & 0 & 0 & 0 \\ 0 & 1 & 0 & 0 \\ 0 & 0 & 0 & 1 \end{bmatrix}, \qquad N = \begin{bmatrix} 0 \\ 0 \\ 1 \end{bmatrix}, \qquad Q = \begin{bmatrix} 0 \\ 0 \\ 1 \\ 0 \end{bmatrix}$$

The (parametric) uncertainty ξ is given by

$$\xi = [0 \quad \Delta a_{32} \quad \Delta a_{34}]y$$

where $\Delta a_{32} = 0.5$ and $\Delta a_{34} = 2$.

The following parameters were chosen for the design of the observer associated with the method described in Sect. 4.5. The filter matrix from (6.66) was chosen as $A_f = 20I_3$. In the notation of (6.67)–(6.68) the augmented system is described by

$$A_a = \begin{bmatrix} -9.9477 & -0.7476 & 0.2632 & 5.0337 & 0 & 0 & 0 \\ 52.1659 & 2.7452 & 5.5532 & -24.4221 & 0 & 0 & 0 \\ 26.0922 & 2.6361 & -4.1975 & -19.2774 & 0 & 0 & 0 \\ 0 & 0 & 1.0000 & 0 & 0 & 0 & 0 \\ 20.0000 & 0 & 0 & 0 & -20.0000 & 0 & 0 \\ 0 & 20.0000 & 0 & 0 & 0 & -20.0000 & 0 \\ 0 & 0 & 0 & 20.0000 & 0 & 0 & -20.0000 \end{bmatrix}$$

$$B_a = \begin{bmatrix} 0.4422 & 0.1761 \\ 3.5446 & -7.5922 \\ -5.5200 & 4.4900 \\ 0 & 0 \\ 0 & 0 \\ 0 & 0 \\ 0 & 0 \end{bmatrix}, \qquad C_a = \begin{bmatrix} 0 & 0 & 0 & 0 & 1 & 0 & 0 \\ 0 & 0 & 0 & 0 & 0 & 1 & 0 \\ 0 & 0 & 0 & 0 & 0 & 0 & 1 \end{bmatrix},$$

$$M_a = \begin{bmatrix} 0 \\ 0 \\ 0 \\ 0 \\ 0 \\ 0 \\ 20 \end{bmatrix}, \qquad Q_a = \begin{bmatrix} 0 \\ 0 \\ 1 \\ 0 \\ 0 \\ 0 \\ 0 \end{bmatrix}$$

The observer design method proposed in Sect. 4.5 was adopted. The tuning parameter for the linear component of the observer, $D_{1,a}$ from Sect. 4.5.1, was chosen as $100I_3$ and $\gamma_{o,a}$ was chosen to be unity (where the subscripts 'a' indicate the parameters are for the observer associated with sensor fault reconstruction). The

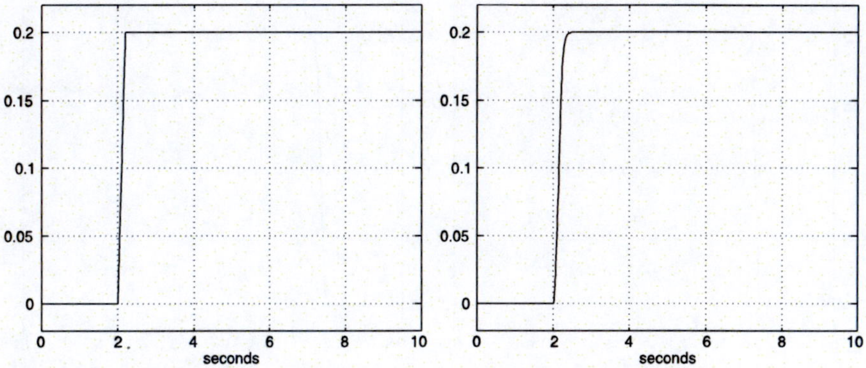

Fig. 6.11 The left subfigure is a fault on the third sensor, and the right subfigure is the reconstruction of the fault for the noise free simulation

optimisation routine yields a value of $\gamma_a = 1.6697 \times 10^{-4}$. The optimal choice of weighting matrix for the equivalent control

$$W_{sc,a} = [0.1951 \quad 0.0002 \quad -0.0500]$$

and the Lyapunov matrix associated with the unit vector component is

$$P_{o,a} = \begin{bmatrix} 0.0316 & 0.0010 & -0.0074 \\ 0.0010 & 0.0001 & -0.0005 \\ -0.0074 & -0.0005 & 0.0035 \end{bmatrix}$$

The matrix which specifies the sliding motion is

$$L_a = \begin{bmatrix} -1.6362 & 0.1615 & 0 \\ 4.7423 & 0.2166 & 0 \\ -0.3648 & -1.0119 & 0 \\ -0.9112 & 0.0034 & 0 \end{bmatrix}$$

and finally

$$E_{2,a} = \begin{bmatrix} 1.9247 & 0.1526 & -0.0011 \end{bmatrix}$$

For this simulation, $\rho_a = 50$ and $\delta_a = 1 \times 10^{-5}$.

Figure 6.11 shows the observer faithfully reconstructing the sensor fault, rejecting the effect of the uncertainty. In Fig. 6.12 the same scenario is used except that the sensor signals were subject to white noise of standard deviation of 5×10^{-4}. As before, the augmented observer replicates the sensor fault, except with noise overlaying the reconstruction.

6.4 Reconstruction in Non-minimum Phase Systems

This section also considers a method for the analysis and design of sliding mode observers for sensor fault reconstruction. The proposed scheme addresses one of

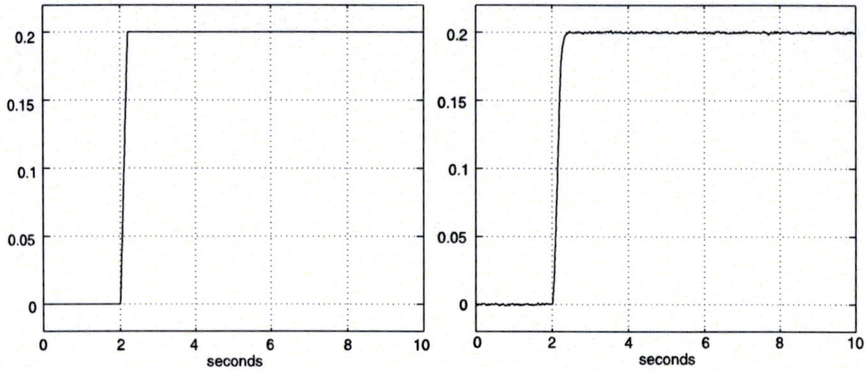

Fig. 6.12 The *left* subfigure is a fault on the third sensor, and the *right* subfigure is the reconstruction of the fault

the restrictions inherent in the earlier sliding mode estimation approaches for sensor faults which effectively require the open-loop system to be stable. For open-loop unstable systems, examples can be found, with certain combinations of sensor faults, for which the earlier sliding mode (and unknown input linear) observer schemes cannot be employed, to reconstruct faults. The method proposed in this section overcomes these limitations.

Consider a dynamical system affected by sensor faults described by

$$\dot{x}(t) = Ax(t) + Bu(t) \tag{6.73}$$

$$y(t) = Cx(t) + Nf_o(t) \tag{6.74}$$

where $A \in \mathbb{R}^{n\times n}$, $B \in \mathbb{R}^{n\times m}$, $C \in \mathbb{R}^{p\times n}$ and $N \in \mathbb{R}^{p\times r}$, and the matrices C and N have full row and column rank respectively. Also assume that the triple (A, B, C) is a minimal realisation of the fault-free input/output behaviour of the system.

Without loss of generality, it can be assumed that the outputs of the system have been reordered (and scaled if necessary) so that the matrix N has a structure

$$N = \begin{bmatrix} 0 \\ I_r \end{bmatrix} \tag{6.75}$$

The objective is to design a sliding mode observer in order to *reconstruct* the faults $f_o(t)$ using only measurements of $y(t)$ and $u(t)$. Suppose the signal $f_o(t)$ is smooth and so assume

$$\varphi(t) := \dot{f}_o(t) \tag{6.76}$$

In this section it is assumed that the sensor faults are incipient and so $\|\varphi(t)\|$ is small in magnitude, but over time the effects of the fault compound, and become significant. Equations (6.73) and (6.76) can be combined to give a system with states $x_a := \mathrm{col}(x, f_o)$ in the form

$$\begin{bmatrix} \dot{x}(t) \\ \dot{f}_o(t) \end{bmatrix} = \underbrace{\begin{bmatrix} A & 0 \\ 0 & 0 \end{bmatrix}}_{A_a} \begin{bmatrix} x(t) \\ f_o(t) \end{bmatrix} + \underbrace{\begin{bmatrix} B \\ 0 \end{bmatrix}}_{B_a} u(t) + \underbrace{\begin{bmatrix} 0 \\ I_r \end{bmatrix}}_{M_a} \varphi(t) \tag{6.77}$$

$$y(t) = \underbrace{[C \quad N]}_{C_a} \begin{bmatrix} x(t) \\ f_o(t) \end{bmatrix} \tag{6.78}$$

Equations (6.77) and (6.78) represent an unknown input problem for the triple (A_a, M_a, C_a) driven by the unmeasurable signal $\varphi(t)$. If a good estimate of $x_a(t)$ can be computed, then $f_o(t)$ can be estimated as the last r states of $x_a(t)$.

From (6.78), and based on the structure of N in (6.75),

$$C_a = [C \quad N] = \begin{bmatrix} C_1 & 0 \\ C_2 & I_r \end{bmatrix} \tag{6.79}$$

where $C_1 \in \mathbb{R}^{(p-r)\times n}$ and $C_2 \in \mathbb{R}^{r\times n}$. Notice that the triple (A_a, M_a, C_a) is inherently relative degree one since $C_a M_a = N$ and $\text{rank}(N) = r$ by assumption.

Lemma 6.6 *The triple* (A_a, M_a, C_a) *is minimum phase if and only if* (A, C_1) *is detectable.*

Proof Consider the Rosenbrock system matrix associated with (A_a, M_a, C_a):

$$R(\mathrm{s}) = \begin{bmatrix} \mathrm{s}I - A & 0 & 0 \\ 0 & \mathrm{s}I & -I_r \\ C_1 & 0 & 0 \\ C_2 & I_r & 0 \end{bmatrix} \tag{6.80}$$

The invariant zeros of (A_a, M_a, C_a) are given by the values of $\mathrm{s} \in \mathbb{C}$ where $R(\mathrm{s})$ loses normal rank. It is clear from (6.80) that

$$\text{rank}\, R(\mathrm{s}) = \text{rank} \begin{bmatrix} \mathrm{s}I - A & 0 \\ C_1 & 0 \\ C_2 & I_r \end{bmatrix} + r$$

and so $R(\mathrm{s})$ loses rank if and only if

$$\text{rank} \begin{bmatrix} \mathrm{s}I - A \\ C_1 \end{bmatrix} < n$$

It follows from the PBH rank test that the invariant zeros of (A_a, M_a, C_a) are the unobservable modes of (A, C_1). Consequently (A_a, M_a, C_a) is minimum phase if and only if (A, C_1) is detectable.

□

Lemma 6.7 *The pair* (A_a, C_a) *is observable if* (A, C_1) *does not have an unobservable mode at zero.*

Proof From the PBH test and the definition of A_a and C_a in (6.77) and (6.78), the pair (A_a, C_a) is observable if and only if

$$\text{rank} \left[\begin{array}{c|c} \mathrm{s}I - A & 0 \\ 0 & \mathrm{s}I_r \\ \hline C_1 & 0 \\ C_2 & I_r \end{array}\right] = n + r, \quad \text{for all } \mathrm{s} \in \mathbb{C} \tag{6.81}$$

For $\mathrm{s} \neq 0$

$$\begin{bmatrix} \mathrm{s}I - A & 0 \\ 0 & \mathrm{s}I_r \\ C_1 & 0 \\ C_2 & I_r \end{bmatrix} \begin{bmatrix} \eta_1 \\ \eta_2 \end{bmatrix} = 0 \Rightarrow \eta_2 = 0 \Rightarrow \begin{bmatrix} \mathrm{s}I - A \\ C_1 \\ C_2 \end{bmatrix} \eta_1 = 0 \Rightarrow \eta_1 = 0 \quad (6.82)$$

since (A, C) is observable, and so for $\mathrm{s} \neq 0$, the rank of the PBH matrix in (6.81) is $n + r$. When $\mathrm{s} = 0$,

$$\operatorname{rank} \begin{bmatrix} \mathrm{s}I - A & 0 \\ 0 & \mathrm{s}I_r \\ C_1 & 0 \\ C_2 & I_r \end{bmatrix}_{\mathrm{s}=0} = \operatorname{rank} \begin{bmatrix} -A & 0 \\ C_1 & 0 \\ C_2 & I_r \end{bmatrix} = \operatorname{rank} \begin{bmatrix} -A \\ C_1 \end{bmatrix} + r \quad (6.83)$$

Consequently (6.81) holds if and only if

$$\operatorname{rank} \begin{bmatrix} -A \\ C_1 \end{bmatrix} = n$$

A sufficient condition for this is that (A, C_1) does not have an unobservable mode at $\mathrm{s} = 0$. □

Corollary 6.1 *If the open-loop system in* (6.73) *is stable the pair* (A_a, C_a) *is observable.*

Assume without loss of generality that C from (6.74) is given as

$$C = \begin{bmatrix} C_1 \\ C_2 \end{bmatrix} = \begin{bmatrix} 0 & 0 & I_{p-r} \\ 0 & I_r & 0 \end{bmatrix} \quad (6.84)$$

For any system with C of full row rank, this canonical form can be achieved by a change of coordinates in (6.73)–(6.74). Change coordinates in the augmented system in (6.77) and (6.78) according to $x_a \mapsto T x_a$ where

$$T = \begin{bmatrix} I_n & 0 \\ C_2 & I_r \end{bmatrix} \quad (6.85)$$

The system triple in the new coordinates is $(T A_a T^{-1}, T M_a, C_a T^{-1})$ where

$$T A_a T^{-1} = \begin{bmatrix} I_n & 0 \\ C_2 & I_r \end{bmatrix} \begin{bmatrix} A & 0 \\ 0 & 0 \end{bmatrix} \begin{bmatrix} I_n & 0 \\ -C_2 & I_r \end{bmatrix} = \begin{bmatrix} A & 0 \\ C_2 A & 0 \end{bmatrix} \quad (6.86)$$

and

$$C_a T^{-1} = \begin{bmatrix} C_1 & 0 \\ C_2 & I_r \end{bmatrix} \begin{bmatrix} I_n & 0 \\ -C_2 & I_r \end{bmatrix} = \begin{bmatrix} C_1 & 0 \\ 0 & I_r \end{bmatrix} = [0 \quad I_p] \quad (6.87)$$

from the definition of C_1 in (6.84). It is also easy to check that

$$T M_a = M_a = \begin{bmatrix} 0 \\ I_r \end{bmatrix} \quad (6.88)$$

where M_a is defined in (6.77). In the original x_a coordinates, the states corresponding to f_o are given by the last r components i.e.

$$f_o(t) = C_f x_a(t) \tag{6.89}$$

where

$$C_f := [0_{r\times n} \quad I_r] \tag{6.90}$$

After the change of coordinates $x_a \mapsto Tx_a$ the new matrix relating the states to the fault signals f_o is

$$C_f T^{-1} = [0 \quad I_r]\begin{bmatrix} I & 0 \\ -C_2 & I_r \end{bmatrix} = [0_{r\times(n-p)} \quad -I_r \quad 0_{r\times(p-r)} \quad I_r] \tag{6.91}$$

using C_2 as defined in (6.84).

Remark 6.3 Although the problem tackled here is similar to the one considered in Sect. 6.2.3 and 6.2.4, the approach is different. The work in Sect. 6.2.4 employed observers in cascade, and both consider filtered output measurements as the basis of the observer design. The net effect is that, in both cases, the sensor signal estimation problem becomes an unknown input problem. This unknown input is then reconstructed using the concept of equivalent output error injection. In this section, the robustness properties of sliding mode observers will be exploited. In this respect, the approach taken here is more akin to the unknown input approaches [52, 219] whereby the fault signal to be estimated is augmented with the plant state vector, then the augmented state vector is robustly estimated using an observer.

6.4.1 Main Results

This section will consider a system, arising from the augmented sensor fault system (6.77)–(6.78), of the form

$$\dot{x}_a(t) = A_a x_a(t) + B_a u(t) + M_a \varphi(t) \tag{6.92}$$

$$y(t) = C_a x_a(t) \tag{6.93}$$

where the faults $f_o(t) = C_f x_a(t)$ and C_f is defined in (6.90). Without loss of generality (following the series of transformations described above) the matrices A_a, M_a, C_a and C_f are assumed to have the forms given in (6.86), (6.87), (6.88) and (6.91), respectively. Write

$$A_a = \begin{bmatrix} A_{11} & A_{12} \\ A_{21} & A_{22} \end{bmatrix} = \left[\begin{array}{c|c} A_{11} & A_{12} \\ \hline A_{211} & \\ A_{212} & A_{22} \end{array}\right] \tag{6.94}$$

where $A_{11} \in \mathbb{R}^{(n+r-p)\times(n+r-p)}$. Define A_{211} as the top $p-r$ rows of A_{21}. By construction, the unobservable modes of (A_{11}, A_{211}) are the invariant zeros of

(A_a, M_a, C_a) (see Lemma 4.3 in Sect. 4.3). Also define $M_2 \in \mathbb{R}^{p\times r}$ as the bottom p rows of M_a so from (6.88)

$$M_2 = \begin{bmatrix} 0_{(p-r)\times r} \\ I_r \end{bmatrix} \tag{6.95}$$

Assumption 1 *Assume that the system triple (A, B, C) is such that the new pair (A, C_1) resulting from the reordering and partitioning of the outputs as shown in (6.77)–(6.79), does not have any unobservable modes at the origin.*

Remark 6.4 It follows from Assumption 1 and Lemma 6.6 that the pair (A_a, C_a) is observable. Using the results of Lemma 6.6, Assumption 1 is equivalent to the assumption that (A_a, C_a) is observable. It is then straightforward to show using the PBH test that the pair (A_{11}, A_{21}) from the partition in (6.94) is also observable.

6.4.1.1 Observer Analysis

For the system in (6.77)–(6.78) a sliding mode observer of the form

$$\dot{\hat{x}}(t) = A_a\hat{x}(t) + B_a u(t) - G_l e_y(t) + G_n \nu \tag{6.96}$$

will be considered. In (6.96) the discontinuous output error injection term

$$\nu = -\rho(t, y, u)\frac{P_o e_y}{\|P_o e_y\|} \quad \text{if } e_y \neq 0 \tag{6.97}$$

where $e_y(t) := C_a\hat{x}(t) - y(t)$ is the output estimation error and P_o is a symmetric positive definite matrix. The matrix G_l is a traditional Luenberger observer gain used to make $(A_a - G_l C_a)$ stable. The function $\rho(\cdot)$ must be an upper bound on the uncertainty and the faults; for details see Sect. 4.5.

An appropriate gain G_n for the nonlinear injection term ν in (6.96) is

$$G_n = \begin{bmatrix} -L \\ I_p \end{bmatrix} \quad \text{where } L = [\, L_1 \quad L_2 \,] \tag{6.98}$$

and $L_1 \in \mathbb{R}^{(n+r-p)\times(p-r)}$ and $L_2 \in \mathbb{R}^{(n+r-p)\times r}$ represent design freedom [82, 256]. In particular the gain L must be chosen so that $A_{11} + LA_{21}$ is stable.

If $e := \hat{x} - x_a$ is the estimation error, then from (6.92) and (6.96)

$$\dot{e}(t) = (A_a - G_l C_a)e(t) - M_a\varphi + G_n\nu \tag{6.99}$$

where φ is defined in (6.76) and represents the derivative of the sensor fault signal. For an appropriate choice of $\rho(t, y, u)$ in (6.97), it can be shown using arguments similar to those used in Sect. 4.5 that an ideal sliding motion takes place on

$$\mathcal{S} = \{e : C_a e = 0\}$$

in finite time. During the ideal sliding motion $e_y = \dot{e}_y = 0$ and the equivalent output error injection term

$$\nu_{eq} = -(C_a G_n)^{-1}(C_a A_a e - C_a M_a \varphi) \tag{6.100}$$

Substituting from (6.100) into (6.99), it follows the sliding motion is governed by the system

$$\dot{e}(t) = \big(A_a - G_n(C_aG_n)^{-1}C_aA_a\big)e(t) - \big(M_a - G_n(C_aG_n)^{-1}C_aM_a\big)\varphi(t) \quad (6.101)$$

Ideally the effect of the unknown disturbance φ on the state estimation, particularly on the states which correspond to estimates of f_o, needs to be minimised.

The effect of φ on the estimate of f_o is given by $C_f e(t)$, where $e(t)$ evolves according to (6.101) since $\hat{f}_o - f_o = C_f e(t)$ if $\hat{f}_o(t) := C_f\hat{x}(t)$. Therefore, the impact of φ on the estimate of f_o can be expressed as $G(\mathrm{s})\varphi$ where

$$G(\mathrm{s}) := \left[\begin{array}{c|c} (A_a - G_n(C_aG_n)^{-1}C_aA_a) & (M_a - G_n(C_aG_n)^{-1}C_aM_a) \\ \hline C_f & 0 \end{array}\right] \quad (6.102)$$

For accurate estimation of the faults f_o, the transfer function matrix $G(\mathrm{s})$ must be 'small', and for complete decoupling $G(\mathrm{s}) = 0$. Here, the $\mathcal{H}_\infty$ norm of $G(\mathrm{s})$ will be minimised by choice of G_n.

Partition the state error vector e from (6.99) conformably with the canonical form in (6.94) as $\mathrm{col}(e_1, e_y)$. One way to identify the reduced order sliding motion is to perform a further change of coordinates according to the nonsingular matrix

$$T_L = \begin{bmatrix} I_{n+r-p} & L \\ 0 & I_p \end{bmatrix} \quad (6.103)$$

so that

$$e = (e_1, e_y) \mapsto (e_1 + Le_y, e_y) \equiv (\tilde{e}_1, e_y) =: \tilde{e} \quad (6.104)$$

It can be easily verified that in the coordinate system in (6.104), during the sliding motion, the error system i.e. (the reduced order sliding motion) can be written as

$$\dot{\tilde{e}}_1(t) = \big(A_{11} + L_1A_{211} + L_2A_{212}\big)\tilde{e}_1(t) + L_2\varphi \quad (6.105)$$

$$\dot{e}_y(t) = e_y(t) = 0 \quad (6.106)$$

The gain matrices L_1 and L_2 must ensure $A_{11} + LA_{211} + L_2A_{212}$ is stable for the sliding motion to be stable. Therefore the effect of φ on the estimation $\hat{f}_o$ is given by $C_f e = \tilde{C}_f\tilde{e}$ where $\tilde{C}_f = C_f T_L^{-1}$ and C_f is given in (6.90). It can be verified

$$\tilde{C}_f = [0_{n-p\times r} \quad I_r \quad \star] \quad (6.107)$$

where $\star$ represents a matrix which plays no part in the subsequent analysis. During the sliding motion,

$$\tilde{C}_f\tilde{e} = [0_{n-p\times r} \quad I_r \mid \star]\begin{bmatrix} \tilde{e}_1 \\ e_y \end{bmatrix} = \underbrace{\begin{bmatrix} 0_{n-p\times r} & I_r \end{bmatrix}}_{C_e}\tilde{e}_1 \quad (6.108)$$

since $e_y \equiv 0$ during sliding. Consequently,

$$G(\mathrm{s})\varphi = \tilde{G}(\mathrm{s})\varphi \quad (6.109)$$

where

$$\tilde{G}(\mathrm{s}) := \left[\begin{array}{c|c} A_{11} + L_1A_{211} + L_2A_{212} & L_2 \\ \hline C_e & 0 \end{array}\right] \quad (6.110)$$

and C_e is defined in (6.108). As argued in Remark 6.4, the pair (A_{11}, A_{21}) is observable, and so from the partition of A_{21} in (6.94) to obtain A_{211} and A_{212}, it follows that there exist matrices L_1 and L_2 so that $A_{11} + L_1 A_{211} + L_2 A_{212}$ is stable.

Proposition 6.2 *If (A_a, M_a, C_a) from (6.92)–(6.93) is minimum phase, then a sliding mode observer of the form in (6.96) exists such that $\hat{f}_o = C_f x_a \to f_o$ as $t \to \infty$.*

Proof If the system (A_a, M_a, C_a) from (6.92)–(6.93) is minimum phase, the pair (A_{11}, A_{211}) is detectable since it can be shown that the unobservable modes of (A_{11}, A_{211}) correspond to the invariant zeros of (A_a, M_a, C_a) (Lemma 4.3 in Sect. 4.3). Consequently there exists an L_o such that $(A_{11} + L_o A_{211})$ is stable. Therefore, the selection $L_1 = L_o$ and $L_2 = 0$ is a feasible choice which makes $A_{11} + L_1 A_{211} + L_2 A_{212} = A_{11} + L_o A_{211}$ stable. Since $L_2 = 0$, equation (6.105) collapses to $\dot{\tilde{e}}_1(t) = (A_{11} + L_o A_{211})\tilde{e}_1(t)$. Asymptotic tracking of the states takes place since $(A_{11} + L_o A_{211})$ is stable and therefore $\tilde{e}_1(t) \to 0$ as $t \to \infty$. It follows $\hat{f}_o(t) - f(t) = C_f e(t) \to 0$ since $e(t) \to 0$, and the fault is estimated asymptotically. □

Proposition 6.3 *If the plant system matrix A from (6.73) is stable, then a sliding mode observer of the form in (6.96) exists such that $\hat{f}_o = C_f \hat{x}_a \to f_o$ as $t \to \infty$.*

Proof If the plant system matrix A from (6.73) is stable, then (A, C_1) is detectable and from Lemma 6.6, (A_a, M_a, C_a) is minimum phase. Therefore from Proposition 6.2, $\hat{f}_o = C_f \hat{x}_a \to f_o$ since $e(t) \to 0$. □

Remark 6.5 If A from (6.73) is unstable then for certain fault conditions, (A, C_1) may be unobservable and perfect reconstruction is not possible. An example of this is discussed in Sect. 6.4.2 in the sequel. Furthermore if (A, C_1) is undetectable then from Lemma 6.6, (A_a, M_a, C_a) is non-minimum phase. Then as argued in [87] classical unknown input observers UIOs also cannot be employed to reject the unknown input $\varphi(t)$: see for example [52, 219]. The next subsection considers the ramifications of this.

6.4.1.2 Observer Design

The observer described in this section embodies the same design philosophies as those proposed in the earlier sections of this chapter. In this section, however, the major difference is the results are no longer predicated on the assumption that (A_a, M_a, C_a) is minimum phase, and as a consequence, the fault signal cannot be perfectly replicated even if no uncertainty is present. The full block structure in (6.98) considered in this section allows the triple (A_a, M_a, C_a) to be non-minimum phase thus broadening the class of systems for which the results are applicable.

Define a Lyapunov matrix for the error system in (6.99) to have the form

$$P = \begin{bmatrix} P_{11} & P_{12} \\ P_{12}^{\mathrm{T}} & P_{22} \end{bmatrix} \tag{6.111}$$

where $P_{11} \in \mathbb{R}^{(n+r-p)\times(n+r-p)}$ is symmetric positive definite. Let $G_l \in \mathbb{R}^{(n+r)\times p}$ be any matrix which satisfies

$$P\underbrace{(A_a - G_l C_a)}_{A_0} + (A_a - G_l C_a)^{\mathrm{T}} P < 0 \tag{6.112}$$

Here, the design of the linear gain G_l for the sliding mode observer from (6.96) will be chosen to satisfy

$$\begin{bmatrix} PA_0 + A_0^{\mathrm{T}} P & P(G_l D - B_d) & E^{\mathrm{T}} \\ (G_l D - B_d)^{\mathrm{T}} P & -\gamma_0 I_{p+r} & 0 \\ E & 0 & -\gamma_0 I_r \end{bmatrix} < 0 \tag{6.113}$$

The matrices $B_d \in \mathbb{R}^{(n+r)\times(p+r)}$, $D \in \mathbb{R}^{p\times(p+r)}$ in (6.113) are defined as

$$B_d := [0 \quad M_a] \tag{6.114}$$

$$D := [D_1 \quad 0] \tag{6.115}$$

where $D_1 \in \mathbb{R}^{p\times p}$, M_a is defined in (6.88), and

$$E := \begin{bmatrix} C_e & M_2^{\mathrm{T}} \end{bmatrix} \tag{6.116}$$

where C_e is defined in (6.108) and M_2 is defined in (6.95). From the Bounded Real Lemma, if (6.113) holds, then $\|\tilde{G}_a(\mathrm{s})\|_\infty < \gamma_0$, where the transfer function matrix $\tilde{G}_a(\mathrm{s}) := E(\mathrm{s}I - A_0)^{-1}(G_l D - B_d)$. This represents an $\mathcal{H}_\infty$ filtering problem [297] associated with the linear part of the observer from (6.96) obtained from setting $\rho = 0$. The matrix D_1 in (6.115) represents design freedom used to trade-off the speed of response of the observer versus the magnitude of the gain matrix G_l. As argued in Sect. 4.5, inequality (6.113) is feasible if and only if

$$\begin{bmatrix} PA_a + A_a^{\mathrm{T}} P - \gamma_0 C_a^{\mathrm{T}} (DD^{\mathrm{T}})^{-1} C_a & -PB_d & E^{\mathrm{T}} \\ -B_d^{\mathrm{T}} P & -\gamma_0 I_{(p+r)} & 0 \\ E & 0 & -\gamma_0 I_r \end{bmatrix} < 0 \tag{6.117}$$

in which case

$$G_l = \gamma_0 P^{-1} C_a^{\mathrm{T}} \left(DD^{\mathrm{T}}\right)^{-1} C_a \tag{6.118}$$

is a choice of the Luenberger gain. Let

$$PA_a + A_a^{\mathrm{T}} P := \begin{bmatrix} X_{11} & X_{12} \\ X_{12}^{\mathrm{T}} & X_{22} \end{bmatrix} \tag{6.119}$$

where P is defined in (6.111) and $X_{11} \in \mathbb{R}^{(n+r-p)\times(n+r-p)}$. It follows that

$$X_{11} = P_{11}A_{11} + P_{12}A_{21} + (P_{11}A_{11} + P_{12}A_{21})^{\mathrm{T}} \tag{6.120}$$

From (6.114)

$$PB_d = P[0 \quad M_a] = \begin{bmatrix} 0 & P_{122} \\ 0 & P_{222} \end{bmatrix} \tag{6.121}$$

where P_{122} and P_{222} are the last r columns of P_{12} and P_{22} from (6.111), respectively. Using (6.119) and (6.121), inequality (6.117) can be written as

$$\begin{bmatrix} X_{11} & X_{12} & 0 & -P_{122} & C_e^{\mathrm{T}} \\ X_{12}^{\mathrm{T}} & X_{22} - \gamma_0^{\mathrm{T}}(DD^{\mathrm{T}})^{-1} & 0 & -P_{222} & M_2 \\ 0 & 0 & -\gamma_o I_p & 0 & 0 \\ -P_{122}^{\mathrm{T}} & -P_{222}^{\mathrm{T}} & 0 & -\gamma_o I_r & 0 \\ C_e & M_2^{\mathrm{T}} & 0 & 0 & -\gamma_o I_r \end{bmatrix} < 0 \tag{6.122}$$

A necessary condition for this inequality to hold is

$$\begin{bmatrix} X_{11} & -P_{122} & C_e^{\mathrm{T}} \\ -P_{122}^{\mathrm{T}} & -\gamma_0 I_r & 0 \\ C_e & 0 & -\gamma_0 I_r \end{bmatrix} < 0 \tag{6.123}$$

If $L := P_{11}^{-1} P_{12}$ then $P_{11} L_2 = P_{122}$ and (6.123) can be re-written as

$$\begin{bmatrix} P_{11}(A_{11} + LA_{21}) + (A_{11} + LA_{21})^{\mathrm{T}} P_{11} & -P_{11} L_2 & C_e^{\mathrm{T}} \\ \star & -\gamma_0 I_r & 0 \\ \star & \star & -\gamma_0 I_r \end{bmatrix} < 0 \tag{6.124}$$

which is the Bounded Real Lemma associated with the transfer function $\tilde{G}(\mathrm{s}) = C_e(\mathrm{s}I - (A_{11} + LA_{21}))^{-1} L_2$ and implies $\|\tilde{G}(\mathrm{s})\|_\infty < \gamma_0$.

Formally the design problem is

For a given matrix D_1 and scalar γ_0, minimise γ with respect to P, subject to

$$\begin{bmatrix} X_{11} & -P_{122} & C_e^{\mathrm{T}} \\ -P_{122}^{\mathrm{T}} & -\gamma I_r & 0 \\ C_e & 0 & -\gamma I_r \end{bmatrix} < 0 \tag{6.125}$$

$$P > 0 \tag{6.126}$$

and (6.117).

This is a convex optimisation problem. Standard LMI software can be used to synthesise numerically γ and P. Once P has been determined, L can be determined as $L = P_{11}^{-1} P_{12}$. The observer gain G_l can be determined from (6.118) and G_n is determined from (6.98). As argued in Sect. 4.5 a possible choice of the symmetric positive definite matrix P_o associated with the unit vector term in (6.97) is $P_o = P_{22} - P_{12} P_{11}^{-1} P_{12}$.

6.4.1.3 System Uncertainty

Suppose the system in (6.73) is subject to uncertainty:

$$\dot{x}(t) = Ax(t) + Bu(t) + Q\xi(t, x, u) \tag{6.127}$$

$$y(t) = Cx(t) + Nf_o(t) \tag{6.128}$$

where $\xi(\cdot)$ represents a bounded unknown disturbance. The term $Q\xi(t, x, u)$ is assumed to capture the mismatch between the model about which the observer is designed, and the real plant which is to be monitored. Therefore the augmented system in (6.77)–(6.78) becomes

$$\dot{x}_a(t) = A_a x_a(t) + B_a u(t) + M_a \varphi(t) + Q_a \xi(t, x, u) \tag{6.129}$$

$$y(t) = C_a x_a(t) \tag{6.130}$$

where the term $Q_a\xi(t, x, u)$ represents the effect of additive bounded uncertainty. Again the fault to be reconstructed is given by $f_o = C_f x_a$. The idea now is to represent (6.129) as

$$\dot{x}_a(t) = A_a x_a(t) + B_a u(t) + [M_a \quad Q_a] \begin{bmatrix} \varphi(t) \\ \xi(t, x, u) \end{bmatrix} \tag{6.131}$$

and to minimise the effect of (ξ, φ) on the reconstruction of f_o. In this new optimisation problem, the disturbance matrix B_d from (6.114) must be augmented and becomes

$$\bar{B}_d = [0 \quad M_a \quad Q_a] \tag{6.132}$$

and the matrix D from (6.115) becomes

$$\bar{D} = [D_1 \quad 0] \tag{6.133}$$

The new optimisation problem becomes

For a given matrix D_1 and γ_0, minimise with respect to γ and P, inequalities (6.125), (6.126) and

$$\begin{bmatrix} PA_a + A_a^{\mathrm{T}}P - \gamma_0 C_a^{\mathrm{T}}(\bar{D}\bar{D}^{\mathrm{T}})^{-1}C_a & -P\bar{B}_d & E^{\mathrm{T}} \\ -\bar{B}_d^{\mathrm{T}}P & -\gamma_0 I & 0 \\ E & 0 & -\gamma_0 I \end{bmatrix} < 0 \tag{6.134}$$

Again this represents a convex optimisation problem and LMI solvers can be employed to synthesise γ and P. Note that Q_a needs to be pre-scaled appropriately so that ξ and φ are of the same order—or suitably weighted to reflect the relative importance of rejection of uncertainty compared to the effect of the fault derivative.

6.4.2 Examples

6.4.2.1 ADMIRE Simulations

An unstable fighter aircraft will now be used to demonstrate the theory which has been developed in the earlier sections. The ADMIRE model represents a rigid small fighter aircraft with a delta-canard configuration based on a real fighter aircraft. The ADMIRE aircraft model has been used by several researchers (e.g., [122]) and within the Group for Aeronautical Research and Technology in Europe (GARTEUR) AG11 and AG12. Details of the model can be found in [97]. The linear model used here has been obtained at a low speed flight condition of Mach 0.22 at an altitude of 3000 m [122] and is similar to the one in [122]. The state, input and output distribution matrices are

$$A = \begin{bmatrix} -0.5432 & 0.0137 & 0 & 0.9778 & 0 \\ 0 & -0.1179 & 0.2215 & 0 & -0.9661 \\ 0 & -10.5128 & -0.9967 & 0 & 0.6176 \\ 2.6221 & -0.0030 & 0 & -0.5057 & 0 \\ 0 & 0.7075 & -0.0939 & 0 & -0.2127 \end{bmatrix}, \quad (6.135)$$

$$B = \begin{bmatrix} 0.0069 & -0.0866 & -0.0866 & 0.0004 \\ 0 & 0.0119 & -0.0119 & 0.0287 \\ 0 & -4.2423 & 4.2423 & 1.4871 \\ 1.6532 & -1.2735 & -1.2735 & 0.0024 \\ 0 & -0.2805 & 0.2805 & -0.8823 \end{bmatrix},$$

$$C = \begin{bmatrix} 0 & 0 & 1 & 0 & 0 \\ 0 & 0 & 0 & 0 & 1 \\ 0 & 0 & 0 & 1 & 0 \end{bmatrix} \quad (6.136)$$

The states are angle of attack (AoA) (rad), sideslip angle (rad), roll rate (rad/s), pitch rate (rad/s) and yaw rate (rad/s). The outputs are roll rate (rad/s), yaw rate (rad/s) and pitch rate (rad/s). The control surfaces represent the deflections (rad) of the canard, right elevon, left elevon and rudder, respectively. The linearised model is obtained from [122]. The obtained linear model is open-loop unstable, which is a typical characteristic of fighter aircraft to allow high manoeuvrability. It is assumed that the sensor for the pitch rate is prone to faults. This system is an example where the fault estimation scheme in Sects. 6.2.3 and 6.2.4 will not work because it can be shown that if $N = [0\ 0\ 1]^{\mathrm{T}}$ in (6.74), then the associated augmented system (A_a, M_a, C_a) is non-minimum phase with a zero at $\{1.0769\}$. Note that the C matrix has been reordered to comply with the requirements in (6.75) where the sensors that are prone to faults are in the lower part of the C matrix. However, the approach proposed in Sect. 6.4.1 is applicable for this particular system. The design parameters for the observer were chosen as $\gamma_0 = 10$ from (6.113) and $D_1 = I_3$ from (6.115). The LMI solver yields a γ and P such that $\|\tilde{G}(s)\|_\infty = 1.2212$. The nonlinear gain in (6.97) has been chosen as $\rho = 1$.

The simulations in Figs. 6.13 and 6.14 have been obtained from the full nonlinear ADMIRE model with the aircraft undergoing a change in altitude (Fig. 6.13).

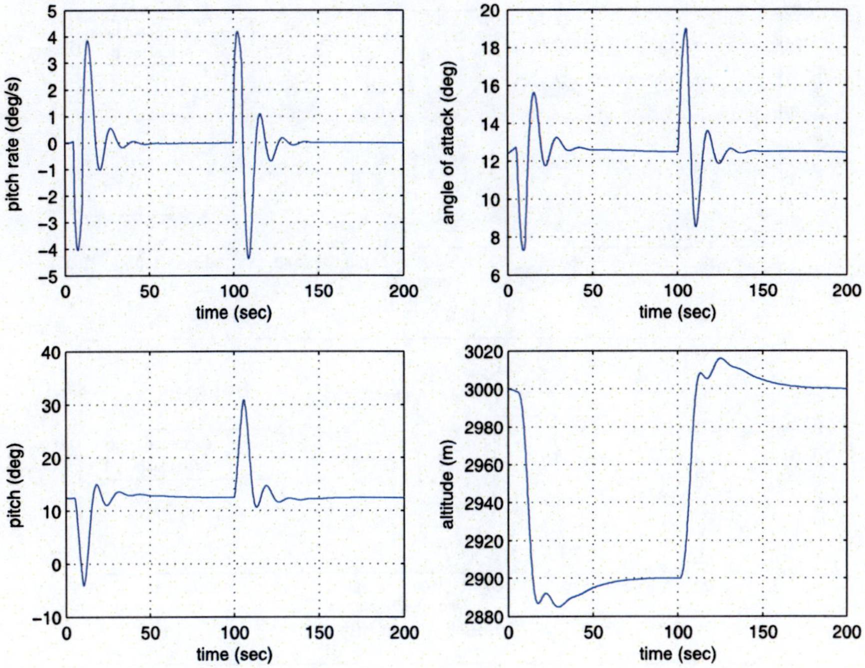

Fig. 6.13 Manoeuvre on ADMIRE full nonlinear model

Figure 6.14 shows the results of the fault reconstruction using different sensor fault shapes, to show the effectiveness of the method.

6.4.2.2 Large Transport Aircraft Simulations

The large transport aircraft model used in this section is based on a linearisation from the FTLAB747 software running under MATLAB®. FTLAB747 has been developed over the years for the study of fault tolerant control and FDI schemes. This software has recently been used in the GARTEUR FM-AG16 as a benchmark model for investigating state of the art FTC schemes. Details of the software can be found in [80, 183, 234]. A linearisation has been obtained around an operating condition of 263,000 kg, 92.6 m/s true airspeed, and an altitude of 600 m at 25.6% of maximum thrust and at a 20 deg flap position. The linearised model is given by

$$A = \begin{bmatrix} -0.6284 & 0 & -0.0000 & 1.0064 & -0.0000 \\ 0.0000 & -0.0999 & 0.1014 & 0 & -0.9887 \\ -0.0000 & -1.6478 & -1.0579 & 0 & 0.1718 \\ -0.5831 & -0.0000 & 0.0000 & -0.5137 & 0.0000 \\ 0.0000 & 0.2767 & -0.1186 & 0 & -0.2066 \end{bmatrix} \tag{6.137}$$

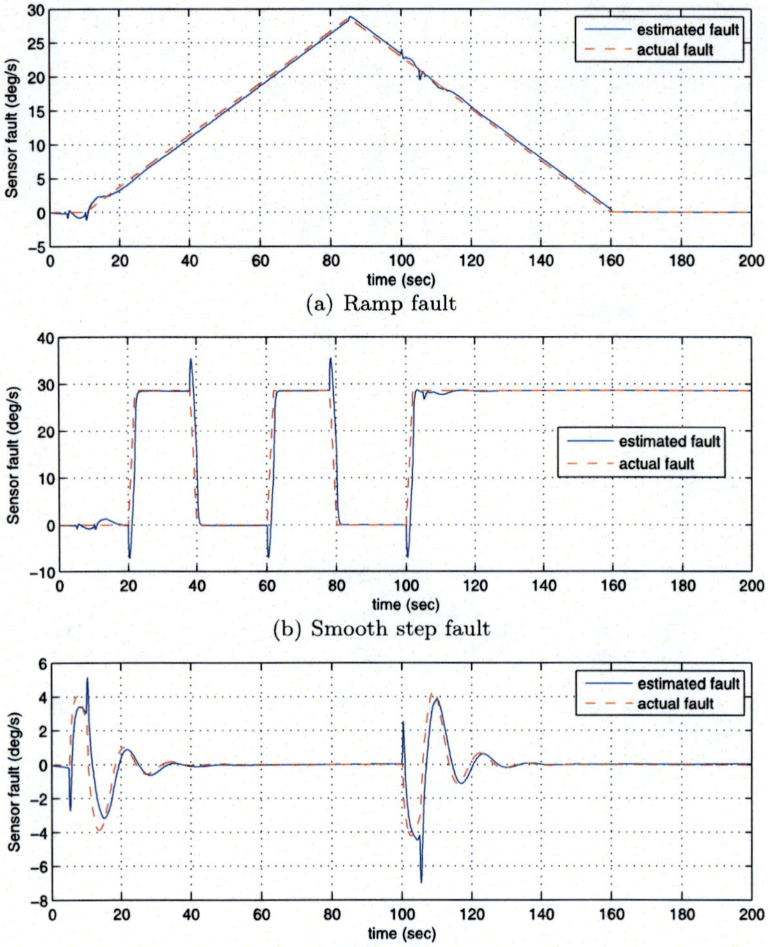

(a) Ramp fault

(b) Smooth step fault

(c) Total failure

Fig. 6.14 Sensor fault reconstruction on the pitch rate sensor on ADMIRE full nonlinear model

$$B = \begin{bmatrix} -0.0116 & -0.0116 & -0.0352 & 0 \\ 0 & 0 & 0 & 0.0174 \\ -0.3118 & 0.3118 & 0 & 0.1187 \\ -0.0617 & -0.0617 & -0.6228 & 0.0000 \\ -0.0276 & 0.0276 & 0 & -0.2478 \end{bmatrix}, \qquad (6.138)$$

$$C = \begin{bmatrix} 0 & 0 & 1 & 0 & 0 \\ 0 & 0 & 0 & 0 & 1 \\ 0 & 0 & 0 & 1 & 0 \end{bmatrix}$$

The output distribution matrix is the same as in (6.136). The states from the FTLAB747 software have been reordered and now represent $x = [\alpha\ \beta\ p\ q\ r]^{\mathrm{T}}$ as in

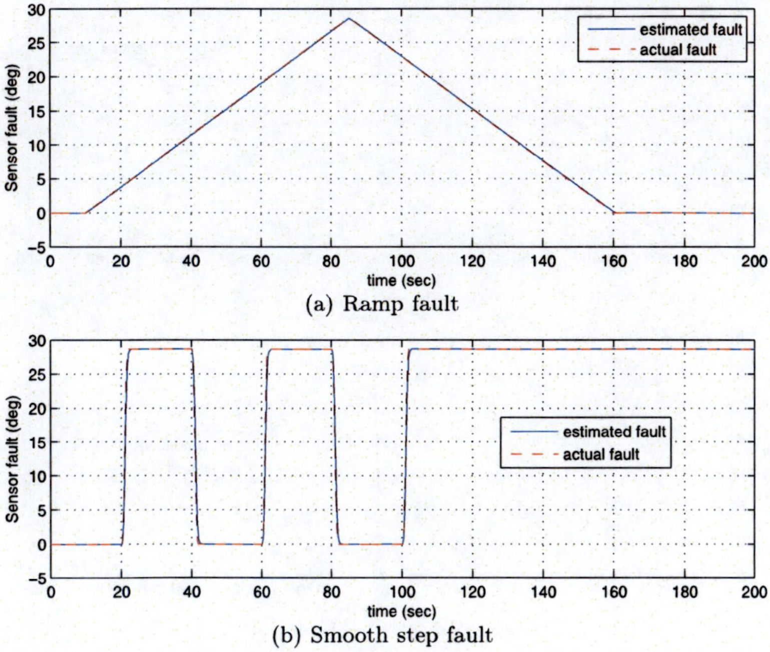

Fig. 6.15 Sensor fault reconstruction on the pitch rate (q) sensor

the ADMIRE example. The control surfaces are $\delta = [\delta_{ar}\ \delta_{al}\ \delta_e\ \delta_r]^{\mathrm{T}}$, which represent the deflections (rad) of the right and left aileron, elevator and rudder, respectively. The linear model given above is open-loop stable, which is a typical characteristic of passenger transport aircraft. Again, it is assumed that the sensor for the pitch rate (q) is prone to faults.

Without Uncertainty The design parameters for the observer were chosen as, $\gamma_0 = 1.1$ from (6.113) and $D_1 = I_3$ from (6.115) to yield $\|\tilde{G}(\mathrm{s})\|_\infty = 2.9092 \times 10^{-8}$. This result is to be expected since the plant is open-loop stable, and therefore the results from Sects. 6.2.3 and 6.2.4 could be used to achieve perfect reconstruction. Here the fact that $\|\tilde{G}(\mathrm{s})\|_\infty = 2.9092 \times 10^{-8}$ indicates almost perfect fault reconstruction. The reduced order eigenvalues for the observer are given by $\{-0.5710 \pm 0.7639i, -0.7457\}$. The nonlinear gain in (6.97) has been chosen as $\rho = 1$. During the simulations the signum function from (6.97) has been approximated using $\frac{P_o e_y}{\|P_o e_y\| + \delta}$ where $\delta = 0.001$.

As in Sect. 6.4.2.1, the aircraft undergoes a change of altitude followed by a banking manoeuvre with a similar sensor fault test. In both tests (Fig. 6.15), the proposed scheme provides satisfactory fault reconstruction for the q sensor.

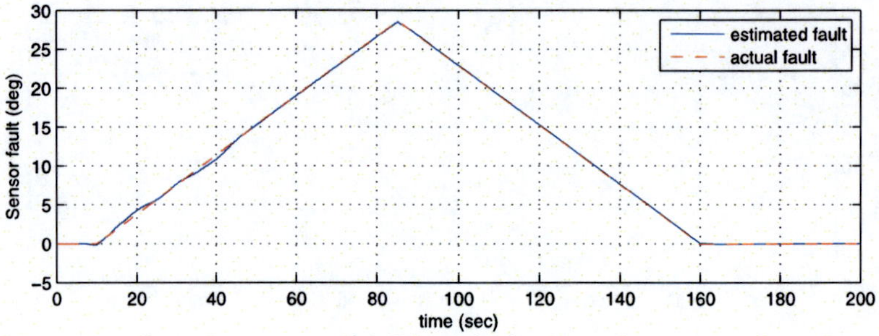

(a) Ramp fault

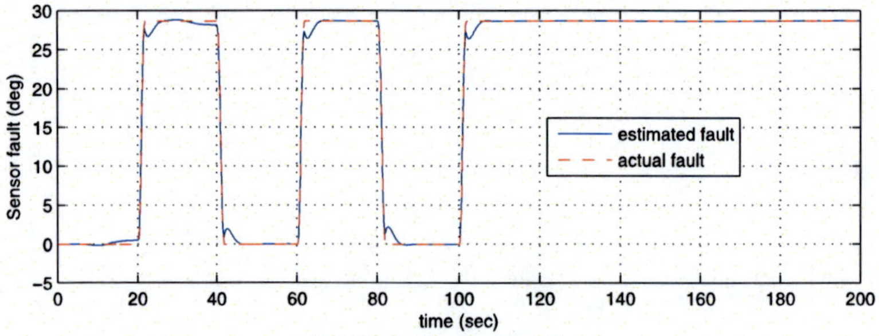

(b) Smooth step fault

Fig. 6.16 Sensor fault reconstruction on the pitch rate (q) sensor

With Uncertainty A similar test as above is repeated, now with the inclusion of uncertainty. The matrix Q in (6.127) is assumed to have a structure given by

$$Q = [1 \quad 0 \quad 0 \quad 0 \quad 0]^{\mathrm{T}}$$

The parametric uncertainty $\xi(\cdot)$ from (6.127) is given by

$$\begin{aligned}\xi(t,x,u) &= [\Delta A_{11} \quad \Delta A_{12} \quad \Delta A_{13} \quad \Delta A_{14} \quad \Delta A_{15}]x(t)\\ &= [-0.3142 \quad 0 \quad 0.0000 \quad 0.6039 \quad -0.0000]x(t)\end{aligned}$$

The design parameters for the observer were chosen as, $\gamma_0 = 1.1$ from (6.113) and $D_1 = I_3$ from (6.133) to yield $\|\tilde{G}(s)\|_\infty = 0.2652$. The reduced order eigenvalues for the observer are given by $\{-0.7062 \pm 0.7031i, -0.7467\}$. The nonlinear gain in (6.97) has been chosen as $\rho = 1$. Again during the simulations the signum function from (6.97) has been approximated.

A similar test to Fig. 6.15 has been repeated, now with the inclusion of the uncertainty. Figures 6.16(a) and 6.16(b) show a small degradation in the fault reconstruction at the beginning of the simulations due to the initial manoeuvre. However, generally Fig. 6.16 shows satisfactory fault reconstructions on the q sensor in the presence of uncertainty.

6.5 Summary

This chapter has discussed techniques for reconstructing sensor faults using sliding mode observers. Various techniques and observer configurations have been developed for different situations and systems. The techniques/schemes described are based in the main on the actuator fault reconstruction technique presented in the previous chapter, where by appropriate filtering of the measured plant signals, the filtered versions are considered as the (measurable) output of a system which treats the sensor fault as an 'actuator fault'. As the existence conditions for the actuator fault reconstruction technique described in the earlier chapters have been thoroughly investigated, the existence conditions for each of the sensor fault reconstruction techniques presented here have been easily established.

6.6 Notes and References

The problem formulation in (6.77)–(6.78) constitutes a 'classical' unknown input observer situation. However, the form in (6.77)–(6.78) is very specific, in that it is inherently relative degree one by construction ($\mathrm{rank}(C_a M_a) = r$). There has been extensive recent research into the use of sliding mode observers for unknown input problems—although the focus has been primarily directed at the situation where the relative degree one requirement is not met: see for example [21, 54, 93, 96, 101, 197]. As far as the authors are aware, much less attention has been directed towards the problem of obviating minimum phase limitations [15]. Thus, although the main motivation in this section was to tackle the problem of sensor fault reconstruction, the problem may also be viewed as one of unknown input reconstruction in non-minimum phase systems. It is important to note that classical linear unknown input observers (UIO) also cannot be employed in this situation [52, 87, 219].

Chapter 7
Case Study: Implementation of Sensor Fault Reconstruction Schemes

The previous chapter discussed the synthesis of sliding mode observers for sensor fault reconstruction, which included a rigourous stability analysis for different situations and systems. In this chapter, two examples of implementations of the theory described in the previous chapter (especially from Sect. 6.3) are described.

7.1 Application to a Crane

The crane rig shown in Fig. 7.1 has a rail of traversable length of 1 m for the cart which suspends the pendulum. The length of the pendulum is 0.35 m and the mass, attached to its non-supported end, is 0.5 kg. The cart itself, of mass 2 kg, is accelerated by a DC motor which drives a belt and pulley system. The DC motor is rated at 200 W and is of a 'flat' design (sometimes known as a 'pancake motor'), with a high torque to speed ratio. The DC motor is powered by a servo-amplifier which can provide 38 V at 7 A unregulated supply.

A potentiometer attached to the cart provides a measurement of angular displacement from the vertical of the pendulum. Another potentiometer attached to the belt and pulley system that drives the cart provides a measurement of truck position. A tacho-generator connected to one of the two pulleys gives the truck velocity.

Although for this system well understood nonlinear models can be developed (see for example [99]), for many systems such a task would be too complex and an uncertain linear model with a well understood 'unknown input' uncertainty distribution matrix may be the best that could be expected to be achieved in terms of modelling [51, 207]. Here, the model has been developed through a mixture of experimental measurement and (subspace) identification [171]. In the model it has been assumed that the dynamics of the motor itself are sufficiently fast that they can be neglected and the relationship between the voltage applied to the motor and the resulting torque can be modelled as a fixed gain. The static friction effects associated with the motor and cart have also been investigated. By experimentation, the smallest voltage applied to the motor necessary to cause an acceleration of the cart was

H. Alwi et al., *Fault Detection and Fault-Tolerant Control Using Sliding Modes*, Advances in Industrial Control,
DOI 10.1007/978-0-85729-650-4_7,

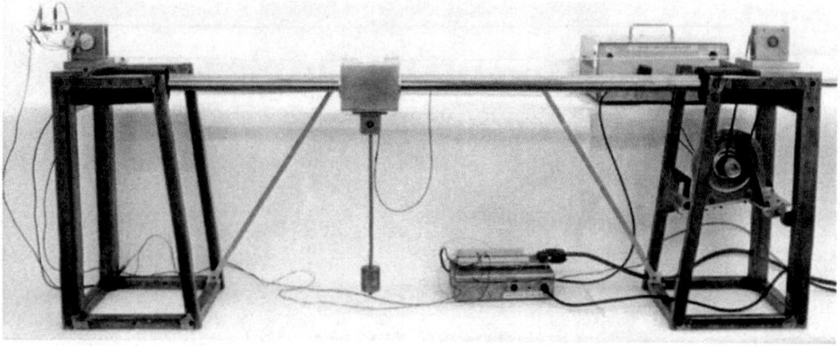

Fig. 7.1 Laboratory crane rig

measured, and the static friction effects modelled as a symmetric dead-zone function $\phi : \mathbb{R} \mapsto \mathbb{R}$ representing the applied motor voltage versus the effective motor voltage.

7.1.1 Modelling of the Crane System

In this case study, the system under consideration is modelled as

$$\dot{x}(t) = Ax(t) + Bu(t) + Q\xi(t, y, u) \tag{7.1}$$

$$y(t) = Cx(t) + Nf_o(t) \tag{7.2}$$

where $A \in \mathbb{R}^{4\times 4}$, $B \in \mathbb{R}^{4\times 1}$, $C \in \mathbb{R}^{3\times 4}$, $N \in \mathbb{R}^{3\times 2}$ and $Q \in \mathbb{R}^4$. The function $f_o : \mathbb{R}_+ \to \mathbb{R}^2$ is unknown but bounded so that

$$\|f_o(t)\| \le \alpha(t) \tag{7.3}$$

where $\alpha : \mathbb{R}_+ \to \mathbb{R}_+$ is a known function. The signal $f_o(t)$ represents sensor faults and N represents a distribution matrix (with columns usually formed from the standard basis for $\mathbb{R}^3$), which indicates which of the sensors providing measurements are prone to possible faults. The map $\xi : \mathbb{R}_+ \times \mathbb{R}^3 \times \mathbb{R} \to \mathbb{R}$ encapsulates any uncertainty or nonlinearities present and is assumed to be unknown but bounded and subject to $\|\xi(t, y, u)\| < \beta$ where the positive scalar β is known.

The data collection and (subsequent) controller implementation has been achieved using MATLAB® and dSPACE®. Taking as states the angular position of the pendulum θ, angular velocity $\dot{\theta}$, plus the position d and velocity $\dot{d}$ of the cart, a fourth order linear model relating the effective motor voltage to the states has been developed. The following state and input distribution matrices have been obtained:

$$A = \begin{bmatrix} 0 & 1.0000 & 0 & 0 \\ -32.6237 & -0.1482 & 0 & 11.6541 \\ 0 & 0 & 0 & 1.0000 \\ 0.9816 & 0.2791 & 0 & -3.1228 \end{bmatrix}, \qquad B = \begin{bmatrix} 0 \\ 12.3537 \\ 0 \\ -9.1153 \end{bmatrix} \tag{7.4}$$

associated with state vector $x = [\theta\ \dot{\theta}\ d\ \dot{d}]^{\mathrm{T}}$. Because of the choice of sensors and the state-space representation, the output distribution matrix

$$C = \begin{bmatrix} 1 & 0 & 0 & 0 \\ 0 & 0 & 1 & 0 \\ 0 & 0 & 0 & 1 \end{bmatrix} \tag{7.5}$$

and (7.4)–(7.5) represents a controllable and observable representation of the crane system. As described above, assumptions and simplifications have been made in the modelling process to obtain the linear representation in (7.4)–(7.5). At the design stage for the FDI scheme, the plant model mismatch will be accounted for by assuming an unknown disturbance signal acts through a known distribution matrix similar to [207].

It will be assumed that the pendulum angular position and the cart velocity sensors are potentially faulty and therefore

$$N = \begin{bmatrix} 1 & 0 \\ 0 & 0 \\ 0 & 1 \end{bmatrix} \tag{7.6}$$

As described in Sect. 6.3, the observer is designed based on the augmented system resulting from augmenting the plant with a new state $z_f \in \mathbb{R}^3$ that is a filtered version of y. The new state $z_f \in \mathbb{R}^3$ is given by

$$\dot{z}_f(t) = -A_f z_f(t) + A_f C x(t) + A_f N f_o(t) \tag{7.7}$$

where $-A_f \in \mathbb{R}^{3\times 3}$ is a stable matrix. The augmented system (A_a, M_a, C_a) is given by

$$\begin{bmatrix} \dot{x}(t) \\ \dot{x}_f(t) \end{bmatrix} = \underbrace{\begin{bmatrix} A & 0 \\ A_f C & -A_f \end{bmatrix}}_{A_a} \begin{bmatrix} x(t) \\ x_f(t) \end{bmatrix} + \underbrace{\begin{bmatrix} B \\ 0 \end{bmatrix}}_{B_a} u(t) + \underbrace{\begin{bmatrix} 0 \\ A_f N \end{bmatrix}}_{M_a} f_o(t)$$
$$+ \underbrace{\begin{bmatrix} Q \\ 0 \end{bmatrix}}_{Q_a} \xi(\cdot) \tag{7.8}$$

$$x_f(t) = \underbrace{[0 \quad I_p]}_{C_a} \begin{bmatrix} x(t) \\ x_f(t) \end{bmatrix} \tag{7.9}$$

where $x_a \in \mathbb{R}^7$ is the augmented state.

For the uncertain system in (7.8)–(7.9) a sliding mode observer of the form

$$\dot{\hat{x}}_a(t) = A_a \hat{x}_a(t) + B_a u(t) - G_l e_y(t) + G_n \nu \tag{7.10}$$

will be considered. In (7.10) the discontinuous output error injection term

$$\nu = -\rho(t, y, u) \frac{P_o e_y}{\|P_o e_y\|} \quad \text{if } e_y(t) \neq 0 \tag{7.11}$$

where $e_y(t) := C_a \hat{x}_a(t) - y(t)$ is the output estimation error and P_o is a symmetric positive definite matrix. Note that this observer is different from the one presented

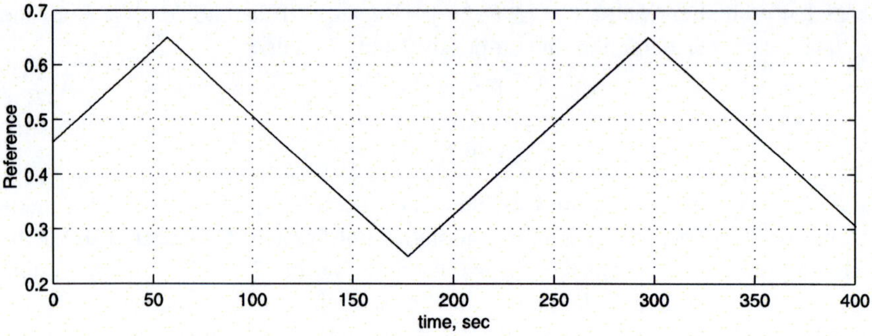

Fig. 7.2 Truck position reference signal

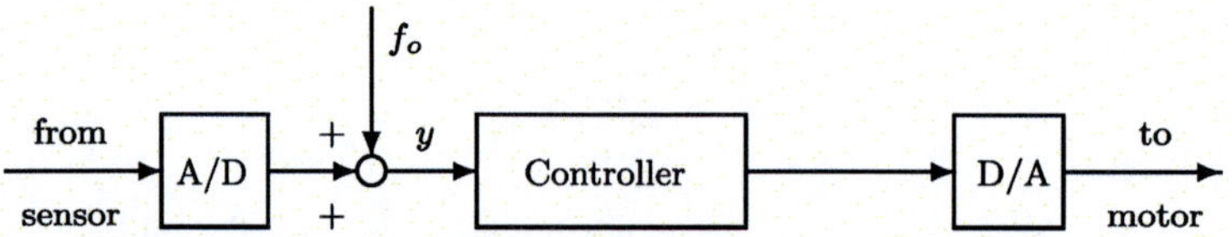

Fig. 7.3 Schematic of the sensor fault implementation

in Sect. 6.3. The above variation is based on the design discussed in Sect. 4.6 where the symmetric positive definite matrix P_o forms part of the unit vector output error injection signal.

7.1.2 Implementation

For the experiments which follow, a simple classical controller was designed based on the measurements of θ and d. It comprises a phase advance compensator with a zero at -0.1 and a pole at -1 driven by θ, and a proportional component driven by the truck position error. A saw-tooth profile was used as a reference signal for the truck to follow. This constitutes a ramp from 0.25 m to 0.65 m and back, over a period of 240 s (Fig. 7.2). As a result the control signal is dynamically changing and the crane system itself is not at rest at an equilibrium position.

The use of dSPACE® allows well defined fault scenarios to be introduced: actuator faults were introduced as sinusoidal disturbances to the control signal between the controller output and the D/A operation. Likewise sensor faults were incorporated as disturbances acting between the A/D operation and the signal used by the controller (Fig. 7.3).

The sine wave injections are not meant to represent realistic faults. They do, however, provide a periodic dynamic signal with a well defined shape against which the quality of the reconstructions can be judged.

The (augmented) uncertainty distribution matrix was chosen as

$$Q_a = \begin{bmatrix} B \\ 0 \end{bmatrix}$$

where B represents the input distribution matrix from (7.1). In this way $\xi(t, y, u)$ in (7.8) represents uncertainty and faults in the actuator channels and so the optimisation process will seek to minimise these effects on the sensor fault reconstructions. Specifying the filter matrix from (7.7) as $A_f = 20I_3$ and choosing $D_{1,a} = I_3$ and $\gamma_{o,a} = 25$, the synthesis procedure yields $\gamma_a = 5.100$. In this case the design parameters $D_{1,a}$ and $\gamma_{o,a}$ do have a significant effect on the resulting optimal γ_a. In this example, it can be verified that (A_a, M_a, C_a) does not possess invariant zeros. Hence the fact that A has an eigenvalue at the origin does not present any difficulties. The observer gains and the scaling matrix for the unit vector injection are given by

$$G_l = \begin{bmatrix} 551.3307 & 4087.7373 & 1334.2722 \\ 3252.3143 & 24113.6678 & 7870.9066 \\ 39.2589 & 291.0776 & 95.0102 \\ 312.3164 & 2315.6111 & 755.8352 \\ 1932.9348 & 25.1044 & 645.5233 \\ 25.1044 & 186.1322 & 60.7551 \\ 645.5233 & 60.7551 & 346.3602 \end{bmatrix},$$

$$G_n = \begin{bmatrix} 0.0000 & 21.9615 & 0 \\ 0.0000 & 129.5513 & 0 \\ 0.0000 & 1.5638 & 0 \\ 0.0000 & 12.4407 & 0 \\ 1.0000 & 0 & -0.0000 \\ 0.0000 & 1.0000 & 0 \\ 0 & -0.0000 & 1.0000 \end{bmatrix}$$

and

$$P_o = \begin{bmatrix} 3.6465 & 1.8313 & -7.1173 \\ 1.8313 & 15.1668 & -6.0735 \\ -7.1173 & -6.0735 & 21.5480 \end{bmatrix}$$

The sensor reconstruction signal is given by $\hat{f}_o = W_{sc,a} T^{\mathrm{T}} \nu_{a,\delta}$ where

$$W_{sc,a} = \begin{bmatrix} 0.0500 & -3.1981 & 0 \\ 0 & -1.1538 & 0.0500 \end{bmatrix}$$

and T is the orthogonal matrix defined in (4.46) in Sect. 4.3 associated with the canonical form for observer design. The signal $\nu_{a,\delta}$ is the smooth (equivalent) injection signal from the augmented observer given by

$$\nu_{a,\delta} = -\rho(y, t, u) \frac{P_o e_y}{\|P_o e_y\| + \delta} \tag{7.12}$$

The gains of the unit vector injection term ρ from (7.12) must be large in this scenario. As argued in the last chapter, the gain must satisfy $\rho > \|A_f N\| \|f_o\|$. Here

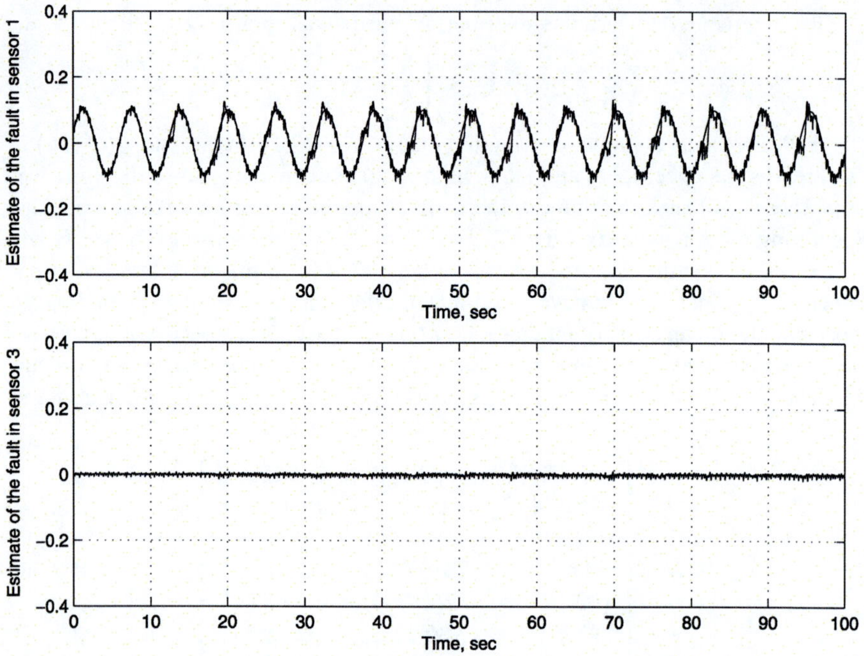

Fig. 7.4 Sensor fault reconstruction: fault in sensor 1

$\|A_f N\| = 20$ and faults in the speed transducer channel up to the order 1 m/s have been explored. In the following series of experiments the scalar $\rho_a = 100$ and $\delta_a = 0.0001$. The first reconstruction signal is associated with sensor 1 (the angular displacement of the pendulum) and the second reconstruction is associated with sensor 3 (the tacho-generator). In the first experiment a sinusoidal disturbance of amplitude 0.1 and frequency 1 rad/s was used to corrupt the measurement signal of angular position. Figure 7.4 shows the reconstruction signals together with the true faults. It can be seen in Fig. 7.4 that the sinusoidal disturbance is well replicated. There is also minimal coupling with the other sensor fault reconstruction.

Figure 7.5 shows the reconstruction signals for a sinusoidal disturbance of amplitude 0.25 on the tachometer signal. It can be seen in Fig. 7.5 that the sinusoidal disturbance is very well replicated and is almost indistinguishable from the true fault. However, there is some coupling on the reconstruction signal for the angular position sensor. In practice therefore a threshold level on the reconstruction signals would need to be set for detection purposes. Heuristically thresholds of the order ± 0.05 would be appropriate. The fact that these signals have a purely physical interpretation enable sensible choices to be made: so for example setting a threshold of ± 0.05 radians on the first reconstruction implies directly that faults in the pendulum angle sensor of up to 3° would be completely ignored before an alarm would be raised. Sensible levels for these thresholds can be computed by examining reconstructions from known fault-free data.

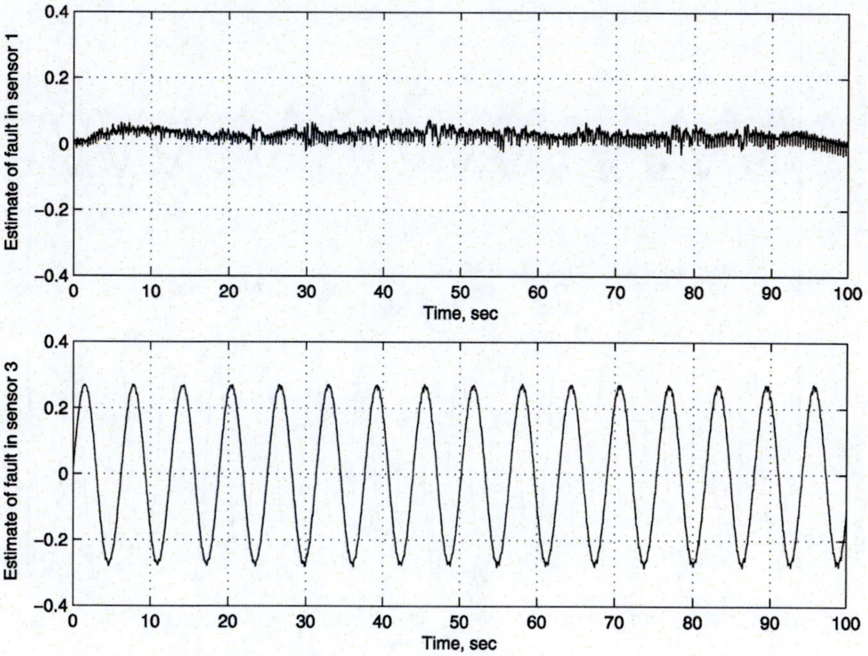

Fig. 7.5 Sensor fault reconstruction: fault in sensor 3

If the probability of the occurrence of simultaneous incipient faults in both sensors was considered to be negligible, then the fact that the coupling between the two reconstruction signals is unidirectional could be exploited. Logical signature information could be used in addition to the absolute sizes of the reconstruction signals and would allow tighter thresholds to be set on the sensor 1 fault reconstruction signal.

Figure 7.6 shows the reconstruction signals $\hat{f}_o$ for an out of phase set of sinusoidal disturbances on both sensors. Again it can be seen in Fig. 7.6 that the sinusoidal disturbance of amplitude 0.1 and 0.75 are well replicated.

7.2 Application to a DC Motor

The theory described in Sect. 6.3 will now be applied to a small DC motor rig which has been used as a safe and practical demonstrator for these ideas. The objective will be to control the speed of rotation of a disk attached (rigidly) to the shaft of the motor subject to an eddy current brake. Control will nominally be achieved in the fault-free case by a simple Proportional+Integral (PI) controller based on the difference between the reference signal and the measurement of angular speed from a tacho-generator. For the purpose of demonstration, however, the measurement from the tacho-generator will be assumed to be prone to faults.

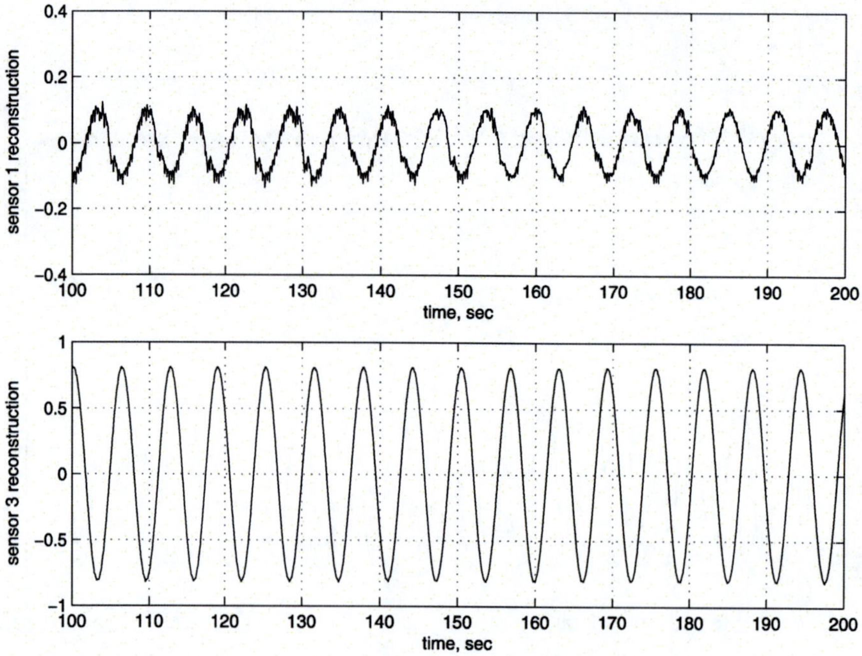

Fig. 7.6 Sensor fault reconstruction: simultaneous faults in sensors 1 and 3

7.2.1 Preliminaries

A slightly modified representation of the problem formulation compared to the one in Sect. 6.3 and in (7.1)–(7.2) will be considered here. Specifically the uncertain system about which the FDI scheme will be designed is given by

$$\dot{x}(t) = Ax(t) + Bu(t) \tag{7.13}$$

$$y(t) = Cx(t) + Nf_o(t) + Q\xi(t, y, u) \tag{7.14}$$

where $A \in \mathbb{R}^{n\times n}$, $B \in \mathbb{R}^{n\times m}$, $C \in \mathbb{R}^{p\times n}$, $N \in \mathbb{R}^{p\times r}$ and $Q \in \mathbb{R}^{p\times h}$ with $n \geq p \geq r$. Assume that the matrices C and N are full row and column rank, respectively, and the function $f_o : \mathbb{R}_+ \to \mathbb{R}^r$ is unknown but bounded so that

$$\| f_o(t) \| \leq \alpha(t) \tag{7.15}$$

where $\alpha : \mathbb{R}_+ \to \mathbb{R}_+$ is a known function. The signal $f_o(t)$ represents (additive) sensor faults and N represents a distribution matrix (with columns usually formed from the standard basis for $\mathbb{R}^p$), which indicates which of the sensors providing measurements are prone to possible faults. The map $\xi : \mathbb{R}_+ \times \mathbb{R}^p \times \mathbb{R}^m \to \mathbb{R}^h$ encapsulates any uncertainty or nonlinearities present and is assumed to be unknown but bounded and subject to $\|\xi(t, y, u)\| < \beta$ where the positive scalar β is known.

The objective is to reconstruct the faults $f_o(t)$ as accurately as possible despite the uncertainty. As argued in the last chapter, an effective way to do this is to first introduce a filter. Consider a new state $x_f \in \mathbb{R}^p$ that is a filtered version of $y(t)$, satisfying

$$\dot{x}_f(t) = -A_f x_f(t) + A_f C x(t) + A_f N f_o(t) + A_f Q \xi(t, y, u) \tag{7.16}$$

where $-A_f \in \mathbb{R}^{p\times p}$ is a stable matrix. Equations (7.13) and (7.16) can be combined to form an augmented state-space system of order $n+p$ given by

$$\begin{bmatrix} \dot{x}(t) \\ \dot{x}_f(t) \end{bmatrix} = \underbrace{\begin{bmatrix} A & 0 \\ A_f C & -A_f \end{bmatrix}}_{A_a} \begin{bmatrix} x(t) \\ x_f(t) \end{bmatrix} + \underbrace{\begin{bmatrix} B \\ 0 \end{bmatrix}}_{B_a} u(t) + \underbrace{\begin{bmatrix} 0 \\ A_f N \end{bmatrix}}_{M_a} f_o(t)$$
$$+ \underbrace{\begin{bmatrix} 0 \\ A_f Q \end{bmatrix}}_{Q_a} \xi(\cdot) \tag{7.17}$$

$$x_f(t) = \underbrace{[0 \quad I_p]}_{C_a} \begin{bmatrix} x(t) \\ x_f(t) \end{bmatrix} \tag{7.18}$$

Define $x_a \in \mathbb{R}^{(n+p)}$ to be the augmented state and note that as earlier, (7.17)–(7.18) treat the 'sensor faults' f_o as 'actuator faults'. By construction $\text{rank}(C_a M_a) = r$ and as shown in the last chapter (Proposition 6.1), the invariant zeros of $(A_a, M_a, C_a) \subseteq \lambda(A)$ and so if the open-loop plant is stable, the system (A_a, M_a, C_a) is minimum phase.

For the uncertain system in (7.17)–(7.18) a sliding mode observer of the form

$$\dot{z}(t) = A_a z(t) + B_a u(t) - G_l e_y(t) + G_n \nu \tag{7.19}$$

will be considered. In (7.19) the discontinuous output error injection term

$$\nu = -\rho(t, y, u) \frac{P_o e_y}{\|P_o e_y\|} \quad \text{if } e_y(t) \neq 0 \tag{7.20}$$

where $e_y(t) := Cz(t) - y(t)$ is the output estimation error and P_o is a symmetric positive definite matrix. The scalar function $\rho(\cdot)$ must be an upper bound on the uncertainty and the faults, and must ensure a sliding motion can be maintained despite the faults and uncertainty.

7.2.2 Description of the Motor Set-up

The rig which has been used in these experiments is based around a 30 W permanent magnet DC motor. The motor is powered by a servo-amplifier which provides a 24 V at 2 A unregulated supply. The motor shaft is connected in line to a rotating eddy current disk brake via a flexible coupling and also to a tacho-generator, the latter providing the measurement of angular speed used for feedback purposes. The servo-amplifier also provides a voltage output which is directly proportional to the load current in the armature coils. These two measurements will be used in the fault tolerant control scheme which is proposed.

The control law and the sliding mode observer for the system have been implemented using dSPACE®. The specific set-up runs MATLAB® 6.1 on a Windows platform with a DS1102 card fixed in one of the expansion ports of the PC and connected directly to the interface bus. The two analogue signals from the sensors have been connected to the first two 16 bit A/D converters in the DSP card. The first 14 bit D/A converter output from the card has been connected to the servo-amplifier.

7.2.3 Modelling

For the purposes of observer design an uncertain linear model representation has been developed. A standard (linear) description of the system has been formed from the state variables i and w which represent the current in the armature circuit and the angular velocity of the motor shaft, respectively. The armature current circuit is assumed to satisfy

$$L_a \frac{di(t)}{dt} + R_a i(t) = u(t) - K_e w(t) \tag{7.21}$$

where L_a and R_a represent the inductance and the resistance, respectively, and K_e represents the gain associated with the back e.m.f. The mechanical equation governing the rotation of the shaft is taken as

$$J \frac{dw(t)}{dt} = K_t i(t) - K_f w(t) \tag{7.22}$$

where J represents the inertia, K_t represents the motor torque constant and K_f the dynamic friction coefficient plus the effect of the eddy current brake. Rather than trying to directly measure or estimate the parameters within the motor model, a system identification approach has been adopted. A continuous-time state-space model, with the structure of the equations given in (7.21)–(7.22), was fitted to the logged data using the methodology described in [171]. Specific known data associated with the motor were used as a starting point for the estimation algorithms [171]. A nominal state-space representation of the form

$$\dot{x}(t) = Ax(t) + Bu(t) \tag{7.23}$$

has been obtained in which

$$A = \begin{bmatrix} -3.6239 & 1049.3257 \\ -2.7147 & -216.2070 \end{bmatrix}, \qquad B = \begin{bmatrix} 0 \\ 445 \end{bmatrix} \tag{7.24}$$

where $x = \mathrm{col}(w, i)$. No attempt was made to explicitly account for the dead-zone nonlinearity resulting from Coulomb friction, which is known to be present in the system.

In the experiments which follow it is assumed that both states are measured. It is also assumed that the tacho-generator signal measuring shaft velocity is prone to error, but that the current sensor is reliable. In terms of the general system architecture in (7.13)–(7.14), the output distribution matrix

$$C = \begin{bmatrix} 0.0239 & 0 \\ 0 & 1 \end{bmatrix} \tag{7.25}$$

which reflects a change from the SI units for the speed and current, respectively, into voltages obtained from the sensors. The sensor fault distribution matrix is

$$N = \begin{bmatrix} 1 \\ 0 \end{bmatrix}$$

The signal $\xi(\cdot)$ in (7.13)–(7.14) represents an additive 'unknown input' and is meant to encapsulate the plant model/mismatch. The distribution matrix Q, however, is assumed to be known. The identification approach does not generate the matrix Q and so it needs to be estimated separately. The value of Q is important since the sub-blocks Q_1 and Q_2 play an important role in the transfer function $\hat{G}(\mathrm{s})$ from (4.123) in Sect. 4.5, and hence in the optimisation procedure to obtain the observer gains. The requirement of obtaining an appropriate value for Q is a well-known problem in the application of robust FDI methodologies which rely on knowledge of the directions associated with the unknown inputs [51]. One of the most practical approaches to estimate the distribution matrix Q is the one proposed in [207]. The approach described in [207] involves estimating in the first instance the quantity $Q\xi$, i.e., the discrepancies between the output of the plant and the model. Then, essentially, principal component analysis is used to obtain the distribution matrix Q [207]. A sinusoidal input has been injected into both the model and the motor rig and Fig. 7.7 shows the discrepancy in each output between the model and the measured data in terms of voltage. In SI units, this corresponds to an actual discrepancy of approximately ± 0.05 amps for the current and ± 2 rad/s for the measured speed.

In this example, a suitable pragmatic choice for the distribution matrix has been found to be

$$Q = \begin{bmatrix} 1 \\ 1 \end{bmatrix}$$

since both error signals in Fig. 7.7 are visually similar. This has proved to be sufficiently accurate for this example. A complete model of the motor system has now been obtained in the form of (7.13)–(7.14) with all the matrices known. It can be verified that both assumptions A1 and A2 are satisfied and therefore the approach described in Sect. 7.2.1 can now be employed.

7.2.4 Observer Design

The observer design involves the introduction of an output filter. After some design iteration, a value of $A_f = 30I_2$ was selected. This choice of A_f results in numerically well-conditioned solutions for G_l and P_o. This is especially important when

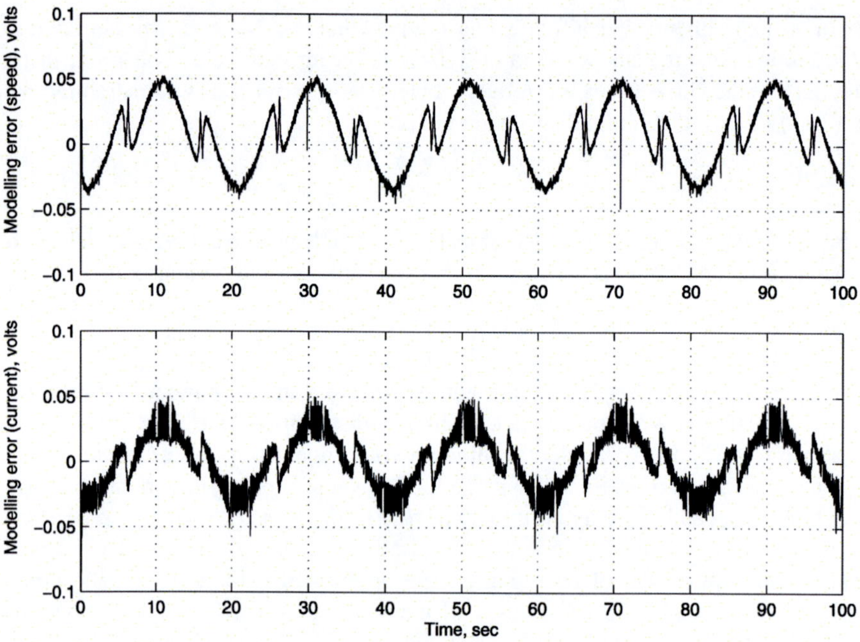

Fig. 7.7 Plant model mismatch in each state-space channel

implementing the observer in real-time using a fixed time-step integration routine (as discussed later in Sect. 7.2.6). The only remaining user defined parameter is a weight used in the optimisation to reflect a trade-off in the state-error tracking performance of the full order observer and the impact of measurement noise (this parameter is represented as $D_1 \in \mathbb{R}^{p\times p}$ in Sect. 4.5. Here the parameter has been chosen as an identity matrix. Once the parameters A_f and D_1 have been specified the algorithm described in Sect. 4.5 yields the following optimal design gains for the observer in (7.19)–(7.20):

$$G_l = \begin{bmatrix} -0.0586 & -0.0591 \\ 0.0054 & 0.0055 \\ 17.5769 & 17.4085 \\ 17.4085 & 17.5690 \end{bmatrix}, \qquad G_n = \begin{bmatrix} 0.0000 & -0.0034 \\ 0.0000 & 0.0003 \\ 1.0000 & 0.0000 \\ 0.0000 & 1.0000 \end{bmatrix}$$

The symmetric positive definite matrix which scales the output error in the injection term (7.20) is

$$P_o = \begin{bmatrix} 3.0545 & -3.0266 \\ -3.0266 & 3.0559 \end{bmatrix}$$

The sensor reconstruction signal is given by $\hat{f}_o = W_{sc,a} T^{\mathrm{T}} \nu_\delta$ where

$$W_{sc,a} = [\,0.0333 \quad -0.0333\,]$$

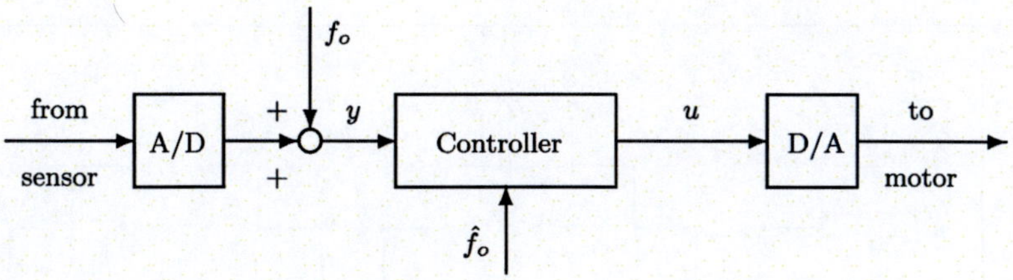

Fig. 7.8 Schematic of the fault implementation

and T is the orthogonal matrix defined in (4.46) in Sect. 4.3. The signal ν_δ is the smooth (equivalent) injection signal from the augmented observer given by

$$\nu_\delta = -\rho(y,t,u)\frac{P_o e_y}{\|P_o e_y\| + \delta} \tag{7.26}$$

The $\mathcal{H}_\infty$ norm of the transfer function matrix $\hat{G}(\mathrm{s})$ in (4.123) is 5.9126×10^{-4}. In the implementation results which follow, the modulation function in (7.26) is $\rho = 4.0$ and the smoothing parameter is $\delta = 0.0001$. From a theoretical standpoint the gain ρ must be large enough to bound the uncertainty and the fault signals, whilst ideally the term δ should be small. This quantity, plus the gain G_l, will determine the accuracy of the estimate $\hat{f}_o$. The online performance will be reduced further by the fact that a fixed time-step integration routine must be employed for the real-time implementation. The final choice of the smoothing coefficient has been arrived at following off-line testing with a fixed time-step integration routine to ensure this value presents no difficulty to the integration routines (too small a value will register as a 'discontinuity' and cause numerical problems).

7.2.5 Implementation

A dSPACE® set-up has been used to implement the (simple) speed feedback controller and the sliding mode observer used to monitor the system. The dSPACE® set-up allows easy and repeatable realisation of the required faults without the need to directly affect the hardware. The fault has been implemented at a 'software level' as shown in Fig. 7.8.

For control purposes a simple PI controller has been designed using only the measured angular velocity w. The input voltage to the motor amplifier is given by

$$u = F(\mathrm{s})(w_r - w) \tag{7.27}$$

where

$$F(\mathrm{s}) = \frac{(\mathrm{s}+15)}{2\mathrm{s}} \tag{7.28}$$

and the reference speed signal is w_r. This controller will be used to provide a benchmark performance level for the fault-free closed-loop system. The objective will be to recover the performance of this controller in the presence of sensor faults.

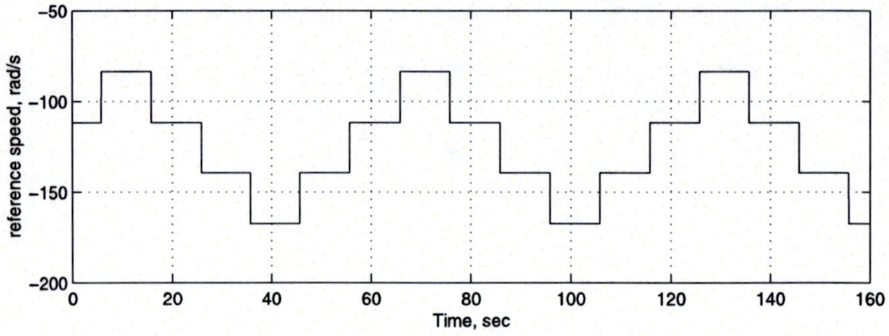

Fig. 7.9 Speed reference signal

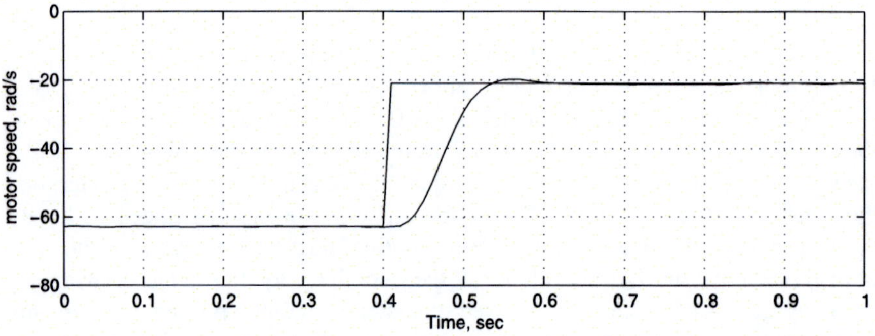

Fig. 7.10 Nominal PI performance

Remark 7.1 The approach that is used to recover the performance in the presence of sensor faults is independent of the controller law and could be 'retro-fitted' to any output feedback controller driven by the error signal between the measured speed w and the reference w_r.

7.2.6 *Results*

In the experiments, the reference signal w_r comprises a series of steps driving the angular speed from approximately 80–160 rad/s and back as shown in Fig. 7.9.

7.2.6.1 Real-Time Results

The control law described in (7.28) provides the nominal performance in the fault-free case. A typical response to one of the step changes in reference speed is given below (Fig. 7.10).

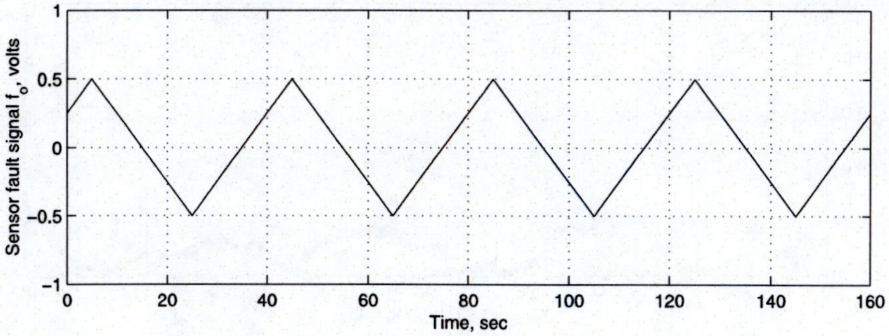

Fig. 7.11 Sensor fault signal

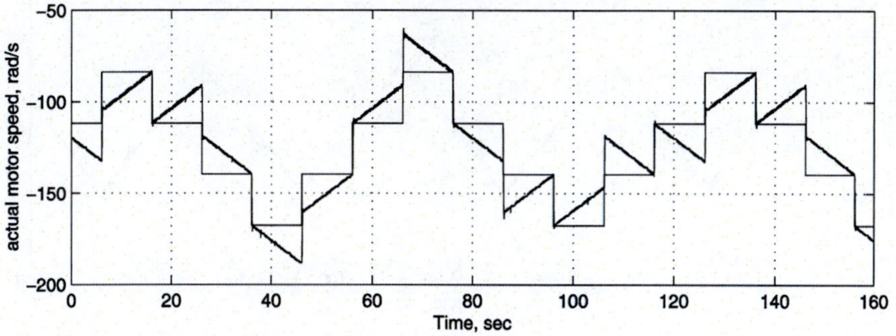

Fig. 7.12 Real motor speed tracking error (no correction)

In these experiments the sampling interval used is 0.001 s. The nominal PI controller gives a rise-time of approximately 0.15 s to a step change input and so the sampling interval gives approximately 150 sample points during the transient response. As argued by [13], this represents an adequate sampling time from a control theory perspective. For the controller in (7.28) the RMS value of the speed tracking error over the 160 s profile in Fig. 7.9 is 2.1415 rad/s. This will be used as the benchmark performance which the sliding mode scheme will attempt to recover in the presence of a sensor fault.

In the following experiments (unless otherwise stated) the speed sensor signal which is used in the controller online calculations has been corrupted by a symmetric saw-tooth signal of amplitude 0.5 volts which, in SI units, corresponds to a peak error of 20.9440 rad/s (Fig. 7.11).

This models a slow drift error in the sensor. As such it represents a difficult to detect incipient fault [51] which is typically used to test the efficacy of FDI/FTC schemes [51, 110, 205]. If the nominal PI controller is used in the presence of this speed sensor fault, whilst the measured output appears to track the reference signal, the actual tracking error is shown in Fig. 7.12. This corresponds to a RMS tracking error of 12.5584 rad/s and significantly degraded performance.

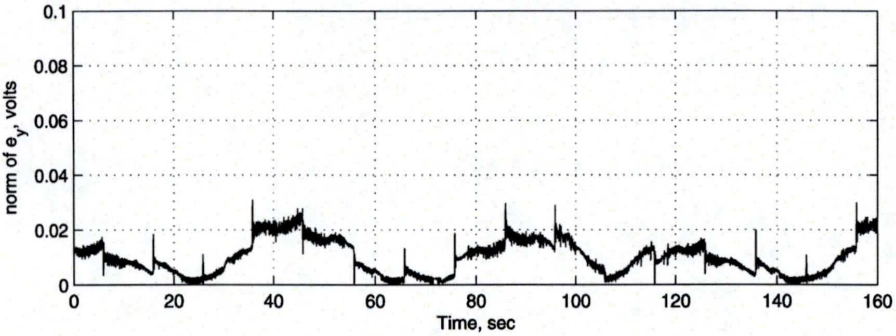

Fig. 7.13 Output estimation error

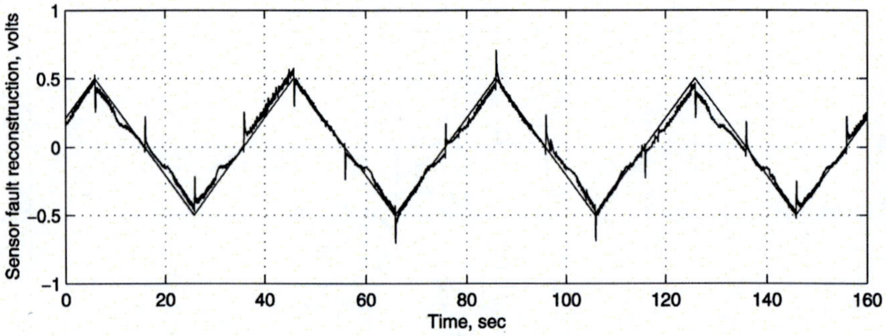

Fig. 7.14 Reconstruction of the fault signal

The fault tolerant control scheme as described in Fig. 7.8 has been implemented to attempt to recover the fault-free performance in the presence of the sensor fault in Fig. 7.11. Specifically the control law in (7.27) has been modified to be

$$u = F(s)(w_r - w + \hat{f}_o) \tag{7.29}$$

This will be discussed more formally in the sequel. Figure 7.13 is concerned with the online sliding mode observer response and represents a plot of $\|e_y(t)\|$ versus time.

This quantity is indicative of whether a sliding motion is taking place (and also in this situation of course, the accuracy with which the output of the observer tracks the measured speed and current signals). As a result of the fixed step integration, a reduction in the predicted tolerance of 0.002 can be observed in Fig. 7.13. Figure 7.14 represents the reconstruction signal $\hat{f}_o$ compared with, in this case, the known fault signal f_o from Fig. 7.11. A very good reconstruction is obtained.

Using the sliding mode estimation scheme, Fig. 7.15 shows the extent to which the actual speed tracks the reference signal w_r (which can be calculated in this situation because the exact value of the fault is known). It can be seen that now very good tracking is still maintained despite the presence of the fault (Figs. 7.15 and 7.16). The RMS speed tracking error now is 2.8869 rad/s. This is not as good

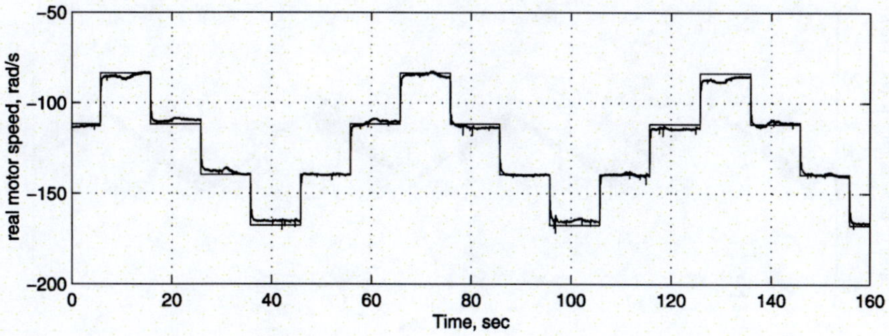

Fig. 7.15 Real motor speed v reference

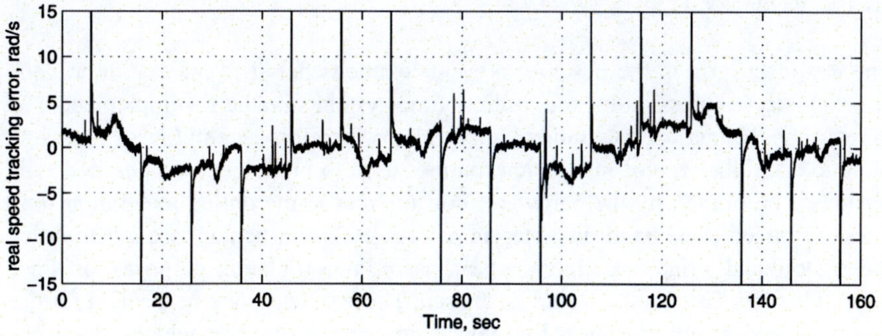

Fig. 7.16 Real motor speed tracking error

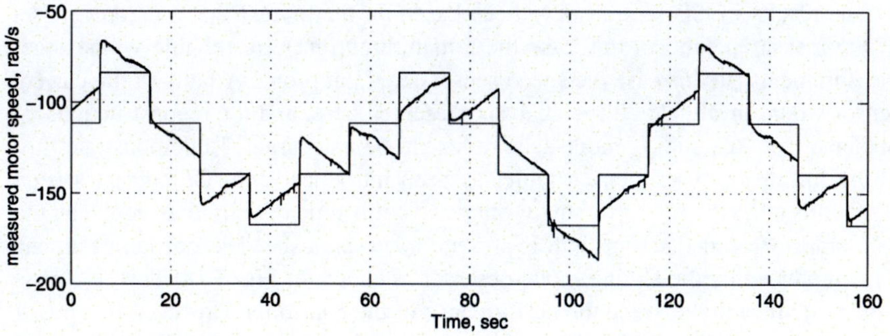

Fig. 7.17 Measured motor speed

as the nominal performance but is significantly better than in the unaccommodated case in Fig. 7.12.

For interest, the measured output is shown in Fig. 7.17. Obviously, apparently poor tracking is being demonstrated here because of the influence of the faults.

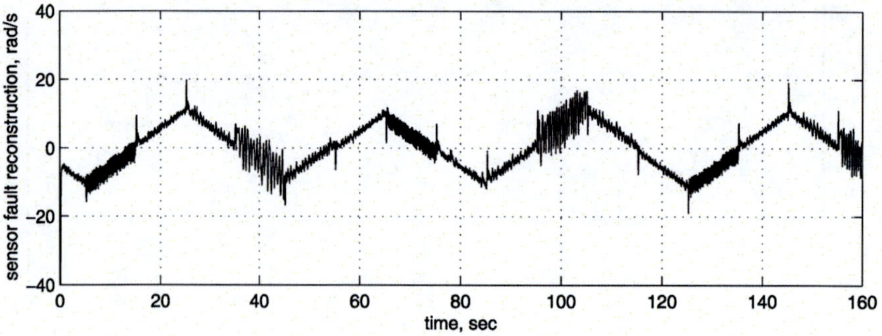

Fig. 7.18 Sensor fault reconstruction (with a fault on current)

7.2.6.2 Remarks

The nonlinear term in the observer prevents explicit analytical integration to generate an exact expression for the discrete time system suitable for implementation in real-time. Numerical integration methods must therefore be employed. The sampling rate and the type of integration routines used to implement the observer will obviously be limited by the hardware. For this reason the online integration routines are forced to be fixed time-step in nature. In this particular case a fifth order 'ode5 Dormand–Prince' routine has been used. For simulation purposes, off-line, a variable time-step routine would be chosen. These schemes are best suited for the 'stiff systems' resulting from the nonlinear term in the observer which effectively has high gain characteristics. Because of the 'stiff' nature of the observer system, low order integration Euler methods are known to encounter difficulties [46], hence the choice of a high order integration routine.

The results in this section have been based on the pretext that a certain subset of the measurements (in this case the armature current) are reliable whilst other measurements (in this case the speed sensor) are fault prone. A fault on the current sensor would invalidate all the analysis presented here, and the reconstructions $\hat{f}_o$ obtained from the sliding mode scheme would be inaccurate. To demonstrate this, in the following experiment, a fault has been introduced into the current sensor. Again this takes the form of a saw-tooth wave—this time of amplitude equivalent to 0.25 amps. No speed sensor fault is present, however, a speed sensor fault estimate $\hat{f}_o$ is produced by the sliding mode observer as shown in Fig. 7.18. This of course would significantly degrade the performance of the controller. However, it is possible to design an FDI scheme (using any robust method of choice), utilising only the armature current measurement, to independently flag the onset of a current sensor fault and thus isolate the situation in which the proposed fault tolerant scheme is not valid. The signal in Fig. 7.18 does, however, indicate the presence of a fault but does not by itself provide fault isolation.

In the present scheme the sensor correction signal $\hat{f}_o$ is a permanent feature. This is deliberate since the focus of this work is to demonstrate that sliding mode estimation can be implemented online and reconstructions of the unknown fault signals can be made with sufficient fidelity to maintain good tracking performance. However, since the fault estimates are not perfect, the effect of the reconstruction error will affect the nominal performance of the controller even when the sensor is operating perfectly.

7.3 Summary

In this chapter sliding mode observers for the reconstruction of sensor faults have been implemented (in real-time) for fault detection and isolation, as well as to achieve fault tolerant control. An estimate of the sensor fault, obtained from an online sliding mode FDI scheme has been used to correct the measured output from the sensor. This 'virtual sensor' has then been used in the control algorithm to form the output tracking error signal which is processed to generate the control signal. This idea has been implemented successfully on a laboratory crane and a DC motor rig, which were chosen as demonstrators. The scheme is not specific to such systems and is applicable to a reasonably wide class of engineering systems which can, at least in an operating region of interest, be adequately represented by an uncertain linear system.

7.4 Notes and References

There are many descriptions of implementation work on DC motors especially for FTC and FDI in the literature e.g., [191, 283]. Survey papers such as [136, 294] provide an excellent overview of the current literature. The popularity of DC motors as real-time implementation test platforms (especially in the academic and research community) is due to their convenient laboratory scale size, availability and low cost. This allows newly developed FTC and FDI schemes to be tested cheaply in a safe laboratory environment compared to safety critical systems like an aircraft. In [145], an implementation of model-based condition monitoring of an actuator system driven by a brushless DC motor is presented. This paper presents real-time implementation tests under laboratory conditions for a cabin pressure control outflow valve system for aircraft. Most recently, [103] has provided a survey for various techniques of fault tolerant control and diagnosis for induction motors.

Acknowledgements The authors are extremely grateful to Peter Barwell, Peter Clarke and David Dryden for their considerable assistance in the development and maintenance of the laboratory crane rig. The efforts of Justin Lado Lomoro in terms of the data collection for the crane are acknowledged.

Chapter 8
Adaptive Sliding Mode Fault Tolerant Control

In this chapter the benefit of using SMC, especially when handling actuator and sensor faults/failures, will be demonstrated using a realistic high fidelity nonlinear aircraft model. The switching surface design and the control law are presented. A simple idea to ensure the SMC can also handle total actuator and sensor failures is presented, based on a specific example involving control of the longitudinal dynamics of the aircraft.

8.1 Introduction

In this chapter, sliding mode schemes for FTC are developed and applied to a high fidelity model of a large civil transport aircraft which has been used by other researchers as a test bed for their developments: see for example [105, 177, 183–185]. The design of the sliding mode switching surface for the controller uses LMI methods building on previous work from the sliding mode literature. An adaptive gain is used in the nonlinear part of the control law which reacts to the occurrence of a fault and attempts to keep the switching function as close as possible to zero, thus trying to maintain nominal performance. If the total failure of an actuator is detected a switch is made to a 'back-up' control surface but the linear component of the control law remains unchanged. This controller is then tested in a number of different actuator fault scenarios. The sliding hyperplane is designed to minimise the effect of unmatched uncertainty on the sliding motion arising from actuator failures. A simple adaptive scheme for the nonlinear unit vector scaling gain is also proposed.

The FTLAB747 software running under MATLAB® has been developed for the study of fault tolerant control and FDI schemes. It represents a 'real world' model of a B747-100/200 aircraft, and the technical data and the underlying differential equations have been obtained from NASA [118, 119]. The software was originally initiated at Delft University of Technology by van der Linden (Delft University Aircraft Simulation and Analysis Tool, DASMAT) [259] and Smaili (Flight Lab 747, FTLAB747) [233], and later developed and enhanced for use in terms of fault detection and fault tolerant control by Marcos and Balas [183] (FTLAB747 V6.1/V6.5).

H. Alwi et al., *Fault Detection and Fault-Tolerant Control Using Sliding Modes*, Advances in Industrial Control,
DOI 10.1007/978-0-85729-650-4_8, © Springer-Verlag London Limited 2011

The high fidelity nonlinear model has 77 states incorporating rigid body variables, sensors, actuators and aero-engine dynamics. All the surfaces and engine dynamics are modelled with realistic position and rate limits. The specific aerodynamic coefficients are taken from [119], which have been obtained from extensive wind tunnel experiments, simulations and test flights. The capabilities of this software as a realistic platform to test FTC and FDI schemes is demonstrated by its subsequent use by many researchers. More recently this software has been upgraded and modified to V6.5/7.1/2006b by Smaili et al. [234] and used as a test bed for the GARTEUR AG16 group [80].

Sliding mode methods have been applied to high performance prototype aircraft such as the tailless aircraft [226, 269] in which redundant actuators have been purposely built into the aircraft for performance and fault tolerance reasons. This is one of the motivations behind the research in this book: to develop sliding mode methods to exploit all the actuators available in large passenger transport aircraft to achieve fault tolerance especially when total actuator failures occur.

8.2 Actuator Fault Tolerant Control

This section will concentrate on the design of a fault tolerant controller to handle actuator faults. Consider the nth order linear time invariant system with m inputs subject to uncertainty given by

$$\dot{x}_p(t) = A_p x_p(t) + B_p u(t) - B_p K(t) u(t) + Q_p \xi(t, x_p) \tag{8.1}$$

where $A_p \in \mathbb{R}^{n \times n}$, $B_p \in \mathbb{R}^{n \times m}$ and $Q_p \in \mathbb{R}^{n \times q}$. Equation (8.1) may be thought of as arising from a linearisation process about a specific operating condition. The matrix $K(t) = \mathrm{diag}(k_1(t), \ldots, k_m(t))$ is composed of scalar functions $k_i(t)$ which satisfy $0 \leq k_i(t) < 1$. These model a decrease in effectiveness of a particular actuator: if $k_i(t) = 0$, the ith actuator is working perfectly whereas if $k_i(t) > 0$, some level of fault is present. Since by assumption $k_i(t) < 1$, this excludes the possibility of the actuators failing completely (although this issue will be addressed in detail separately later in the chapter). Without loss of generality it can be assumed that the input distribution matrix B_p has full rank and the pair (A_p, B_p) is controllable. The function $\xi(t, x_p)$ is assumed to be unknown but bounded and represents uncertainty in the system. Here, it is assumed to satisfy

$$\left\|\xi(t, x_p)\right\| < \mathrm{c}_1 \left\|x_p(t)\right\| + \mathrm{c}_2 \tag{8.2}$$

where c_1 and c_2 are known constants. This uncertainty structure has been considered in Section 3.6 in [85]. Only longitudinal control is considered: all lateral and directional states have been set to trim values. The controller is designed for an 'up and away' [105] flight envelope and the main objective is to obtain good tracking of flight path angle (FPA) and true airspeed (V_{tas}). The nominal (fault-free) sliding mode controller has first been designed using a linear model obtained

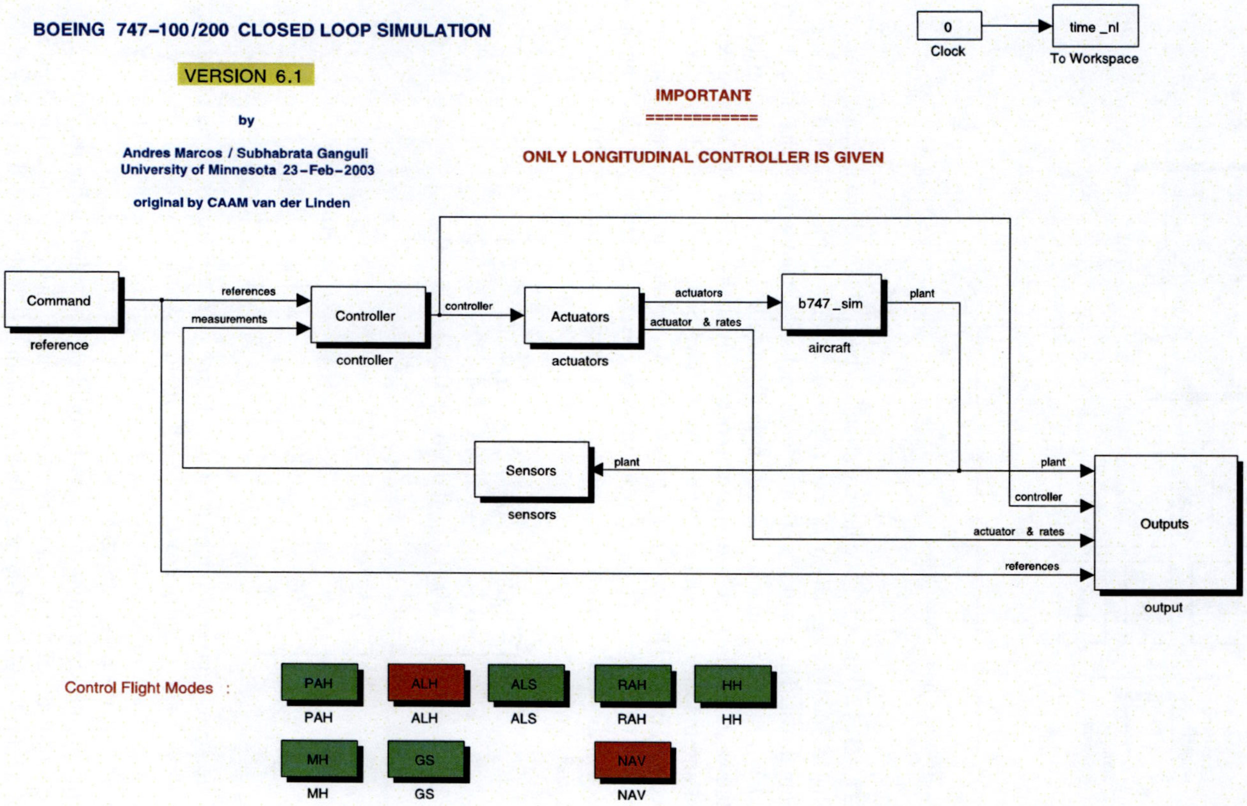

Fig. 8.1 FTLAB747 SIMULINK® model

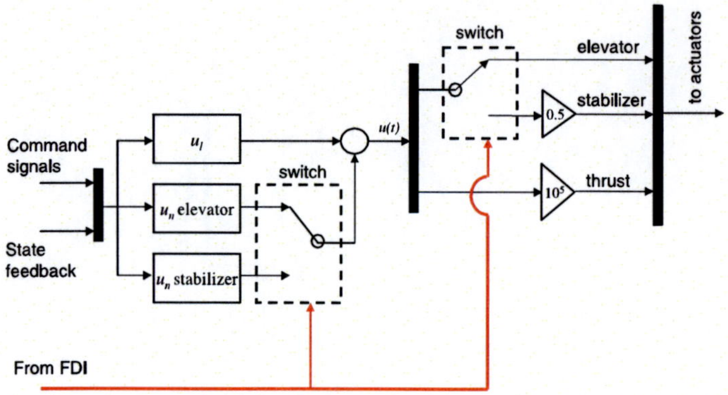

Fig. 8.2 Controller interconnection

from FTLAB747 (Fig. 8.1). The linearisation has been obtained around an operating condition of 300,000 kg, 184 m/s true airspeed, and an altitude of 4000 m at half maximum thrust. The result is a sixth order model associated with pitch rate q, true airspeed V_{tas}, angle of attack α, pitch angle θ, altitude h_e and horizontal position along the earth x-axis x_e. For design purposes, only the first four states have been retained and the four individual engine thrusts have been aggregated to produce a single control input. The two other inputs represent elevator deflection and horizontal stabiliser deflection. In the following state-space representation, the three inputs have been individually scaled which results in a system and input distribution matrix pair (A_p, B_p) with

$$A_p = \begin{bmatrix} -0.6803 & 0.0002 & -1.0490 & 0 \\ -0.1463 & -0.0062 & -4.6726 & -9.7942 \\ 1.0050 & -0.0006 & -0.5717 & 0 \\ 1 & 0 & 0 & 0 \end{bmatrix} \tag{8.3}$$

$$B_p = \begin{bmatrix} -1.5539 & 0.0154 \\ 0 & 1.3287 \\ -0.0398 & -0.0007 \\ 0 & 0 \end{bmatrix}, \qquad b_s = \begin{bmatrix} -1.5760 \\ 0 \\ -0.0398 \\ 0 \end{bmatrix} \tag{8.4}$$

where the states represent pitch rate (rad/s), true airspeed (m/s), angle of attack (rad) and pitch angle (rad), respectively. The inputs associated with B_p are elevator deflection (rad) and total thrust (N) (scaled by 10^5), and b_s is the distribution matrix associated with the horizontal stabiliser.

During normal operation, the aircraft would be controlled using the thrust and elevator, however, in the event of an elevator failure, the horizontal stabiliser can be used as 'back-up' (see Fig. 8.2). In this situation b_s will be used to replace the first column of B_p when the 'back-up' controller is activated (this will be discussed later). When implementing the controller on the nonlinear model a simple gain block (10^5 for thrust and 0.5 for horizontal stabiliser [105]) is used to recover the signal sent to the actuator (see Fig. 8.2). The controlled output distribution matrix is

$$C_c = \begin{bmatrix} 0 & 0 & -1 & 1 \\ 0 & 1 & 0 & 0 \end{bmatrix} \tag{8.5}$$

which represents flight path angle (FPA)[1] and true airspeed (V_{tas}). This linear model will be used to design the controller scheme which will be described in the sections which follow.

8.2.1 Sliding Mode Controller Design

Integral action (as discussed in Sect. 3.5.1) will be included to add a tracking facility for the two controlled outputs FPA and V_{tas}. The uncertain faulty system from (8.1) has been augmented with integral action states $x_d \in \mathbb{R}^m$ satisfying

$$\dot{x}_d(t) = y_c(t) - C_c x_p(t) \tag{8.6}$$

where the differentiable signal $y_c(t)$ satisfies

$$\dot{y}_c(t) = \Gamma\big(y_c(t) - Y_d\big) \tag{8.7}$$

In (8.7) the design matrix $\Gamma \in \mathbb{R}^{m\times m}$ is stable and Y_d is a piecewise constant demand vector. Augmenting the states from (8.1) with the integral action states and defining $x = \text{col}(x_d, x_p)$ it follows that

$$\dot{x}(t) = Ax(t) + Bu(t) + B_d y_c(t) - BK(t)u(t) + Q\xi(t,x) \tag{8.8}$$

where

$$A = \begin{bmatrix} 0 & -C_c \\ 0 & A_p \end{bmatrix}, \quad B = \begin{bmatrix} 0 \\ B_p \end{bmatrix}, \quad B_d = \begin{bmatrix} I_m \\ 0 \end{bmatrix}, \quad Q = \begin{bmatrix} 0 \\ Q_p \end{bmatrix} \tag{8.9}$$

Since the pair (A_p, B_p) is controllable, if (A_p, B_p, C_c) does not have any invariant zeros at the origin, then (A, B) is controllable (see Sect. 3.5.1). For the analysis which follows, define an augmented version of b_s from (8.4) as

$$B_s = \begin{bmatrix} 0 \\ b_s \end{bmatrix} \tag{8.10}$$

Although B_s does not directly appear in (8.8), it represents the distribution matrix associated with (8.8) when the horizontal stabiliser is employed as a 'back-up' control surface if a total failure in the elevator occurs. Define a switching function

$$s(t) = Sx(t) \tag{8.11}$$

[1]Flight path angle is the difference between pitch angle and angle of attack [36].

as a linear combination of the states, where $S \in \mathbb{R}^{m\times(n+m)}$. If a control law can be developed which forces the closed-loop trajectories onto the surface $s(t)=0$ in finite time (despite faults), and constrains the states to remain there, then an ideal sliding motion is said to have been attained. Suppose the matrix S is designed so that the square matrix SB is nonsingular (in practice this is easily accomplished since B is full rank and S is a free parameter). Then, as discussed in Sect. 3.5.1, the ideal sliding motion is given by

$$\begin{aligned}\dot{x}(t) = \big(I - B(SB)^{-1}S\big)Ax(t) + \big(I - B(SB)^{-1}S\big)Q\xi(t,x)\\ + \big(I - B(SB)^{-1}S\big)B_d y_c(t)\end{aligned} \tag{8.12}$$

for all $t \geq t_s$ where $Sx(t_s)=0$. It can be seen from (8.12) that the sliding motion is a control independent free motion which depends on the choice of sliding surface. If Q_p belongs to the range-space of the matrix B_p then $(I - B(SB)^{-1}S)Q = 0$ and the sliding motion is independent of the uncertainty. Several approaches have been proposed in the literature for the design of the matrix S (see Sect. 3.4) and without loss of generality, the surface can always be designed so that $SB = I_m$.

The proposed control law comprises two components; a linear component to stabilize the nominal linear system; and a discontinuous component. Specifically

$$u(t) = u_l(t) + u_n(t) \tag{8.13}$$

where the linear component is given by

$$u_l(t) = \underbrace{-(SB)^{-1}(SA - \Phi S)}_{L_x}x(t) \underbrace{- (SB)^{-1}SB_d}_{L_d} y_c(t) \tag{8.14}$$

where $\Phi \in \mathbb{R}^{m\times m}$ is any stable design matrix and $u_n(t)$ is a discontinuous component which is a function of $s(t)$.

In this chapter, the choice of the nonlinear term $u_n(t)$ is facilitated by choosing S such that $SB = I_m$, which effectively decouples the components of the sliding surface and associates with each element of S a particular control input. Componentwise, the proposed control structure has the form

$$u_i(t) = u_{l_i}(t) - \big(\rho_i(t) + \eta_i\big)\,\mathrm{sgn}\big(s_i(t)\big), \quad i = 1,\dots,m \tag{8.15}$$

where the η_i are positive constants,[2] $u_{l_i}(t)$ is the ith component of $u_l(t)$ and $s_i(t)$ is the ith component of $s(t) = Sx(t)$. It is easy to see from (8.14) that $u_{l_i}(t)$ is bounded by $|u_{l_i}(t)| < l_1\|x(t)\| + l_2$ where l_1 and l_2 are known positive constants. The gains $\rho_i(\cdot)$ in each of the control channels are defined as

$$\rho_i(t) = r_i(t)\big(\bar{r}_{(i,1)}\big\|x(t)\big\| + \bar{r}_{(i,2)}\big) \tag{8.16}$$

[2]The η_i could be chosen as functions of the state, large enough to bound the uncertainty in the fault-free case when $K(t)=0$.

where the fixed constants are

$$\bar{r}_{(i,1)} := \big(l_1 + \|S_i Q\|\mathrm{c}_1\big), \qquad \bar{r}_{(i,2)} := \big(l_2 + \|S_i Q\|\mathrm{c}_2\big) \tag{8.17}$$

and c_1 and c_2 are from (8.2). The variables $r_i(t)$ are adaptive gains which vary according to

$$\dot{r}_i(t) = \alpha_i\big(\bar{r}_{(i,1)}\|x(t)\| + \bar{r}_{(i,2)}\big)D_\epsilon\big(|s_i(t)|\big) - \beta_i r_i(t), \quad r_i(0) = 0 \tag{8.18}$$

where the α_i and β_i are positive design constants. The function $D_\epsilon : \mathbb{R} \mapsto \mathbb{R}$ is the dead-zone function

$$D_\epsilon(s) = \begin{cases} 0 & \text{if } |s| < \epsilon \\ s & \text{otherwise} \end{cases} \tag{8.19}$$

where ϵ is a positive scalar. Here, ϵ is set to be small and helps define a boundary layer about the surface $\mathcal{S} = \{x(t) : Sx(t) = 0\}$ inside which an acceptably close approximation to ideal sliding takes place. Provided the states evolve (with time) inside the boundary layer, no adaptation of the switching gains takes place. If a fault occurs, which starts to make the sliding motion degrade so that the states evolve outside the boundary layer, i.e., $|s_i(t)| > \epsilon$, then the dynamic coefficients $r_i(t)$ increase in magnitude (according to (8.18)) to force the states back into the boundary layer around the sliding surface.

Remark 8.1 In a fault-free situation it is not necessary and indeed is not advisable to have a large gain on the switched terms $u_n(t)$—therefore ideally the term $\rho(\cdot)$ should adapt to the onset of a fault and react accordingly. This adaptation scheme differs from the one in [270] and is more akin to the scheme from [278].

The choice of the design parameters η_i, α_i, β_i and ϵ depends on the closed-loop performance specifications and requires some design iteration. In general, the η_i need to be chosen as the nominal (no fault) gains for the nonlinear component of the control law (8.15) to ensure that sliding occurs in the fault-free system. The parameter ϵ is chosen to be small to form a boundary layer about $\mathcal{S}$, but not too small to cause 'false alarms' and unnecessary increases in $\rho_i(t)$. Thus ϵ dictates how sensitive the adaptive gains $r_i(t)$ are to changes in $s(t)$. The gain α_i dictates the rate at which $r_i(t)$ increases in reaction to faults: a large value for α_i indicates a fast increase of $r_i(t)$. On the other hand β_i dictates the rate at which $r_i(t)$ decreases to the nominal gain η_i when the fault has been rectified. A relationship between ϵ, η_i, α_i and β_i will be determined in the proof of the proposition which follows. The choice of these design parameters will be discussed further in Sect. 8.3. The following proposition shows that the gain functions are bounded and motion inside a boundary layer around $\mathcal{S}$ is obtained.

Proposition 8.1 *Consider the potentially faulty augmented system represented by (8.8) with the control law in (8.15); then each of the components $r_i(t)$ remain bounded and the switching states $s(t)$ enter a boundary layer around $\mathcal{S}$ in finite time.*

Proof Consider $\bar{k} = \max\{k_1(t) \dots k_m(t)\}$. Notice that by assumption $\bar{k} < 1$. From the decoupled structure which results from $SB = I_m$, it follows that

$$\dot{s}_i = -\phi_i s_i - \big(1 - k_i(t)\big)\big(\rho_i(t) + \eta_i\big)\operatorname{sgn}(s_i) - k_i(t)u_{l_i}(t) + S_i Q\xi(t,x) \tag{8.20}$$

where it has been assumed that $\Phi = \operatorname{diag}(-\phi_1, \dots, -\phi_m)$ and the ϕ_i are positive scalars. Therefore

$$s_i\dot{s}_i \le -\phi_i s_i^2 - (1-\bar{k})\big(\rho_i(t) + \eta_i\big)|s_i| + s_i\big(S_i Q\xi(t,x) - k_i(t)u_{l_i}(t)\big) \tag{8.21}$$

Using (8.17), and the fact that $\bar{k} = 1 - (1 - \bar{k})$, by construction

$$\begin{aligned} \big|\big(S_i Q\xi(t,x) - k_i(t)u_{l_i}(t)\big)\big| &\le \big|S_i Q\xi(t,x)\big| + \bar{k}\big|u_{l_i}(t)\big| \\ &\le \big(\bar{r}_{(i,1)}\big\|x(t)\big\| + \bar{r}_{(i,2)}\big) - (1-\bar{k})\big|u_{l_i}(t)\big| \end{aligned} \tag{8.22}$$

for $i = 1, \dots, m$, since from (8.2), $\|\xi(t,x)\| < \mathrm{c}_1\|x(t)\| + \mathrm{c}_2$. Define a scalar

$$\zeta := 1/(1-\bar{k}) > 0 \tag{8.23}$$

and component-wise Lyapunov functions

$$V_i = \frac{1}{2}\bigg(s_i^2 + \frac{1}{\alpha_i}(1-\bar{k})\big(r_i(t) - \zeta\big)^2\bigg) \tag{8.24}$$

where α_i is the positive scalar from (8.18). Clearly $V_i(\cdot)$ is positive definite with respect to s_i, the adaptive gain errors $r_i(t) - \zeta$, and is radially unbounded. Taking derivatives

$$\dot{V}_i = s_i\dot{s}_i + \frac{1}{\alpha_i}(1-\bar{k})\big(r_i(t) - \zeta\big)\dot{r}_i(t) \tag{8.25}$$

then substituting from (8.16), (8.18), (8.21) and (8.22) into the above and using the fact that $(1-\bar{k})\zeta = 1$, it follows that

$$\begin{aligned} \dot{V}_i \le{} & -\phi_i s_i^2 - |s_i|(1-\bar{k})\big(\eta_i + \big|u_{l_i}(t)\big|\big) \\ & - |s_i|(1-\bar{k})\big(\bar{r}_{(i,1)}\big\|x(t)\big\| + \bar{r}_{(i,2)}\big)\big(r_i(t) - \zeta\big) \\ & + \frac{1}{\alpha_i}(1-\bar{k})\big(r_i(t) - \zeta\big)\big(\alpha_i\big(\bar{r}_{(i,1)}\big\|x(t)\big\| + \bar{r}_{(i,2)}\big)D_\epsilon\big(\big|s_i(t)\big|\big) - \beta_i r_i(t)\big) \end{aligned} \tag{8.26}$$

If $|s_i| > \epsilon$ then $D_\epsilon(|s_i|) = |s_i|$ and so substituting in (8.26) and simplifying terms yields

$$\dot{V}_i \le -\phi_i s_i^2 - |s_i|(1-\bar{k})\big(\eta_i + \big|u_{l_i}(t)\big|\big) - \frac{\beta_i}{\alpha_i}(1-\bar{k})\big(r_i(t) - \zeta\big)r_i(t) \tag{8.27}$$

Notice by construction $\bar{k} < 1$ and $r_i(t) \geq 0$. Further manipulation of (8.27), and using (8.23), yields

$$\dot{V}_i \leq -\phi_i s_i^2 - |s_i|(1-\bar{k})\big(\eta_i + |u_{l_i}(t)|\big) - \frac{\beta_i}{\alpha_i}(1-\bar{k})\left(\frac{1}{2}\zeta - r_i(t)\right)^2 + \frac{\beta_i}{4\alpha_i(1-\bar{k})} \tag{8.28}$$

since expanding the quadratic term on the right-hand side of (8.28) gives (8.27). If $|s_i| > \epsilon$, then $|s_i|(1-\bar{k})\eta_i \geq (1-\bar{k})\epsilon\eta_i$. The quantities $\epsilon, \eta_i, \alpha_i$ and β_i are design parameters and so if they are chosen to satisfy

$$(1-\bar{k})\epsilon\eta_i \geq \frac{\beta_i}{4\alpha_i(1-\bar{k})} \tag{8.29}$$

then

$$\dot{V}_i \leq -\phi_i s_i^2 - |s_i|(1-\bar{k})|u_{l_i}(t)| - \frac{\beta_i}{\alpha_i}(1-\bar{k})\left(\frac{1}{2}\zeta - r_i(t)\right)^2 \leq 0 \tag{8.30}$$

If $|s_i| < \epsilon$ then $D_\epsilon(|s_i|) = 0$ and so substituting in (8.26) and simplifying terms yield

$$\dot{V}_i \leq -\phi_i s_i^2 - |s_i|(1-\bar{k})\big(\eta_i + |u_{l_i}(t)|\big) - \frac{\beta_i}{\alpha_i}(1-\bar{k})\big(r_i(t) - \zeta\big)r_i(t) - |s_i|(1-\bar{k})\big(\bar{r}_{(i,1)}\|x(t)\| + \bar{r}_{(2,i)}\big)\big(r_i(t) - \zeta\big) \tag{8.31}$$

Notice again, by construction, $\bar{k} < 1$ and $r_i(t) \geq 0$ and therefore for $|s_i| < \epsilon$ and $r_i(t) > \zeta$, it follows that $\dot{V}_i < 0$. Define a rectangle in $\mathbb{R}^2$ as

$$\mathcal{R}_i = \big\{(s_i, r_i) \mid |s_i| \leq \epsilon,\ 0 \leq r_i \leq \zeta\big\} \tag{8.32}$$

Also define $\mathcal{R}_+ = \{(s_i, r_i) \mid r_i \geq 0\}$. By construction of the adaptive gains, $r_i(t) \geq 0$ for all time and so the trajectory of $(s_i(t), r_i(t)) \in \mathcal{R}_+$ for all time, and so outside $\mathcal{R}_i \cap \mathcal{R}_+ = \mathcal{R}_i$, from (8.28) and (8.31), $\dot{V}_i < 0$. Let $\mathcal{V}_{r,i}$ denote the truncated ellipsoid

$$\mathcal{V}_{r,i} = \big\{(s_i, r_i) \mid V_i(s_i, r_i) \leq r\big\} \cap \mathcal{R}_+$$

where $V_i(\cdot)$ is defined in (8.24). Because $\mathcal{R}_i$ in (8.32) is a compact set, there exists a unique $r_{i,0} > 0$ such that $r_{i,0} = \min\{r \in \mathbb{R}_+ \mid \mathcal{R}_i \subset \mathcal{V}_{r,i}\}$ and in fact $r_{i,0} = \frac{1}{2}(\epsilon^2 + \frac{\zeta}{\alpha_i})$.

As shown in Fig. 8.3, since $\mathcal{R}_i \subset \mathcal{V}_{r_{i,0}}$, it follows that outside $\mathcal{V}_{r_{i,0}}$ the derivative of the Lyapunov function $\dot{V}_i < 0$ and so $\mathcal{V}_{r_{i,0}}$ is an invariant set which is entered in finite time t_0. Since $\mathcal{V}_{r_{i,0}}$ is entered in finite time, $V_i(s_i, r_i) \leq r_{i,0}$ for all $t > t_0$ which implies $|s_i| \leq \sqrt{2r_{i,0}}$ for all time $t > t_0$, and hence s_i enters and remains in a boundary layer of size $\sqrt{2r_{i,0}}$ around the ideal sliding surface $\mathcal{S}$. □

Fig. 8.3 Level set of the Lyapunov functions V_i

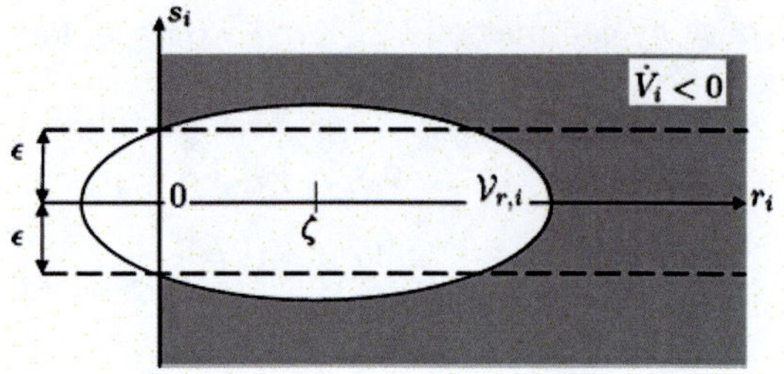

From the arguments above, for an appropriate choice of α_i, β_i and ϵ, close approximation to ideal sliding can be maintained even in the presence of faults. The reduced order sliding motion is then governed by (8.12). The motion depends on the uncertainty, but using arguments similar to those in Section 3.6 in [85], for a small enough c_1, ultimate boundedness of the states $x(t)$ can be proved.

Remark 8.2

- If $\epsilon = 0$ and $\beta_i = 0$ then ideal sliding can be guaranteed since it follows from (8.27) that $\dot{V}_i(s) \leq -\phi_i s_i^2 - |s_i|(1-\bar{k})(\eta_i + |u_{l_i}(t)|)$. This means ideal sliding can be attained and maintained in finite time. However, this adaptive scheme has disadvantages in practice since the gains $r_i(t)$ may become unbounded in the presence of noise [270].
- The adaptive gains act as a measure of severity of the actuator fault. Once the adaptive gain $\rho_i(t)$ from (8.16) exceeds a predetermined maximum value $\rho_{\max,i}$, a very severe fault or failure can be declared and a 'backup' control strategy can be initiated if required.
- From (8.23), as $\bar{k} \to 1$, ζ becomes infinitely large. In the case of total failure ($k_i(t) = 1 \Rightarrow \bar{k} = 1$), an alternative control strategy must be employed.

8.2.2 Sliding Mode Hyperplane Design

The first step in sliding mode controller design is the selection of the sliding surface matrix S. One methodology is the quadratic cost function approach which was discussed in Sect. 3.4.1. A modification of this approach is considered here to take into account the occurrence of failures. The design approach adopted here is described specifically for the aircraft system. However, its underlying philosophy is generic and could be adopted in other systems.

First consider the problem of designing a sliding surface matrix S for the nominal linear system associated with (8.24). Assume there are no faults (i.e., $K(t) = 0$) and there is no reference demand ($y_c(t) = 0$).[3] Also for the purpose of design, ig-

[3]As argued in Section 7.3.3 in [85], since $y_c(t) \to Y_d$ the effect of the demand signal can be removed by a change of coordinates which considers the system states relative to their steady-state values.

nore the uncertainty term. For this nominal linear system, as discussed previously in Sect. 3.4.1, consider the problem of minimising the quadratic performance index

$$J = \frac{1}{2}\int_{t_s}^{\infty} x(t)^{\mathrm{T}} \mathbf{Q} x(t)\, dt \tag{8.33}$$

where $\mathbf{Q}$ is a symmetric positive definite matrix and t_s is the time at which the sliding motion commences. Define a change of coordinates

$$z(t) = T_r x(t) \tag{8.34}$$

where T_r is an orthogonal matrix, so that the system in (8.8) is in regular form: i.e.,

$$T_r A T_r^{\mathrm{T}} = \begin{bmatrix} A_{11} & A_{12} \\ A_{21} & A_{22} \end{bmatrix}, \qquad T_r B = \begin{bmatrix} 0 \\ B_2 \end{bmatrix} \tag{8.35}$$

where $A_{11} \in \mathbb{R}^{n \times n}$, $B_2 \in \mathbb{R}^{m \times m}$. Also assume that (in regular form) the matrix $T_r \mathbf{Q} T_r^{\mathrm{T}}$ associated with equation (8.33) has a block diagonal structure $T_r \mathbf{Q} T_r^{\mathrm{T}} = \mathrm{diag}(\mathbf{Q}_1^{\mathrm{T}}\mathbf{Q}_1, \mathbf{Q}_2^{\mathrm{T}}\mathbf{Q}_2)$ where $\mathbf{Q}_2^{\mathrm{T}}\mathbf{Q}_1 = 0$ and the matrix $\mathbf{Q}_2^{\mathrm{T}}\mathbf{Q}_2 \in \mathbb{R}^{m \times m}$ is nonsingular. It follows that

$$J = \frac{1}{2}\int_{t_s}^{\infty} z_1(t)^{\mathrm{T}} \mathbf{Q}_1^{\mathrm{T}}\mathbf{Q}_1 z_1(t) + z_2(t)^{\mathrm{T}} \mathbf{Q}_2^{\mathrm{T}}\mathbf{Q}_2 z_2(t)\, dt \tag{8.36}$$

where $z = \mathrm{col}(z_1, z_2)$ with $z_1 \in \mathbb{R}^n$. As a result of the assumption of regular form, under nominal fault-free operation, the differential equation constraint (8.8), whilst sliding, may be written as

$$\dot{z}_1(t) = A_{11} z_1(t) + A_{12} z_2(t) \tag{8.37}$$

where the ‘virtual control’ z_2 satisfies

$$\mathcal{K} z_1 + z_2 = 0 \tag{8.38}$$

Here (8.38) represents the hyperplane equation $Sx = 0$ where

$$S = S_2[\mathcal{K} \quad I_m] T_r$$

and $S_2 \in \mathbb{R}^{m \times m}$ and is nonsingular. Substituting for z_2 from (8.38) in (8.37) gives an autonomous reduced order sliding motion. The matrix $\mathcal{K} \in \mathbb{R}^{m \times n}$ must be chosen to make $(A_{11} - A_{12}\mathcal{K})$ stable. This is always possible since (A_{11}, A_{12}) is controllable if (A, B) is controllable. As argued in [33] the optimal cost is given by $J = z_1(t_s)^{\mathrm{T}} P_c z_1(t_s)$ where P_c is the symmetric positive definite solution to the Riccati equation

$$P_c A_{11} + A_{11}^{\mathrm{T}} P_c - P_c A_{12} \big(\mathbf{Q}_2^{\mathrm{T}}\mathbf{Q}_2\big)^{-1} A_{12}^{\mathrm{T}} P_c + \mathbf{Q}_1^{\mathrm{T}}\mathbf{Q}_1 = 0 \tag{8.39}$$

and $z_1(t_s)$ is the value of the state component z_1 at the time at which sliding occurs. Furthermore the optimal choice of

$$\mathcal{K} = (\mathbf{Q}_2^{\mathrm{T}}\mathbf{Q}_2)^{-1} A_{12}^{\mathrm{T}} P_c$$

This problem can be posed as an LMI optimisation: Minimise trace(X^{-1}) subject to

$$\begin{bmatrix} A_{11}X + XA_{11}^{\mathrm{T}} - A_{12}N - N^{\mathrm{T}}A_{12}^{\mathrm{T}} & (\mathbf{Q}_1 X - \mathbf{Q}_2 N)^{\mathrm{T}} \\ \mathbf{Q}_1 X - \mathbf{Q}_2 N & -I \end{bmatrix} < 0 \qquad (8.40)$$
$$X > 0$$

where $X \in \mathbb{R}^{n \times n}$ and is symmetric and $N := \mathcal{K}X$. As argued on page 114 in [33], any solution to (8.40) satisfies $X^{-1} \geq P_c$. Consequently trace$(X^{-1}) \geq$ trace(P_c) and hence the minimisation process results in $X^{-1} = P_c$.

In the 'back-up' case, the input distribution matrix is perturbed by the change in actuator. Now the new input distribution matrix $\tilde{B}$ (say) is formed from replacing the first column from B in (8.9) associated with the elevator, with B_s in (8.10) which is associated with the horizontal stabiliser. In the regular form coordinates

$$T_r \tilde{B} = \begin{bmatrix} B_1 \\ B_2 R \end{bmatrix}$$

where $B_1 \in \mathbb{R}^{n \times m}$, $R \in \mathbb{R}^{m \times m}$ and B_2 is given in (8.35). Provided a sliding motion can be maintained with the new actuator set, in the regular form coordinates, the uncertain reduced order motion can be represented as

$$\dot{z}_1(t) = (A_{11} - A_{12}\mathcal{K})z_1(t) + Q_1\xi + B_1 u_{eq}(t) \qquad (8.41)$$

where $u_{eq}(t)$ is the equivalent control signal necessary to maintain a sliding motion on $\mathcal{S}$ and Q_1 represents the top n rows of $T_r Q$ i.e., the unmatched uncertainty distribution matrix in the regular form coordinates. The signal $u_{eq}(t)$ will be a function of the states z_1 and will include the effects of any additional mismatched disturbances resulting from the failure (such as turning moments generated from stuck actuators). The objective is to minimise the effect of $u_{eq}(t)$ and ξ on the nominal performance of the system in (8.41) in an $\mathcal{L}_2$ sense. Under the constraint that a common Lyapunov function for both the quadratic cost problem and the $\mathcal{L}_2$ gain problem is sought, from the Bounded Real Lemma, the $\mathcal{L}_2$ gain between $u_{eq}(t)$ and ξ and the state $z_1(t)$ is less than γ if

$$\begin{bmatrix} A_{11}X + XA_{11}^{\mathrm{T}} - A_{12}N - N^{\mathrm{T}}A_{12}^{\mathrm{T}} & [\, B_1 \;\; Q_1 \,] & X \\ [\, B_1 \;\; Q_1 \,]^{\mathrm{T}} & -\gamma I & 0 \\ X & 0 & -\gamma I \end{bmatrix} < 0 \qquad (8.42)$$

The overall optimisation problem used here is

Minimise $(a_1 \operatorname{trace}(Z) + a_2\gamma)$ subject to (8.40), (8.42) and

$$\begin{bmatrix} -Z & I_n \\ I_n & -X \end{bmatrix} < 0 \tag{8.43}$$

The matrix variable Z is a 'slack variable' which satisfies $Z > X^{-1}$ and so $\operatorname{trace}(Z)$ bounds $\operatorname{trace}(X^{-1})$. Here a_1 and a_2 are positive scalars which determine the relative weighting between the quadratic cost and the $\mathcal{L}_2$ problem. This represents a convex optimisation problem in terms of X, Z, N and γ and can be solved using standard LMI packages. The matrix which determines the hyperplane is computed as $\mathcal{K} = NX^{-1}$ and finally (in the original coordinates), the matrix

$$S = S_2[\mathcal{K} \quad I_m]T_r \tag{8.44}$$

The nonsingular matrix S_2 is then chosen to ensure $SB = I_m$. Although the development above is specific to the B747 backup stabiliser scenario, the approach is more flexible and could be used in more general situations.

8.3 Simulation Results

This subsection describes the actuator fault tolerant controller designed for the B747-100/200 aircraft. The controller is designed for longitudinal axis control in the 'up and away' flight envelope [105]. The main objective is to obtain tracking of flight path angle (FPA) and true air speed V_{tas}. The settling time when there is no fault/failure should be approximately 20 s for FPA and 45 s for V_{tas}. If a fault/failure occurs, the tracking requirement is 30 s for FPA with no difference in the V_{tas} tracking. These specifications are adopted from [105].

The weighting matrix for the hyperplane design has been chosen as

$$T_r \mathbf{Q} T_r^{\mathrm{T}} = \begin{bmatrix} 0.5 & 0 & 0 & 0 & 0 & 0 \\ 0 & 0.5 & 0 & 0 & 0 & 0 \\ 0 & 0 & 2 & -1 & 0 & 0 \\ 0 & 0 & -1 & 1 & 0 & 0 \\ 0 & 0 & 0 & 0 & 5 & 0 \\ 0 & 0 & 0 & 0 & 0 & 20 \end{bmatrix} \tag{8.45}$$

The last two elements of $T_r\mathbf{Q}T_r^{\mathrm{T}}$ multiply the z_2 term in (8.36) and hence weight the 'virtual control' term. Thus, by analogy to a more typical LQR framework, they affect the speed of response of the closed-loop system. The last state is weighted heavily to reduce the gains in the engine channels. The first two terms in (8.45) are associated with the integral action states and are less heavily weighted. The non-diagonal term in (8.45) arises from the fact that flight path angle is the quantity of interest. In the following design, the parameters $a_1 = a_2 = 1$ i.e., equal weighting of the quadratic cost performance and the $\mathcal{L}_2$ robustness. In this example, the choice of a_2 is not crucial because the degree of mismatch between B and $\tilde{B}$, represented by $\|B_1\|$, is small. The LMI optimisation software gives a unique solution for $\mathcal{K}$ in (8.44) which results in

$$S = \begin{bmatrix} 0.2163 & -0.0013 & -0.6524 & 0.0077 & 0.3471 & -0.9034 \\ -0.0000 & -0.1192 & 0.0000 & 0.7526 & -0.0005 & -0.0000 \end{bmatrix}$$

Note the original sliding surface matrix S obtained from the optimisation software (8.44) has been scaled using S_2 in order that $SB = I_2$. The poles associated with the sliding motion are $\{-0.6786, -0.3566 \pm 0.3802i, -0.1584\}$. From (8.14) the stable matrix has been chosen as $\Phi = -I_2$ which gives faster poles than those associated with the reduced order sliding motion. This results in

$$L_x = \begin{bmatrix} -0.2163 & 0.0013 & 0.7643 & -0.0086 & -1.0891 & 1.1956 \\ 0.0000 & 0.1192 & 0.1107 & -0.8672 & -3.8542 & 7.3710 \end{bmatrix}$$

$$L_d = \begin{bmatrix} -0.2163 & 0.0013 \\ 0.0000 & 0.1192 \end{bmatrix}$$

The pre-filter matrix from (8.7) has been designed to be

$$\Gamma = \begin{bmatrix} -0.2400 & 0 \\ 0 & -0.1250 \end{bmatrix}$$

This may be viewed as representing the ideal response in the FPA and the V_{tas} channels. Again the FPA response is faster than the V_{tas} response. In the simulations the discontinuity in the nonlinear control term has been smoothed, as in Sect. 3.2.2, by using a sigmoidal approximation where the fixed scalar $\delta = 0.01$. This removes the discontinuity and introduces a further degree of tuning to accommodate the actuator rate limits, especially during actuator fault or failure conditions. The initial fixed gains for the 'back-up' controller (using the horizontal stabiliser) are given by $\rho_{s,1} = 0.4$ and $\rho_{s,2} = 0.05$. Here the smoothing parameter is chosen as $\delta_s = 0.1$. The larger value of δ_s is used to accommodate the smaller positional movement and lower rate limits of the horizontal stabiliser. In this chapter, only the gains in the elevator channel are allowed to adapt: the gains associated with the thrust channel are fixed. When employing the adaptive gain for the controller from (8.15), it was found for this particular example the $\bar{r}_{(i,1)}(t)$ in (8.17) have no significant effect on the closed-loop performance and so $l_1 = \mathrm{c}_1 = 0$ was chosen and therefore $\bar{r}_{(i,1)}(t) \equiv 0$. The parameters, l_2 and c_2 have been chosen as $l_2 = 0.5$ and $\mathrm{c}_2 = 0.9117$ and therefore $\bar{r}_{(i,2)}(t) \equiv 1$. The upper and lower limits for $\rho_1(t)$ have been chosen as $\rho_{\max,1} = 5$ and $\rho_{\min,1} = \eta_1 = 0.2$, respectively. Here $\eta_1 = 0.2$ is chosen to be larger than the uncertainty in the no fault condition. The choice of $\rho_{\max,1}$ dictates how fast a severe or total failure can be detected. Here, $\rho_{\max,1}$ has been chosen large enough to compensate for the worse case fault on the elevator (before the switch to the stabiliser is activated) at a 70% decrease in effectiveness. The adaptation parameters are $\alpha_1 = 600$ and $\beta_1 = 0.02$ and the tolerance $\epsilon = 0.0005$. Appropriate values for α_1, β_1 and ϵ involve some design iteration. The parameter ϵ was chosen to be able to tolerate the variation in $s(t)$ due to normal changes in flight conditions but small enough to enable the adaptive gain to be sensitive to deviations from zero in the switch term s_1 when a fault or severe disturbance occurs. The term α_1 dictates the rate at which $\rho_1(t)$ increases and reacts to the faults. Here, it needs to be large to enable small changes in s_1 to cause significant changes in the gain so that the control system reacts quickly to the onset of a fault. From (8.29), $(1-\bar{k})\epsilon\eta_i = 3.0 \times 10^{-5}$ and $\beta_i/(4\alpha_i(1-\bar{k})) = 2.78 \times 10^{-5}$ and therefore the condition in Proposition 8.1 is satisfied.

The simulations presented in this chapter are all based on the *full nonlinear model*. For the 'up and away' flight condition, the elevator is used to track FPA demands.

8.3.1 Fault-Free Simulations

In this section, simulations are presented for the nominal controller design as described in Sect. 8.3. The simulation covers the entire 'up and away' flight region, from an altitude of 4000 m with a velocity of 184 m/s, to the end of the region at an altitude of 8500 m and with a velocity of 280 m/s. A series of 3 degree FPA and 10 m/s V_{tas} commands are issued during the simulation to take the aircraft through the entire envelope. Figures 8.4(a), 8.4(c) and 8.4(d) show the results associated with the controller designed for the elevator and thrust, whilst Fig. 8.4(b) shows the performance with the horizontal stabiliser as the control surface (together with thrust). In these responses the adaptive gain in each channel has been fixed throughout the simulation. Figure 8.4 shows that both controllers are able to maintain satisfactory tracking performance (using either the elevator or horizontal stabiliser) over the range of the 'up and away' flight region even though these conditions become increasingly distant from the design condition. This indicates the robustness of the single fixed sliding mode controller.

8.3.2 Changes in Effectiveness Gain

This section shows how the controller with an adaptive gain $\rho(t)$, as defined in Sect. 8.2.1, copes with different percentages of faults as modelled in (8.1). This section also demonstrates the fault detection capability of the controller. The simulations have been conducted at an altitude of 4000 m and a V_{tas} of 184 m/s with a 3 degree of FPA (0.5–30 s) step change and a 20 m/s V_{tas} step change at 55 s as command signals. The 'effectiveness gain' has been implemented as a simple but unknown (as far as the controller is concerned) gain between the output of the controller block and the actuator dynamics. These simple tests indicate the effect of a

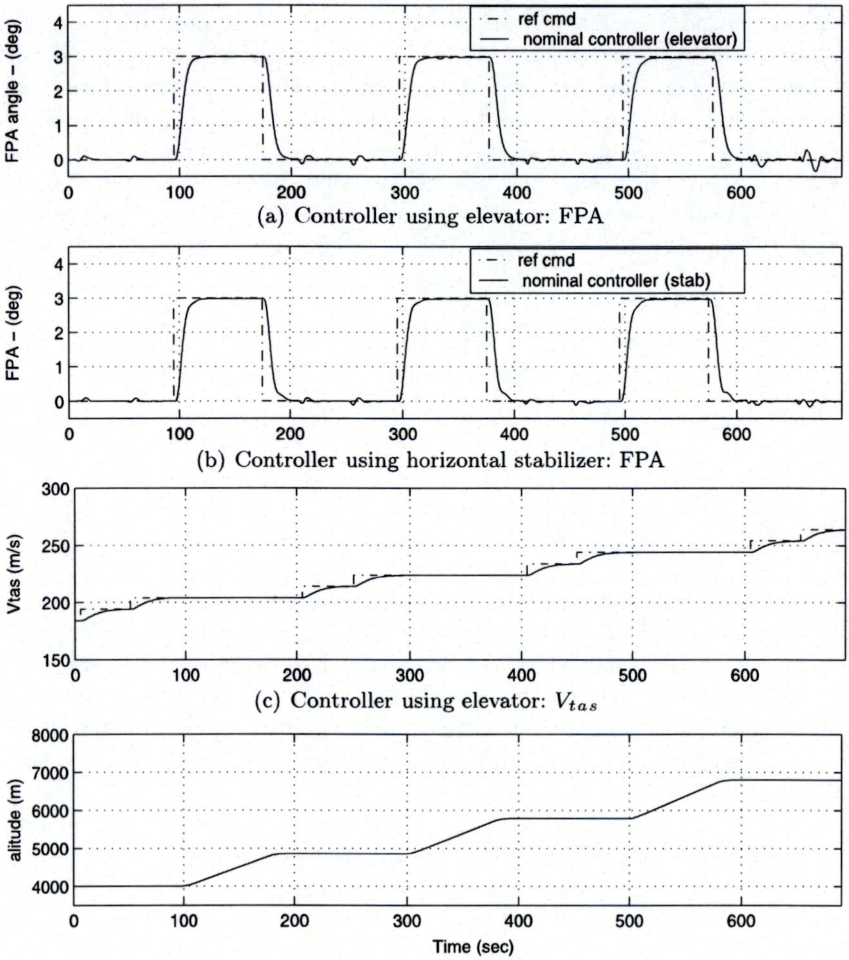

(a) Controller using elevator: FPA

(b) Controller using horizontal stabilizer: FPA

(c) Controller using elevator: V_{tas}

(d) Controller using elevator: altitude

Fig. 8.4 Nominal fault-free performance

loss of efficiency of the elevator due to damage or faults. Figure 8.5 shows comparisons of the adaptive gain controller with $k_1 = 0, 0.5, 0.6, 0.7, 0.7, 0.8, 0.9$ (indicating 0, 50, 60, 70, 80, 90 percent failure, respectively). It can be seen that there is a slight variation in FPA tracking but no visible difference in V_{tas} tracking. Overall FPA tracking is still possible (although with some degradation in performance) even with a 90% fault. This indicates that the adaptive gain controller is robust and has the ability to maintain good tracking even for severe faults. Looking at the signal s_1 (the switching signal associated with the elevator channel) in Fig. 8.5(c), its value is still relatively close to zero even when the percentage of fault increases. The adaptive gain signal shows that at 70% failure, the gain reaches the maximum allowable set gain ($\rho_{\text{max},1} = 5$). At this point it would be possible for a warning signal to be sent

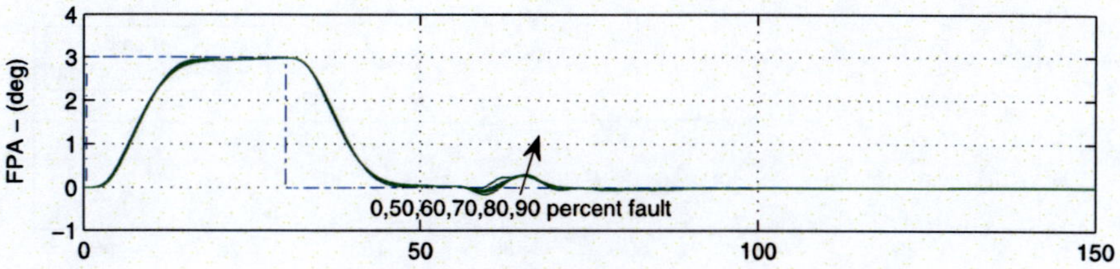

(a) Controller performance: FPA tracking

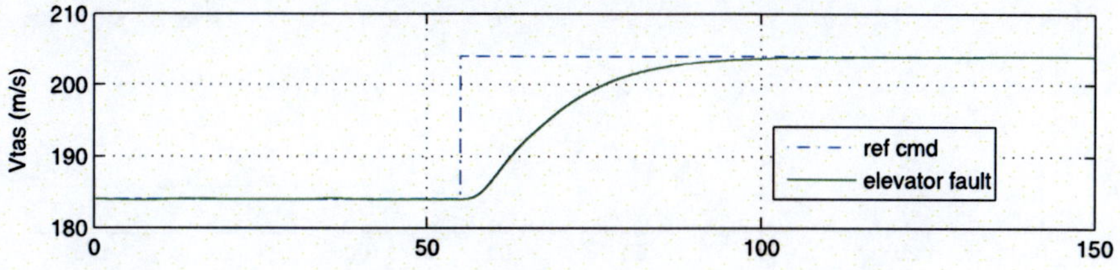

(b) Controller performance: V_{tas} tracking

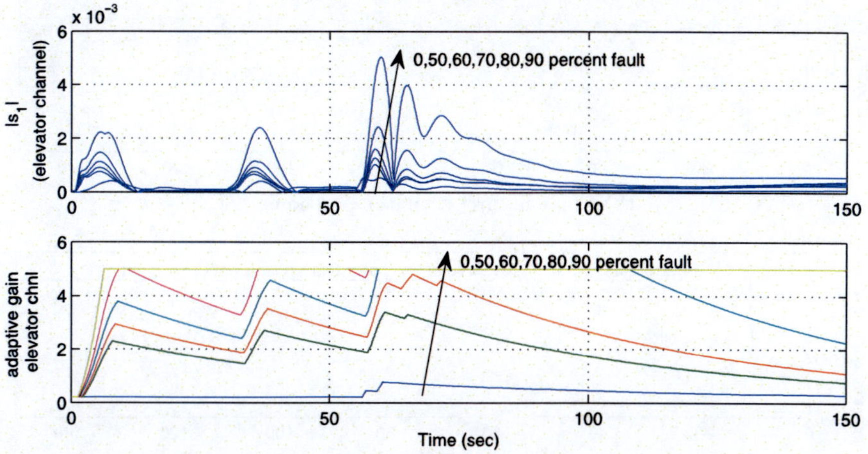

(c) Switching function $|s_1|$ and adaptive gain ρ_1

Fig. 8.5 Responses of different effectiveness gains

to the pilot or an automatic change to the 'back-up' controller could be initiated. Notice that even though the nominal adaptive gain controller is still able to maintain stability up to 90% failure, early failure detection is more desirable to provide advanced warning and to avoid potentially irrecoverable instability.

8.3.3 Total Elevator Failure Simulations

This section shows the results of nonlinear simulations when the elevator develops floating and/or lock type actuator failures. These simulate total failure of the eleva-

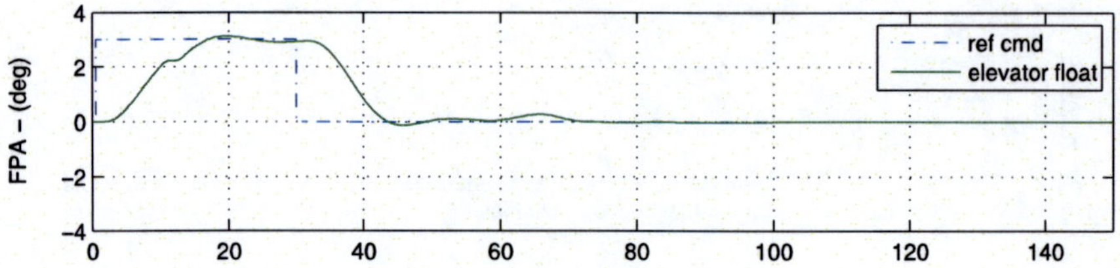

(a) Float failure: FPA response

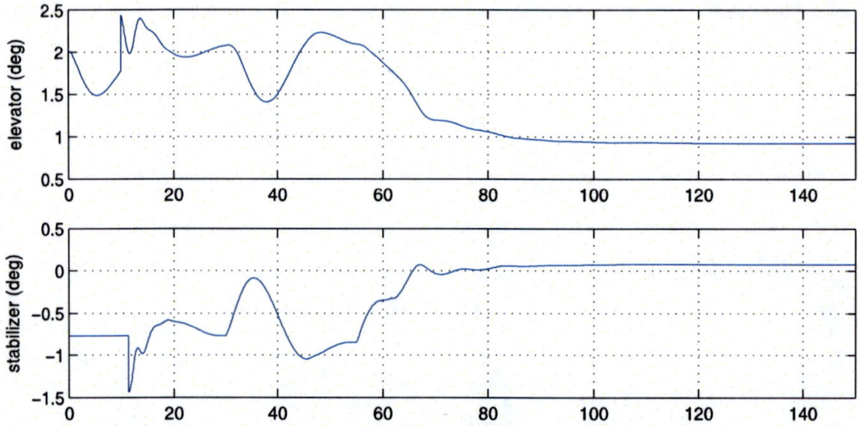

(b) Float failure: actuator positions

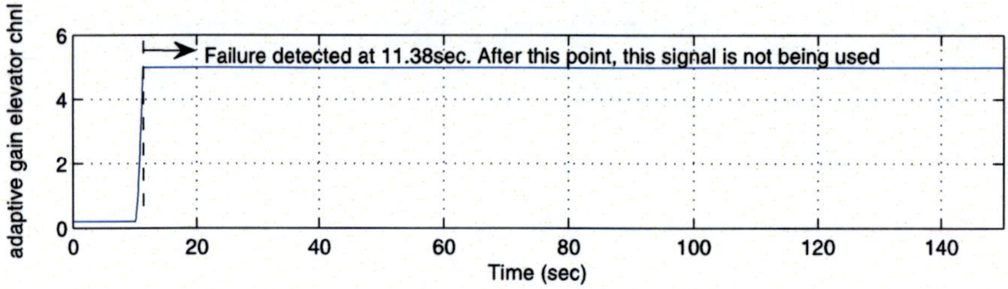

(c) Float failure: The adaptive gain ρ_1

Fig. 8.6 Responses to a float failure

tor and therefore require stabilization of the aircraft using the 'back-up' controller (which uses the horizontal stabiliser). A similar flight scenario is considered. The failure is set to occur during the climb (pitch up) manoeuvre at 10 s for both failure scenarios. To simulate a floating actuator type of failure, the elevator signal is replaced with the angle of attack [105]. This simulates the ineffectiveness of the elevator to provide a moment, and therefore the aircraft is unable to perform a pitch manoeuvre. Figure 8.6(a) shows that FPA tracking performance is slightly degraded and the response is slower. Figure 8.6(c) shows that the failure is detected at 11.38 s when the adaptive gain reaches its maximum value. Some peaks can be seen in the

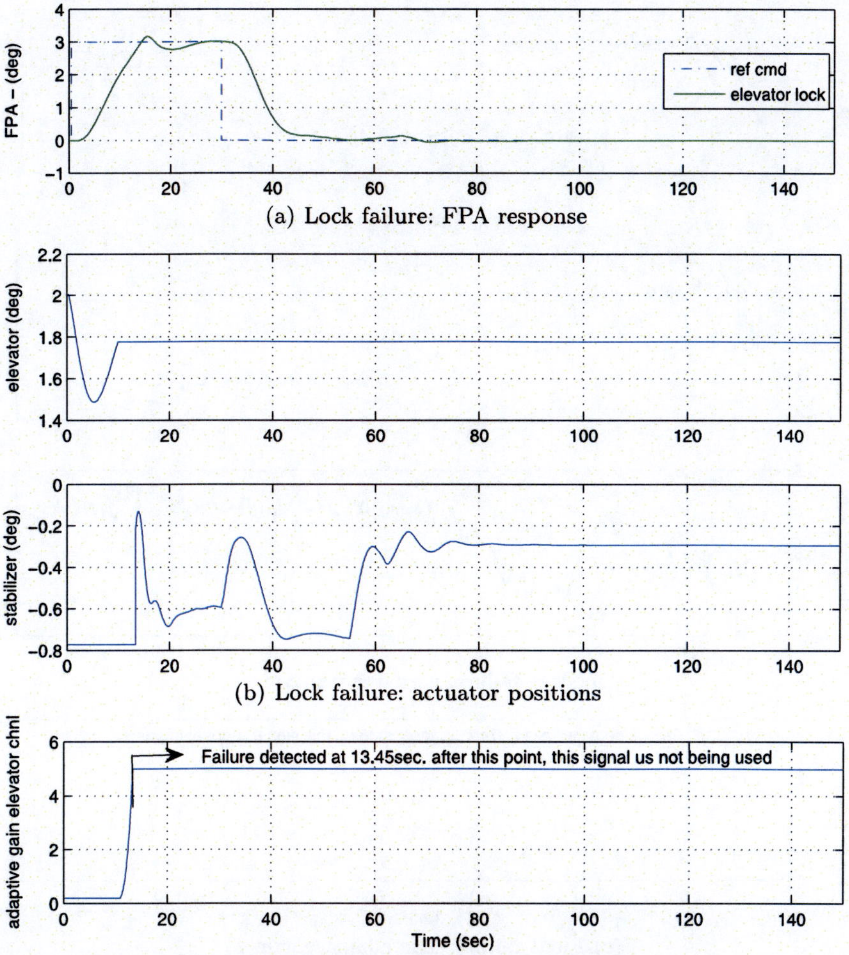

(a) Lock failure: FPA response

(b) Lock failure: actuator positions

(c) Lock failure: The adaptive gain ρ_1

Fig. 8.7 Responses to a lock failure

horizontal stabiliser signal (Fig. 8.6(b)) after activation due to the sudden change of control signal, but this stabilises after a few seconds. Once the controller is switched to the horizontal stabiliser, that surface is used for the remainder of the simulation. Overall performance is satisfactorily maintained after detection of the failure and after the change to the 'back-up' controller. To simulate lock failures, the elevator position is held at its value at 10 s. Figure 8.7(a) shows that, as before, the FPA tracking is slightly degraded. The failure is detected at 13.45 s (Fig. 8.7(c)) and the horizontal stabiliser is activated (Fig. 8.7(b)). A high peak occurs in the horizontal stabiliser signal but disappears after a few seconds (Fig. 8.7(b)). Overall tracking performance is maintained.

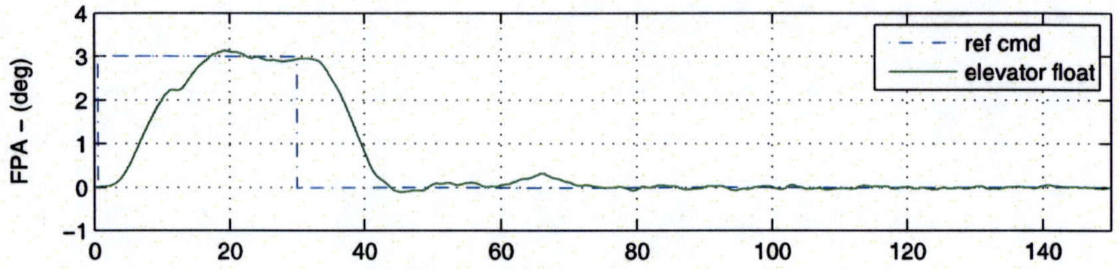

(a) Float failure: FPA response

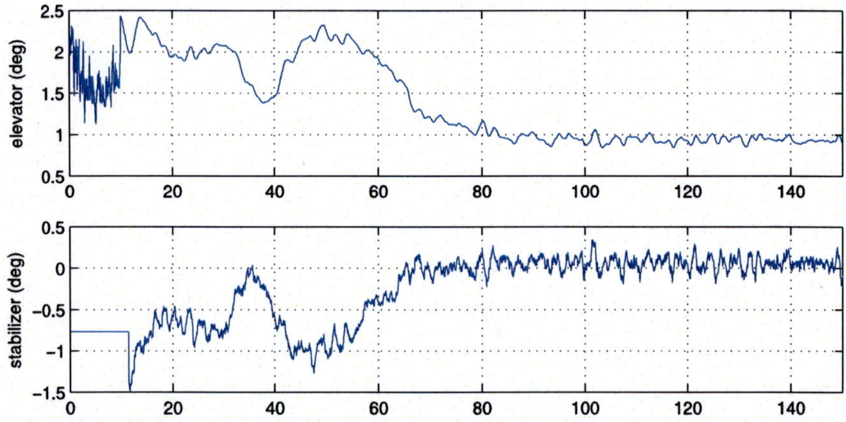

(b) Float failure: actuator positions

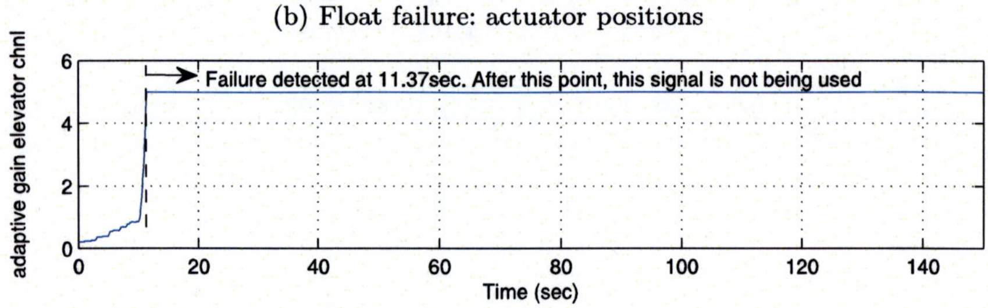

(c) Float failure: The adaptive gain ρ_1

Fig. 8.8 Responses to a float failure with wind and gust disturbances

8.3.4 Total Elevator Failure with Wind and Gusts

The same set of flight conditions and tests has been repeated in the presence of nonzero wind and gust (turbulence) conditions. The nonzero wind condition is set as $-11, -12, 0$ (m/s) for the u, v, w wind axis, respectively. As in [183], the turbulence parameter is set as white noise with a variance of 1. Figures 8.8 and 8.9 show the simulation results for float and lock type failures with wind and turbulence disturbances. As in the previous test, all the plots show that the controller has been able to maintain stability and tracking performance even in the presence of disturbances.

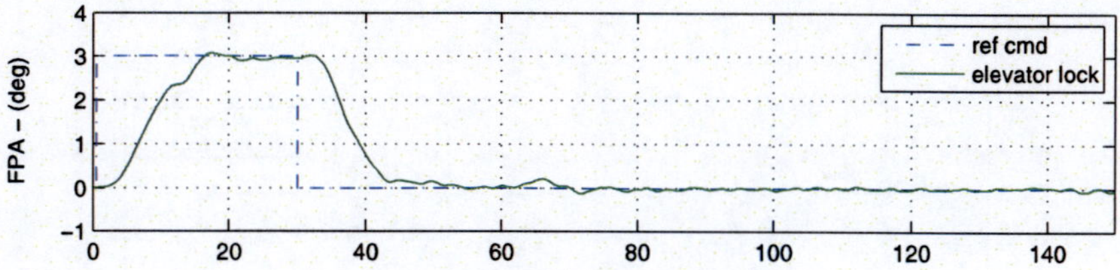

(a) Lock failure: FPA response

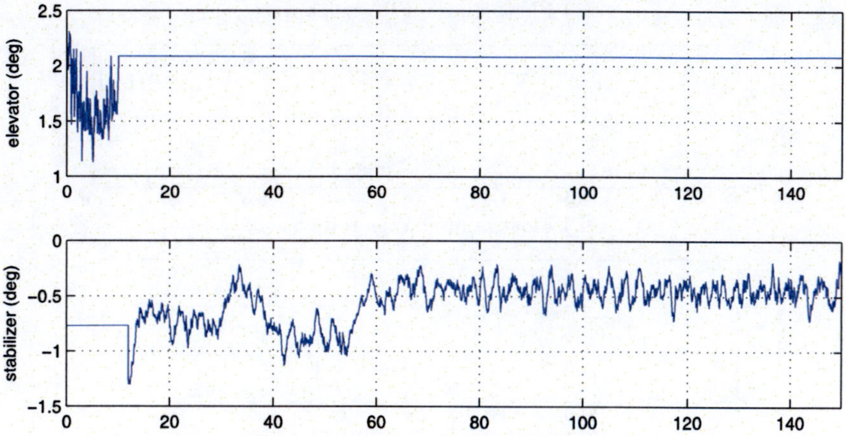

(b) Lock failure: actuator positions

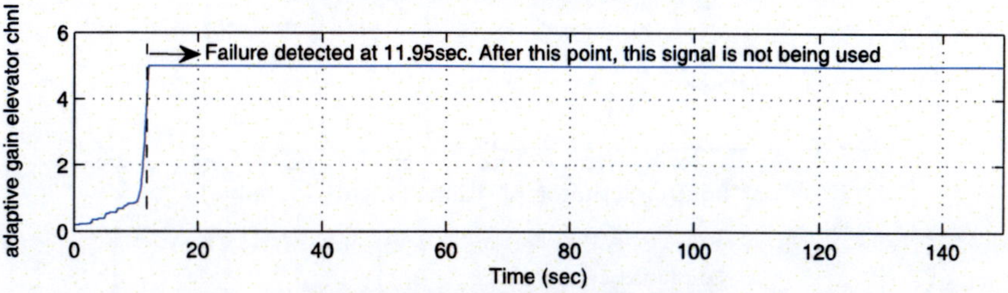

(c) Lock failure: The adaptive gain ρ_1

Fig. 8.9 Responses to a lock failure with wind and gust disturbances

8.3.5 Combined Loss of Effectiveness and Elevator Failure

The simulations have been conducted at an altitude of 4000 m and a V_{tas} of 184 m/s. The reference command requests a change in flight path angle of 3 deg for 20 s followed by a 20 m/s change in speed over a period of 45 s (in two steps). The command sequence for the FPA demand is then reversed after 250 s so that the aircraft is returned to (approximately) the initial flight conditions. These simple tests indicate the effect of a loss of efficiency of the elevator due to damage or faults. Figures 8.10 and 8.11 show the results of nonlinear simulations for various fault conditions: a nominal (no fault) period for the first 150 s, followed by degradation of the elevator effectiveness (150–260 s) and finally total failure (260–400 s). From

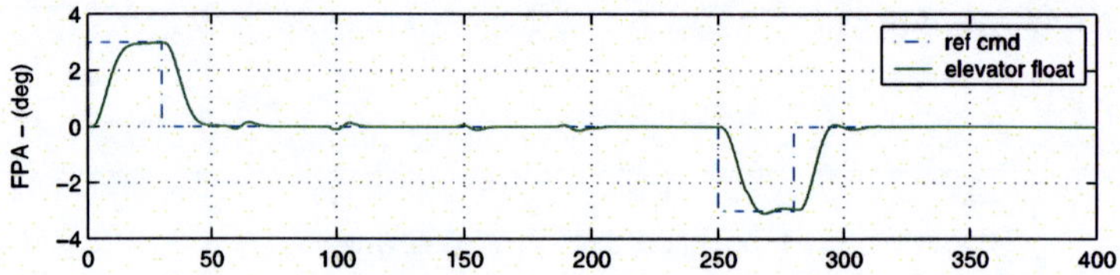

(a) Float failure: FPA response

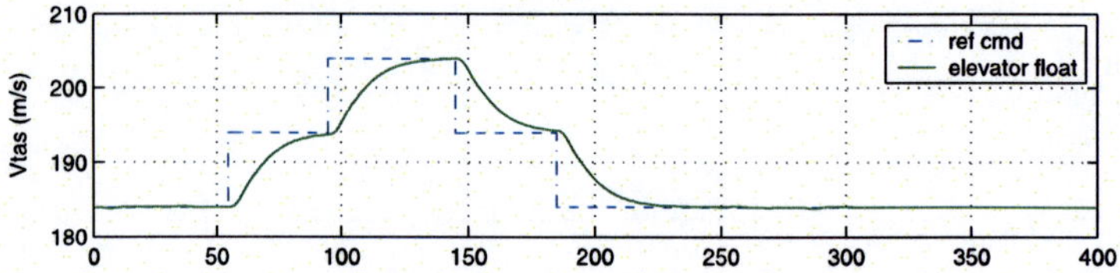

(b) Float failure: Vtas response

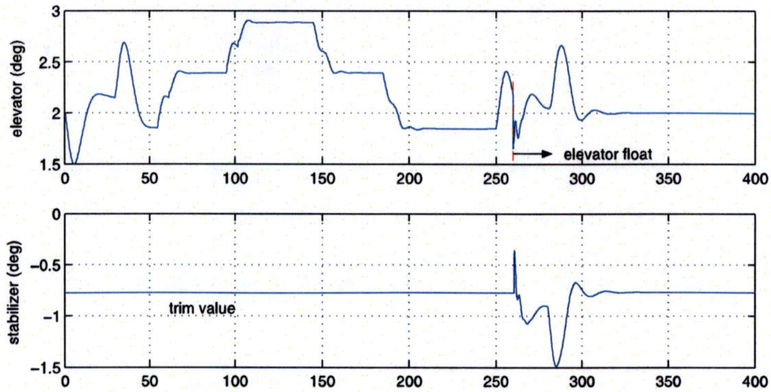

(c) Float failure: actuator positions (with trim value)

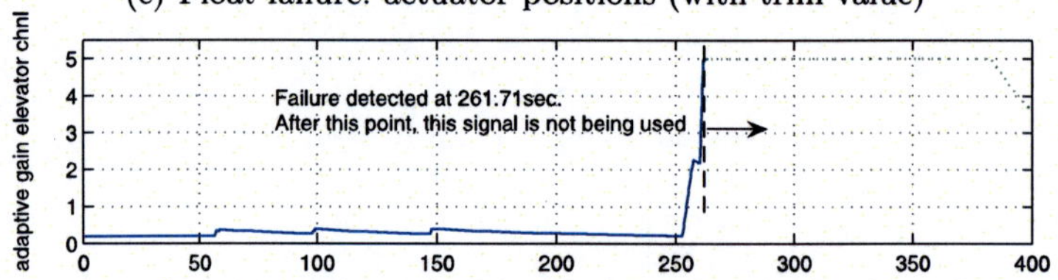

(d) Float failure: The adaptive gain ρ_1

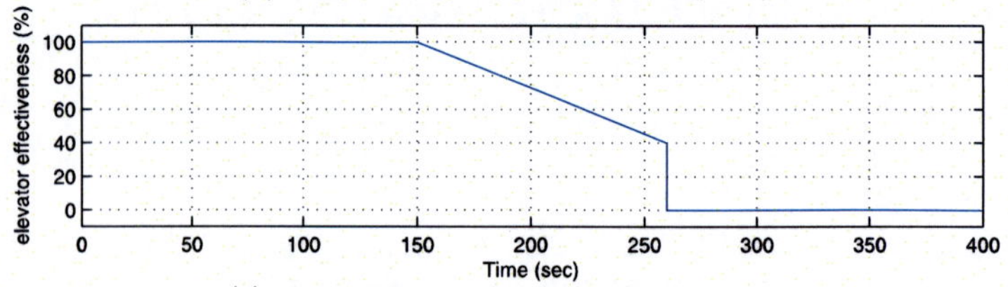

(e) Float failure: elevator effectiveness

Fig. 8.10 Responses to nominal, faults and failures (float failure)

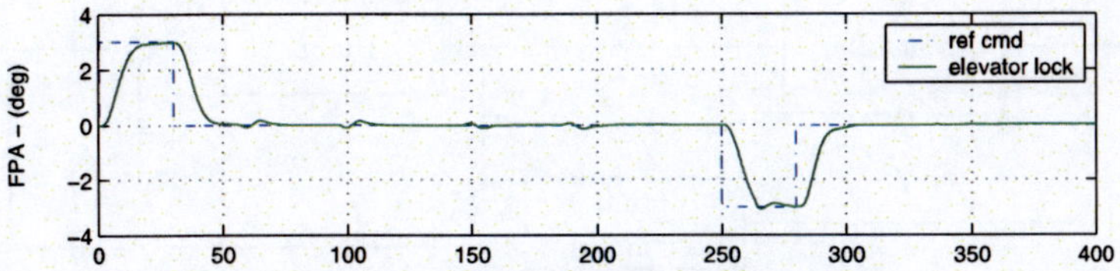

(a) Lock failure: FPA response

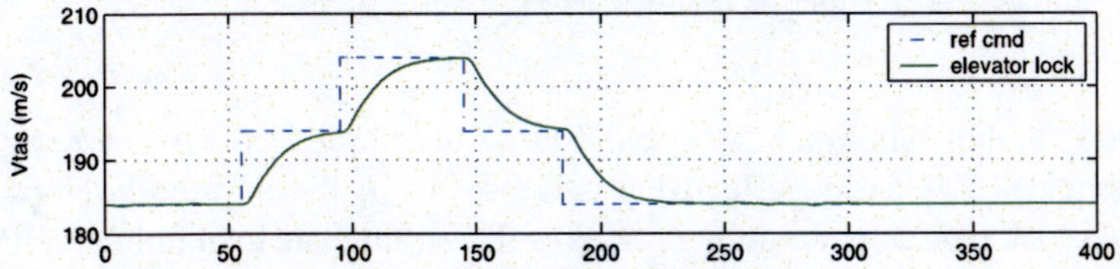

(b) Lock failure: Vtas response

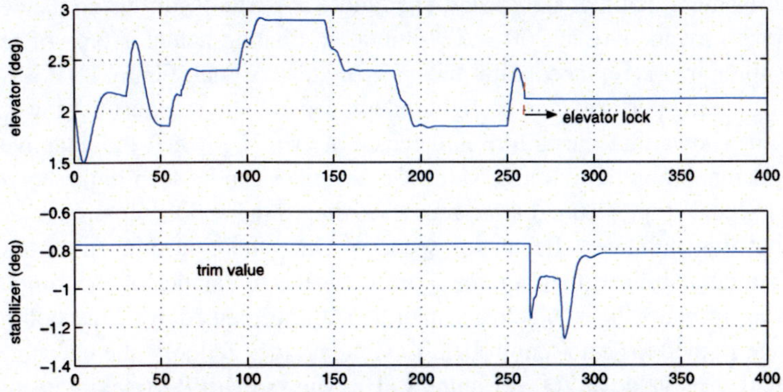

(c) Lock failure: actuator positions (with trim value)

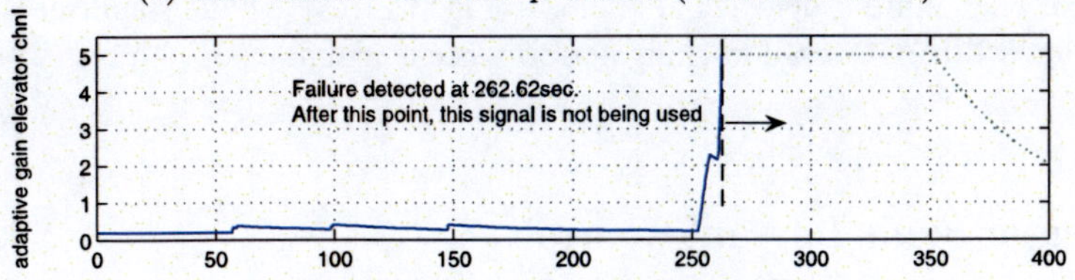

(d) Lock failure: The adaptive gain ρ_1

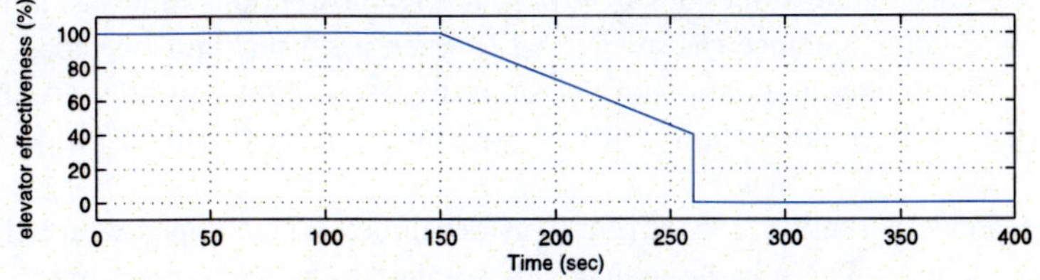

(e) Lock failure: elevator effectiveness

Fig. 8.11 Responses to nominal, faults and failures (lock failure)

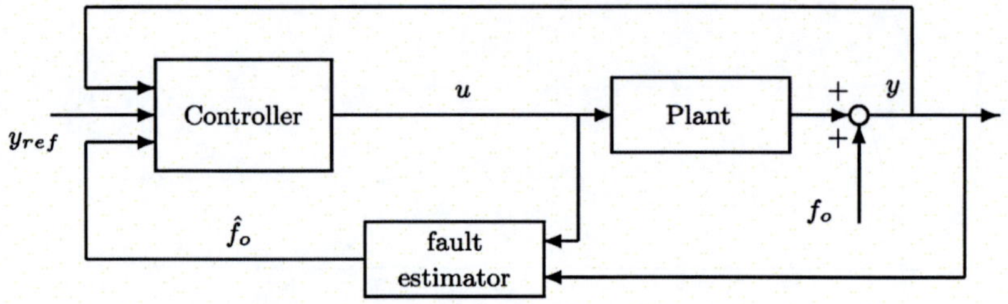

Fig. 8.12 SMC sensor fault tolerant control schemes

150–260 s, as shown in Fig. 8.10(e) and 8.11(e), the elevator effectiveness degrades from (normal) 100% to 40% effectiveness. Subsequently the elevator develops floating and/or lock type actuator failures at 260 s. These simulate total failure of the elevator and therefore require stabilization of the aircraft using the 'back-up' controller (which uses the horizontal stabiliser). The failure is set to occur during the descent (pitch down) manoeuvre at 260 s. To simulate a floating actuator type of failure, the elevator signal is replaced with the angle of attack signal. Figure 8.10(a) shows that FPA tracking performance is slightly degraded and the response is slower. Figure 8.10(d) shows that the failure is detected at 261.71 s when the adaptive gain reaches its maximum permissable value. Some spikes can be seen in the horizontal stabiliser signal (Fig. 8.10(c)) after activation due to the sudden change of control signal, but this stabilises after a few seconds. Once the controller is switched to the horizontal stabiliser, that surface is used for the remainder of the simulation. Overall performance is satisfactorily maintained after detection of the failure and the change to the 'back-up' controller. To simulate lock failures, the elevator position is held at its value at 260 s. Figure 8.11(a) shows that FPA tracking is slightly degraded. The failure is detected at 262.62 s (Fig. 8.11(d)) and the stabiliser is activated (Fig. 8.11(c)). Overall tracking performance is maintained.

8.4 Sensor Fault Tolerant Control

In the following sections, the idea of using fault reconstruction signals as a means of achieving fault tolerance (as in Fig. 8.12) when a sensor fault occurs, will be presented. The model and the controller from the last section will be used. In this section, the fault reconstruction will be based on the robust method proposed in Sect. 6.3.

In the previous section, a controller was developed which copes with actuator faults and failures. The scheme assumes that accurate fault-free measurements of the states are available. Here the possibility of faulty state measurements will be considered. The idea is to use a sliding mode observer to reconstruct the fault signals and to use these signals to correct the measured values before they are used in the control law.

8.4.1 Preliminaries

Consider an uncertain dynamical system affected by sensor faults described by

$$\dot{x}_p(t) = A_p x_p(t) + B_p u(t) + Q_p \xi(t, x_p) \tag{8.46}$$

$$y(t) = x_p(t) + N_p f_o(t) \tag{8.47}$$

where $A_p \in \mathbb{R}^{n\times n}$, $B_p \in \mathbb{R}^{n\times m}$, $N_p \in \mathbb{R}^{n\times r}$ and $Q_p \in \mathbb{R}^{n\times q}$ with $n > r$. In (8.47), it is assumed that all states are available for measurement. Assume that the matrix N_p is full column rank and the function $f_o : \mathbb{R}_+ \to \mathbb{R}^r$ is unknown but bounded so that

$$\|f_o(t)\| \le \hat{a}(t) \tag{8.48}$$

where $\hat{a} : \mathbb{R}_+ \to \mathbb{R}_+$ is a known function. The signal $f_o(t)$ represents (additive) sensor faults and N_p represents a distribution matrix, which indicates which of the sensors providing measurements are prone to possible faults. Notice in this special case, all the states are assumed to be measured by sensors—which is normal for modern civil aircraft.

The objective is to design a sliding mode observer in order to reconstruct the faults $f_o(t)$. As argued in Chapter 6, an effective way to do this is to first introduce a filter. Consider a new state $x_f \in \mathbb{R}^n$ that is a filtered version of $y(t)$ from (8.47), satisfying

$$\dot{x}_f(t) = -A_f x_f(t) + A_f\big(x_p(t) + N_p f_o(t)\big) \tag{8.49}$$

where $-A_f \in \mathbb{R}^{n\times n}$ is a stable matrix. Equations (8.46) and (8.49) can be combined to give a system of order $2n$ with states $x_a = \mathrm{col}(x_p, x_f)$ in the form

$$\dot{x}_a(t) = A_a x_a(t) + B_a u(t) + M_a f_o(t) + Q_a \xi(t, x_p) \tag{8.50}$$

$$x_f(t) = C_a x_a(t) \tag{8.51}$$

for appropriate matrices $A_a \in \mathbb{R}^{(2n)\times(2n)}$, $B_a \in \mathbb{R}^{(2n)\times m}$, $C_a \in \mathbb{R}^{n\times(2n)}$, $M_a \in \mathbb{R}^{(2n)\times r}$ and $Q_a \in \mathbb{R}^{(2n)\times q}$. For the uncertain system in (8.50)–(8.51) a sliding mode observer of the form

$$\dot{\hat{x}}_a(t) = A_a \hat{x}_a(t) + B_a u(t) - G_l e_y(t) + G_n \nu \tag{8.52}$$

will be considered. In (8.52) the discontinuous output error injection term

$$\nu = -\rho_o(t, y, u)\frac{P_o e_y}{\|P_o e_y\|} \quad \text{if } e_y \neq 0 \tag{8.53}$$

where $e_y(t) := C_a \hat{x}_a(t) - x_f(t)$ is the output estimation error and P_o is a symmetric positive definite matrix.

As argued in Sect. 6.3, the gains G_l and G_n are chosen to generate a stable sliding motion on $\mathcal{S}_o = \{e : C_a e = 0\}$ in finite time, where e is the state estimation

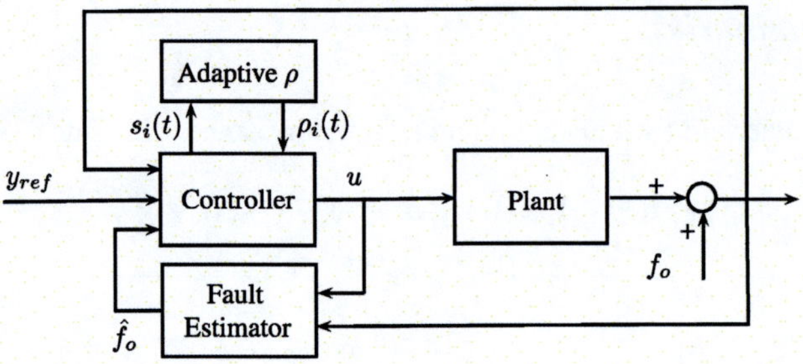

Fig. 8.13 Schematic of the sensor fault implementation

error $(\hat{x}_a - x_a)$. Furthermore, the gains G_l and G_n can be chosen to minimise the impact of $\xi(t, x_p)$ on the reconstruction of f_o obtained as $\hat{f}_o := W_{sc}\nu_{eq}$ in an $\mathcal{L}_2$ sense. In fact only $r \times (n-r)$ elements in W_{sc} are freely assignable since W_{sc} must be chosen to ensure that $W_{sc}C_aM_a = I_r$. The signal ν_{eq} can be approximated to any degree of accuracy, and is computable online as

$$\nu_\delta = -\rho_o(t, y, u)\frac{P_oe_y}{\|P_oe_y\| + \delta_o} \tag{8.54}$$

where δ_o is a small positive scalar. Then if $\hat{f}_o = W_{sc}\nu_\delta$, by straightforward manipulation, it can be shown that the fault reconstruction signal from (8.4.1) satisfies

$$\hat{f}_o(t) = f_o(t) + \hat{G}(\mathrm{s})\xi(t, x_p) \tag{8.55}$$

where $\hat{G}(\mathrm{s})$ is a transfer function matrix which depends on the plant matrices A_a, Q_a, the observer matrix G_n and the weighting matrix W_{sc}.

A general configuration representing the proposed sensor fault tolerant control scheme which will be used in this chapter is shown in Fig. 8.13. In this particular figure, the specific output of the FDI component is the sensor fault estimate $\hat{f}_o$.

8.4.2 Closed-Loop Analysis

In this chapter, the estimated sensor fault $\hat{f}_o$ will be used to correct the measured output signal so that $y - N_p\hat{f}_o$ will be the output of a 'virtual sensor' that will be used in the control law calculations to generate u. Suppose the corrected output measurement is given by $\hat{x}_p$, then

$$\hat{x}_p := y - N_p\hat{f}_o = x_p + N_p(f_o - \hat{f}_o) \tag{8.56}$$

Also the integral action states from (8.6) are corrected so that

$$\dot{x}_d = y_c - C_c\hat{x}_p = y_c - C_cx_p - C_cN_p(f_o - \hat{f}_o) \tag{8.57}$$

After the coordinate change $x \mapsto T_r x = z$ to induce regular form, and assuming for stability analysis purposes that $y_c \equiv 0$, then

$$\underbrace{\begin{bmatrix} \dot{z}_1 \\ \dot{z}_2 \end{bmatrix}}_{\dot{z}} = \underbrace{\begin{bmatrix} A_{11} & A_{12} \\ A_{21} & A_{22} \end{bmatrix}}_{T_r A T_r^{\mathrm{T}}} \underbrace{\begin{bmatrix} z_1 \\ z_2 \end{bmatrix}}_{z} + \underbrace{\begin{bmatrix} 0 \\ B_2 \end{bmatrix}}_{T_r B} u + \underbrace{\begin{bmatrix} Q_1 \\ Q_2 \end{bmatrix}}_{T_r Q} \xi - \underbrace{\begin{bmatrix} B_d^1 \\ B_d^2 \end{bmatrix}}_{T_r B_d C_c N_p} (f_o - \hat{f}_o) \tag{8.58}$$

where $z_2 \in \mathbb{R}^m$. By construction

$$f_o - \hat{f}_o = -\hat{G}(\mathrm{s})\xi(t, x_p) \tag{8.59}$$

Suppose $\hat{G}(\mathrm{s})$ has a state-space realisation $(\hat{A}, \hat{B}, \hat{C}, \hat{D})$ with states $\hat{e} \in \mathbb{R}^{(n-p)}$ which implies $f_o - \hat{f}_o = -\hat{C}\hat{e} - \hat{D}\xi$. During a sliding motion, since $\hat{x}$ is used in place of x in the control law, it follows that $\hat{s}(t) = S\hat{x} = 0$ and so

$$\begin{aligned} S\hat{x} = 0 \quad &\Leftrightarrow \quad S T_r^{\mathrm{T}} T_r x + S\hat{N}_p(f_o - \hat{f}_o) = 0 \\ &\Leftrightarrow \quad S_2[\mathcal{K} \quad I_m]z + S\hat{N}_p(f_o - \hat{f}_o) = 0 \end{aligned} \tag{8.60}$$

where $\hat{N}_p^{\mathrm{T}} = [0_{m\times m} \;\; N_p^{\mathrm{T}}]$ to account for the augmentation of the integral action states. From (8.60), during a sliding motion,

$$z_2(t) = -\mathcal{K}z_1(t) + S_2^{-1}S\hat{N}_p\big(\hat{C}\hat{e}(t) + \hat{D}\xi(t, x_p)\big) \tag{8.61}$$

Consequently, from (8.58), the reduced order sliding motion is governed by

$$\begin{aligned} \dot{z}_1(t) &= (A_{11} - A_{12}\mathcal{K})z_1(t) + Q_1\xi(t, x_p) \\ &\quad + \big(A_{12}S_2^{-1}S\hat{N}_p + B_d^1\big)\big(\hat{C}\hat{e}(t) + \hat{D}\xi(t, x_p)\big) \end{aligned} \tag{8.62}$$

$$\dot{\hat{e}}(t) = \hat{A}\hat{e}(t) + \hat{B}\xi(t, x_p) \tag{8.63}$$

By assumption $\|\xi(t, x_p)\| \leq \mathrm{c}_1\|z_p(t)\| + \mathrm{c}_2$. Consequently, since

$$\big(\mathrm{c}_1\|z_p(t)\| + \mathrm{c}_2\big)^2 \leq 2\mathrm{c}_1^2\|z_p(t)\|^2 + 2\mathrm{c}_2^2$$

it follows from (8.61) that

$$\begin{aligned} \|\xi(t, x_p)\|^2 &\leq 2\mathrm{c}_1^2\big(\|z_1(t)\|^2 + \|z_2(t)\|^2\big) + 2\mathrm{c}_2^2 \\ &\leq 2\mathrm{c}_1^2\big((1 + 4\|\mathcal{K}\|^2)\|z_1(t)\|^2 + 4\|S_2^{-1}S\hat{N}_p\hat{C}\|^2\|\hat{e}(t)\|^2 \\ &\quad + 2\|S_2^{-1}S\hat{N}_p\|^2\|\hat{D}\|^2\|\xi(t, x_p)\|^2\big) + 2\mathrm{c}_2^2 \end{aligned}$$

Let $\alpha^2 := \max\{1 + 4\|\mathcal{K}\|^2,\ 4\|S_2^{-1}S\hat{N}_p\hat{C}\|^2\}$. Using the fact that

$$\|\hat{G}(\mathrm{s})\|_\infty < \hat{\gamma} \Rightarrow \|\hat{D}\| < \hat{\gamma}$$

means

$$\|\xi\|^2 \leq 2\mathrm{c}_1^2(\alpha^2(\|z_1\|^2 + \|\hat{e}\|^2) + 2\|S_2^{-1}S\hat{N}_p\|^2\hat{\gamma}^2\|\xi\|^2) + 2\mathrm{c}_2^2 \tag{8.64}$$

Suppose $4\mathrm{c}_1^2\hat{\gamma}^2\|S_2^{-1}S\hat{N}_p\|^2 < 1$, which will always be satisfied for a small enough $\hat{\gamma}$, then rearranging (8.64) yields

$$\begin{aligned}\|\xi\|^2 &\leq \Bigg(\underbrace{\frac{2\mathrm{c}_1^2\alpha^2}{(1-4\mathrm{c}_1^2\hat{\gamma}^2\|S_2^{-1}S\hat{N}_p\|^2)}}_{\hat{\mathrm{c}}_1^2}\left\|\begin{bmatrix} z_1 \\ \hat{e}\end{bmatrix}\right\|^2 + \underbrace{\frac{2\mathrm{c}_2^2}{(1-4\mathrm{c}_1^2\hat{\gamma}^2\|S_2^{-1}S\hat{N}_p\|^2)}}_{\hat{\mathrm{c}}_2^2}\Bigg) \\ &\leq \left(\hat{\mathrm{c}}_1\left\|\begin{bmatrix} z_1 \\ \hat{e}\end{bmatrix}\right\| + \hat{\mathrm{c}}_2\right)^2 \end{aligned} \tag{8.65}$$

Notice that $\hat{\mathrm{c}}_1 \to 0$ as $\mathrm{c}_1 \to 0$, and so as the plant uncertainty decreases, the uncertainty in (8.62)–(8.63) diminishes. If $\xi(t, x_p) \equiv 0$ then (8.62)–(8.63) is stable since both $A_{11} - A_{12}\mathcal{K}$ and $\hat{A}$ are stable by design. Consequently using Lyapunov arguments similar to those in Section 3.6 in [85], there exists a value of $\mathrm{c}_1 > 0$ for which the system (8.62)–(8.63) retains stability.

8.5 Robust Sensor Fault Reconstruction for the B747

This section describes the development of the fault reconstruction scheme for the *full nonlinear high fidelity model* of the B747-100/200 aircraft from [183]. As in Sect. 8.2, a linearisation of the B747 model about an operating condition of 300,000 kg, 184 m/s true airspeed, and an altitude of 4000 m at half maximum thrust has been obtained. The system matrices (A_p, B_p) are given in (8.3)–(8.4) and the output distribution matrix is given by

$$C_p = \begin{bmatrix} 1 & 0 & 0 & 0 \\ 0 & 1 & 0 & 0 \\ 0 & 0 & 1 & 0 \\ 0 & 0 & 0 & 1 \end{bmatrix}$$

A key aspect of the design is to establish the matrix Q_p from (8.46) which captures the discrepancy between the nonlinear and linear models. A much more accurate model is required here for analytical redundancy purposes than for the controller design. Prior to obtaining the matrix Q_p, the second state (V_{tas}) has been scaled by 0.1 and therefore the plant system triple (A_p, B_p, C_p) is transformed by the matrix given by $T_s = \text{diag}(1, 0.1, 1, 1)$. This has been done so that the magnitude of each of the states is comparable. Uniformly sampled data at 10 Hz were collected from the nonlinear (open-loop) simulation which was excited using a PRBS signal (see Fig. 8.14) of amplitude 1 deg in the elevator channel. An estimate of the derivatives

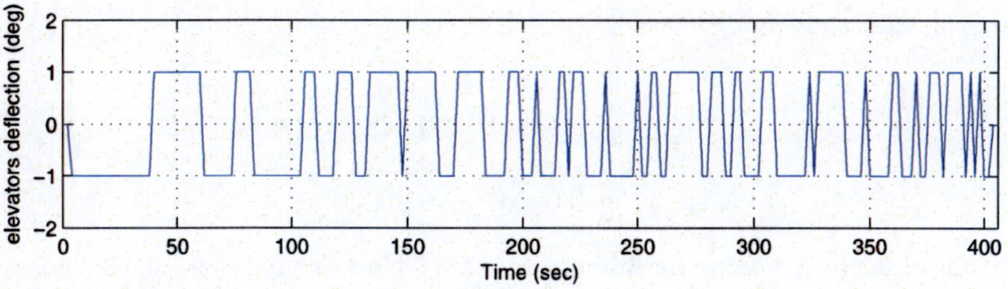

Fig. 8.14 PRBS signal

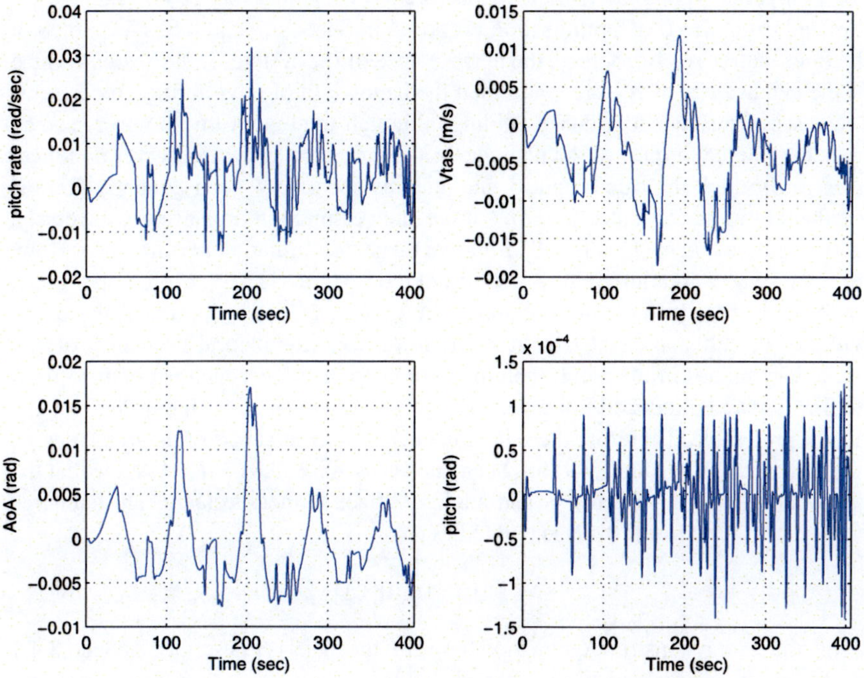

Fig. 8.15 Plots of $e_p = \dot{x} - Ax - Bu$

of each of the state-space vector components was obtained numerically (off-line) and an error vector $e_p := \dot{x} - Ax - Bu$ was computed for each sample.

In terms of the uncertainty model from (8.46) the time series data $\{e_p\} = Q_p\{\xi\}$. Figure 8.15, shows the four components of the signals e_p. Principal component analysis on the signals e_p using singular value decomposition of the matrix $e_p^{\mathrm{T}} e_p$ has been employed to compute Q_p. This is based on the procedure proposed by Chen and Patton [51]. The singular values of the matrix $e_p^{\mathrm{T}} e_p \in \mathbb{R}^{4\times 4}$ are given by $\{3.2332, 1.9011, 0.3644, 0.0001\}$. The first two are significantly larger than the last two and so Q_p has been chosen as the eigenvectors associated with the first two singular values giving

$$Q_p = \begin{bmatrix} -0.8562 & 0.4262 \\ -0.3149 & -0.8786 \\ -0.4049 & -0.2155 \\ 0.0000 & -0.0000 \end{bmatrix} \tag{8.66}$$

Details of the raison d'etre for this appear in [51]. Note that the elements in the last row of Q_p are small compared to the others. This is in accordance with the observation that pitch (the last state) is the integral of pitch rate, and therefore no modelling uncertainty is present. Once the matrix Q_p from (8.50) has been obtained, the observer gains G_l and G_n and the reconstruction weighting matrix W_{sc} can be synthesized using the LMI optimisation proposed in Sects. 6.3 and 4.5. The choice of the filter matrix A_f impacts on the performance of the system. If the absolute value of the eigenvalues of A_f are small then the bandwidth of the filtering properties is decreased. Consequently, during sliding, although the output of the observer may track the filtered outputs of the plant perfectly, the outputs of the observer no longer necessarily track the true output of the plant as accurately—consequently there is a reduction in performance in terms of the state estimation properties. Conversely, large negative eigenvalues for A_f improve the state estimation performance. However, the state estimation performance is not the key criteria here. More importantly, the choice of A_f affects the optimal value of $\hat{\gamma} = \|\hat{G}(s)\|_\infty$. Often if the bandwidth of the filter associated with A_f is lower than the natural frequency of any oscillatory modes in the plant, then the optimal value of $\hat{\gamma}$ which is obtained from the LMIs may be reduced/improved, and consequently smaller eigenvalues for A_f maybe preferable. Therefore the selection of A_f is a crucial part of the initial design iteration. Here A_f from (8.49) has been chosen as $A_f = 0.01 \times I_4$. Assume that the pitch rate q, true air speed V_{tas} and angle of attack α measurements are fault-free and therefore N_p from (8.47) is defined as

$$N_p^{\mathrm{T}} = [0 \quad 0 \quad 0 \quad 1]$$

Using the fault reconstruction method based on the observer described in Sect. 8.4.1 applied to the augmented system, and choosing

$$D_1 = \mathrm{diag}(0.1 \quad 0.11 \quad 0.1 \quad 0.1)$$

and $\hat{\gamma}_o = 0.003$ the following observer gains have been obtained:

$$G_l = \begin{bmatrix} 0.3325 & -4.5479 & -1.3616 & -0.6191 \\ -8.0016 & 74.7746 & -1.1413 & -12.9629 \\ 1.8657 & -0.0352 & 1.4094 & 1.7649 \\ 3.2872 & -11.6213 & 2.6326 & 4.8176 \\ 0.0324 & -0.1128 & 0.0248 & 0.0461 \\ -0.1365 & 1.6228 & -0.0029 & -0.2522 \\ 0.0248 & -0.0024 & 0.1640 & 0.1631 \\ 0.0461 & -0.2084 & 0.1631 & 0.3181 \end{bmatrix}$$

$$G_n = \begin{bmatrix} 8.0739 & -2.2555 & -9.5628 & -0.0000 \\ -81.3037 & 40.4364 & 6.0311 & 0.0000 \\ 85.9276 & 5.9437 & -4.2736 & 0.0000 \\ 99.8244 & -0.2224 & 0.9782 & 0.0000 \\ 1.0000 & 0 & -0.0000 & 0.0000 \\ 0.0000 & 1.0000 & 0.0000 & 0.0000 \\ 0 & 0 & 1.0000 & 0 \\ -0.0000 & 0.0000 & -0.0000 & 1.0000 \end{bmatrix}$$

where

$$P_o = \begin{bmatrix} 15.7240 & 1.0217 & -1.8273 & -0.5313 \\ 1.0217 & 0.2582 & -0.4207 & 0.2724 \\ -1.8273 & -0.4207 & 4.4240 & -2.3378 \\ -0.5313 & 0.2724 & -2.3379 & 2.4352 \end{bmatrix}$$

and the injection scaling matrix from (8.4.1) is

$$W_{sc} = \begin{bmatrix} 0.8295 & 1.0211 & -2.5213 & 100.0000 \end{bmatrix}$$

It can be shown that with this choice of gains

$$\left\| \hat{G}(s) \right\|_\infty = 5.8668 \times 10^{-4}$$

The choice of the design matrix D_1 has been used to fine tune the observer gain G_l, while $\hat{\gamma}_o$ is chosen to be small to ensure that the $\mathcal{H}_\infty$ norm of $\hat{G}(s)$ from (8.55) is small (which means that the fault reconstruction will be less affected by the uncertainty). When trying to ensure that the $\mathcal{H}_\infty$ norm of $\hat{G}(s)$ is small (using a small $\hat{\gamma}_o$), the observer gain G_l might become large and unrealistic for implementation. Therefore in terms of design, there is a trade-off between obtaining a small $\hat{\gamma}$ and a realistic observer gain G_l. The simulation parameters from (8.54) were chosen as $\rho_o = 50$ and $\delta_o = 0.005$. A large ρ_o is required to ensure that sliding still occurs in the presence of uncertainty and faults; and a small δ_o is necessary to closely approximate the discontinuous switching injection signal. The ν_{eq} signal used for the reconstruction is filtered using a first order low pass filter with time constant 0.1 before being scaled by the weighting matrix W_{sc}. This filtering operation is quite in keeping with the notion of the equivalent injection being the low frequency component of ν [256]. In the same way, other observers can be designed to specifically reconstruct faults on the angle of attack and pitch rate measurement signals.

8.6 Sensor Fault Tolerant Control Simulation Results

The effect of feeding faulty sensor signals into the controller has been investigated. For comparison purposes, the performance of the scheme in Fig. 8.13 has been measured using the root mean square (RMS) of the FPA tracking error. As in Sect. 8.3.5, the same flight conditions and controller have been used. The following subsections

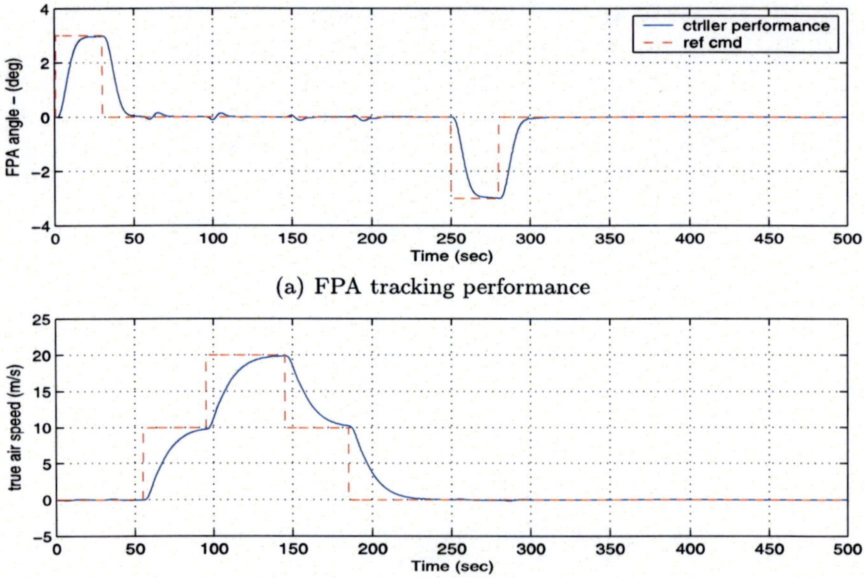

(a) FPA tracking performance

(b) V_{tas} tracking performance

Fig. 8.16 Fault-free controller responses

will discuss the different scenarios for testing the observer-based schemes which have been designed. First a nominal no fault condition is considered. Then the scenario in which a fault occurs and fault reconstruction is not used is discussed, to see the effect of the corrupted signals on the performance of the controller from Sect. 8.3. Finally, the FTC scheme from Fig. 8.13 is employed in the presence of faults, using a sliding mode fault reconstruction scheme. Note that the simulations are done on the full 77 state nonlinear high fidelity model of the B747-100/200.

8.6.1 Fault-Free Simulation

The fault-free performance of the controller is given in Fig. 8.16. The reference command requests a change in flight path angle of 3 deg for 20 s followed by a 20 m/s change in speed over a period of 45 s (in two steps). The command sequence for the FPA demand is then reversed after 250 s so that the aircraft is returned to (approximately) the initial flight conditions. (The results presented in the following figures do not include the trim values for ease of interpretation.) Since the design of the FDI scheme assumes that the measurement for V_{tas} is free from faults, only the FPA tracking error is shown, and the RMS value is used as a measure of the controller performance. Figure 8.17(a) shows the FPA tracking error when no fault is present. The RMS of the error signal is 0.0150.

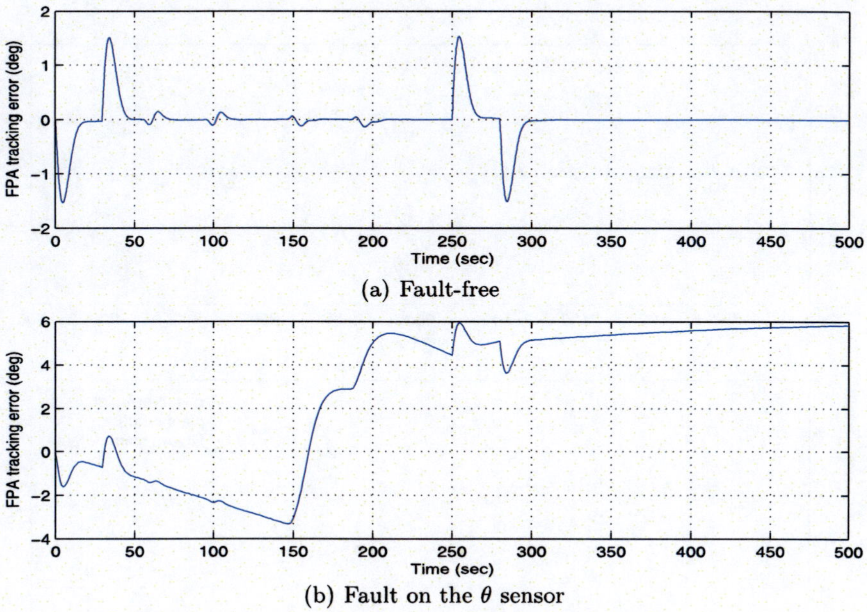

Fig. 8.17 Fault simulation: FPA tracking error: fault-free and with fault (FDI switched off)

8.6.2 Fault Simulations: FDI Switched Off

Figure 8.17(b) shows the FPA tracking error when the corrupted pitch signal (from a faulty sensor) is directly used by the controller. The fault signal was set as a sensor drift represented by a positive ramp signal starting from zero at the beginning of the simulation, with a peak of 5.73 deg at 250 s and then a negative ramp back to zero at 500 s (as in Fig. 8.19(a)). When the corrupted pitch signal is used directly by the controller, Fig. 8.17(b) shows a significant degradation in the performance of the FPA tracking error. A RMS value of 0.1969 is observed compared to 0.0150 in the fault-free case (Fig. 8.17(a)). In the subsequent tests, the FDI scheme will be switched on and the performance of the FTC from Fig. 8.13 will be evaluated when the fault reconstruction signal is used to correct the corrupted sensor signal before being used by the controller. As before, the same set of flight conditions and tests, using the same controller and command references, will be performed in the scenario that there is a fault on the θ measurement.

8.6.3 Fault Simulation: FDI Switched On

Figure 8.18 shows the corrupted plant output with an error on the pitch (θ) measurement. Figure 8.18 also shows that the observer states track the filtered plant outputs closely. The filtering effect is clearly visible in Fig. 8.18 where the filtered plant

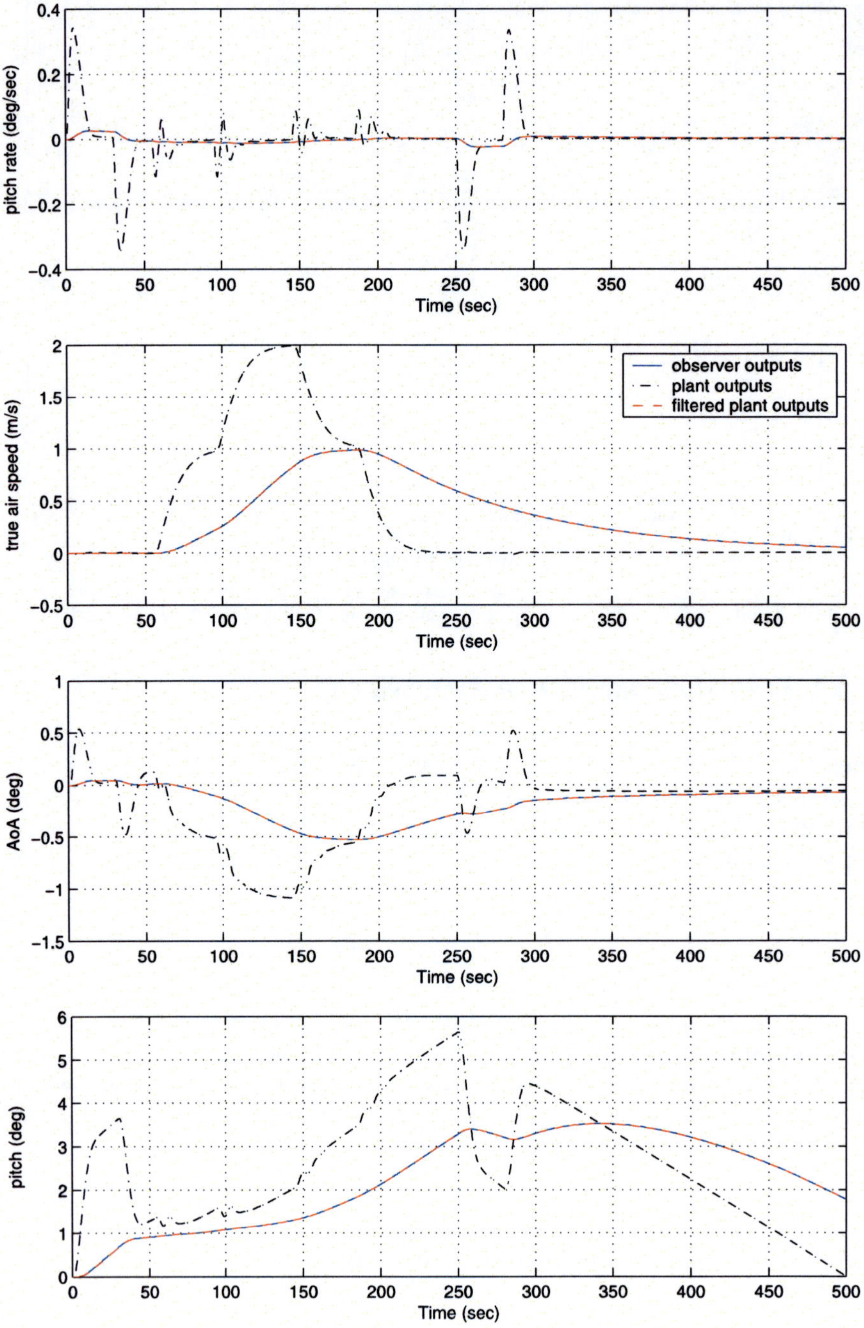

Fig. 8.18 FDI on: fault simulation: corrupted plant outputs

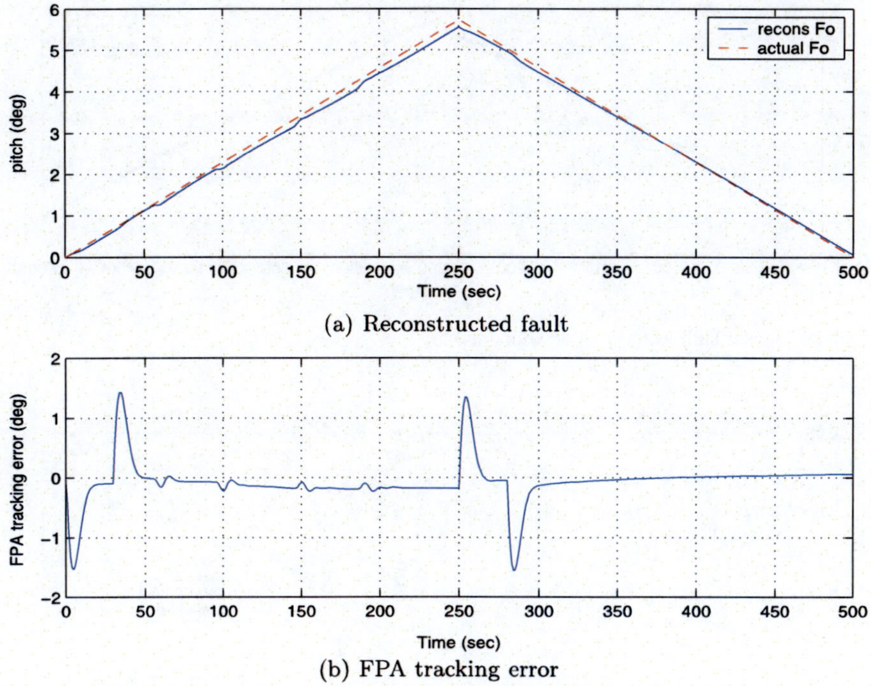

Fig. 8.19 Fault simulation responses: FDI switched on (without trim values)

and observer output are different from the (actual) plant output. Here, the objective of the observer is to provide a good estimate of the faults; however, the plant state estimates are compromised by the low pass filter. Figure 8.19(a) shows the reconstructed θ sensor fault. The corrupted measured pitch signal (Fig. 8.18) is corrected by the fault estimation signal in Fig. 8.19(a) before being used by the controller. The closed-loop performance of the aircraft is given in Fig. 8.19(b). The RMS of 0.0154 in Fig. 8.19(b) is better than the one in Fig. 8.17(b) (0.1969), which shows that the sensor fault reconstruction has enabled the controller to maintain the required performance in the presence of a sensor fault.

8.6.4 Fault Simulations with Sensor Noise

Figure 8.20 shows the simulation results under similar conditions to the previous tests but with the addition of (bounded) sensor noise. The noise has been implemented using a scaling of the band limited white noise block in SIMULINK® with a noise power of 0.01 and sampling time of 0.1. Figure 8.20 shows satisfactory sensor fault reconstruction in the presence of noise.

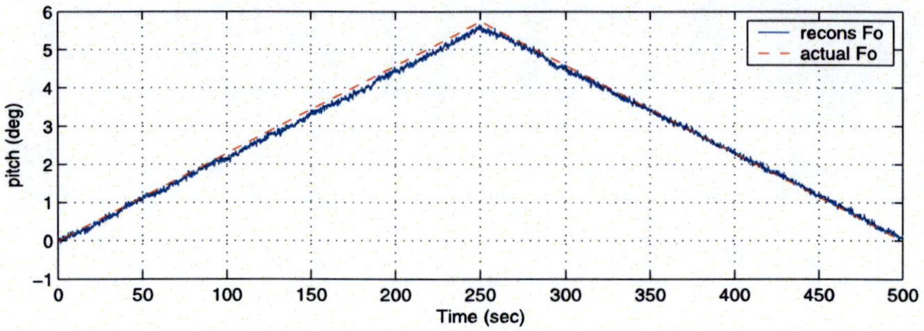

Fig. 8.20 Sensor fault reconstruction with noise

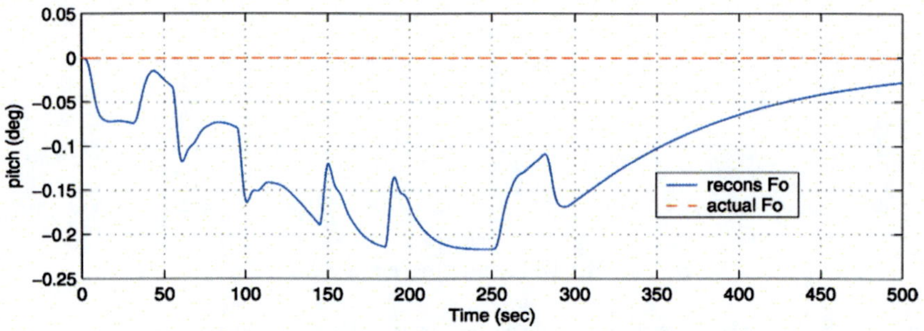

Fig. 8.21 Fault-free simulation: fault reconstruction signal—threshold selection

8.6.5 Threshold Selection

The reconstruction schemes above have been implemented on the full nonlinear model of the B747-100/200 using FTLAB747 V6.5. The observers have been run under the assumption that the measurement of true air speed V_{tas} is fault-free. In the event of a pitch sensor fault, a reconstruction from the FDI provides the fault estimation. Note that in a nominal fault-free scenario, the FDI will not be switched on to correct the signals used in the controller because this would degrade the performance since the reconstructions are not perfect. The simplest approach would be to use the correction signals only if the fault estimate exceeds a predetermined threshold value indicating the presence of a fault. An initial estimate of the required threshold can be obtained using the reconstruction scheme in a fault-free situation. Figure 8.21 shows the fault reconstruction signals when no fault is present. The selected threshold values need to be larger than the 'normal' variations in the fault estimation signals, to avoid false alarms, but not too large to miss faults. In the above simulations, the threshold values can be chosen as ± 0.3 deg for the pitch angle sensor.

8.7 Summary

The first part of this chapter presented a SMC scheme for fault tolerant control of a civil aircraft. Only longitudinal control with a fault and/or failure occurring in the elevator channel has been considered. The controller is based on a state-feedback sliding mode scheme and the gain associated with the nonlinear term was allowed to adaptively increase when the onset of a fault was detected. Compared to other FTC schemes which have been implemented on this model, the controller proposed here is simple and yet is shown to work across the entire 'up and away' flight envelope. It is not scheduled with respect to any variables and its structure remains fixed (except for the adaptive gain associated with the nonlinear switching term). Unexpected deviation of the switching variable from its nominal condition initiates the adaptation mechanism. Total failure can also be detected from the switching function, and in this example has been used to trigger the use of a 'back-up' control surface. A range of realistic fault scenarios was considered and the results of simulations using the full nonlinear aircraft model have been presented.

The second part of the chapter has highlighted the capabilities and benefits of sliding mode observers as a means of fault reconstruction and sensor fault tolerant control. The scheme involved the reconstruction of the sensor faults using a sliding mode observer. The fault reconstructions were then used to correct the measured outputs before they were used in the controller calculations and therefore the controller does not need to be reconfigured.

8.8 Notes and References

The representation of actuator faults/failures in (8.1) indicates a *reduction in the effectiveness* of a particular actuator via the diagonal structure of K. This representation is used by many other researchers such as [32, 148, 251, 296]. Posing the faults/failures as in (8.1), makes the stability analysis and FDI integration much simpler.

In terms of improving the safety of large passenger aircraft, there is already significant literature investigating FTC, see e.g., [35, 105, 177, 184, 298]; including the successful NASA work on propulsion controlled aircraft (PCA) which involved a successful flight test. Apart from [35], all these papers describe work on the same high fidelity nonlinear model of a B747 aircraft based on software called FTLAB747 [183, 233, 234]. In [105], an LPV controller is proposed for FTC, however, only longitudinal control is considered. Marcos and Balas [184] investigated robust integrated controller design and diagnosis for the longitudinal axis. The work in [298] considered both longitudinal and lateral control, but is only applied to the linear model of the aircraft. In [120] an integrated propulsion controlled aircraft has been proposed using a modified version of the FTLAB747 software. However, this paper considers thrust vectoring,[4] which is currently not available in any large passen-

[4]Thrust vectoring refers to the capability of the nozzle of the jet engine to be directed at any angle. Typically most jet engine nozzles are fixed and not movable.

ger transport aircraft. (Although thrust vectoring is currently available in advanced fighter aircraft such as the Saab JAS 39 'Gripen' [97] and Sukhoi Su-30MKM.) Researchers in [222] use SMC on a similar aircraft but not the validated FTLAB747 high fidelity model.

Chapter 9
Fault Tolerant Control with Online Control Allocation

In Chap. 8, SMC was shown to be able to handle actuator faults without requiring any FDI. However, sliding mode methods are still hampered by one drawback; the inability to directly handle total actuator failures.[1] In Chap. 8, a simple example showed that when an elevator fails, the horizontal stabiliser can be used to generate the same pitch manoeuvre. In Chap. 8, the control signal to the elevator was simply rerouted to the stabiliser as the relationship between the stabiliser and the elevator is known. The idea of using a single controller which can redistribute the control signals to 'redundant' actuators is appealing. It avoids the need of having many controllers (or a bank of controllers) for each type of fault or failure as required by many other strategies in the FTC literature. It also removes the requirement of switching between controllers when faults/failures occur and the associated bumpless transfer issues. This chapter will further explore the ideas from Chap. 8 for rerouting control signals, using a strategy called control allocation (CA). CA provides SMC with access to redundant actuators enabling the SMC to handle actuator failures directly, thus opening new and exciting ways of achieving FTC. In this chapter, a novel control allocation strategy is considered whereby the control signal distribution is based on the actuator effectiveness level.

9.1 Introduction

In most safety critical systems, e.g., passenger aircraft [35] and modern fighter aircraft [97], there is actuator redundancy. This gives freedom to design FTC systems to maintain stability and acceptable performance during faults and failures. CA is one approach to manage the actuator redundancy for different control strategies handling actuator faults (see for example [38, 63]). In this chapter, a combination of SMC and CA will be explored to achieve FTC. A rigorous design procedure is developed from a theoretical perspective. The proposed scheme has been tested in

[1] This drawback is also inherent in almost all traditional feedback control paradigms such as LQR, $\mathcal{H}_\infty$ and μ-synthesis.

H. Alwi et al., *Fault Detection and Fault-Tolerant Control Using Sliding Modes*, Advances in Industrial Control,
DOI 10.1007/978-0-85729-650-4_9,

simulation on an aircraft model which has been used in the literature to demonstrate a CA scheme [122]. The control strategy uses the effectiveness level of the actuators and redistributes the control to the remaining actuators when faults/failures occur.

9.2 Controller Design

Consider the nth order linear time invariant system with m inputs given by

$$\dot{x}(t) = Ax(t) + Bu(t) \tag{9.1}$$

where $A \in \mathbb{R}^{n \times n}$ and $B \in \mathbb{R}^{n \times m}$ with $1 \leq m < n$. In most of the CA literature, it is assumed that B can be factorised as

$$B = B_\nu N \tag{9.2}$$

where $B_\nu \in \mathbb{R}^{n \times l}$, $N \in \mathbb{R}^{l \times m}$ and both have rank $l < m$ [122]. Then the 'virtual control input' [122] is defined as

$$\nu(t) := Nu(t) \tag{9.3}$$

where $\nu(t) \in \mathbb{R}^l$ can be interpreted as the total control effort produced by the actuators [122]. Typically the control law $\nu(t)$ is designed based on the pair (A, B_ν) which is assumed to be controllable. Direct manipulation of (9.3) gives

$$u(t) = N^\dagger \nu(t) \tag{9.4}$$

where $N^\dagger \in \mathbb{R}^{m \times l}$ is a right pseudo-inverse of matrix N, so that $NN^\dagger = I_l$. Note that the choice of $N^\dagger$ is not unique. One choice for $N^\dagger$ is obtained from the following minimisation problem:

$$\min_u u^{\mathrm{T}} W^{-1} u \quad \text{subject to } Nu = \nu \tag{9.5}$$

where $W \in \mathbb{R}^{m \times m}$ is a symmetric positive definite diagonal weighting matrix [122, 203]. This minimises *at each time instant* the 'weighted sum of squares' cost, associated with the control vector u. The optimal solution to (9.5) is $u = N^\dagger \nu$ where

$$N^\dagger := WN^{\mathrm{T}}\left(NWN^{\mathrm{T}}\right)^{-1} \tag{9.6}$$

Often in the literature, W from (9.5) and (9.6) is set to the identity [222, 226, 269], which gives the classical Moore–Penrose pseudo-inverse [129]. Another approach is to choose the weighting W to be a diagonal matrix formed from the control surface limits squared [38, 63]. In this way, W scales each control surface based upon deflection limits to equally distribute the control effort [38] (assuming symmetric position limits and no rate limits).

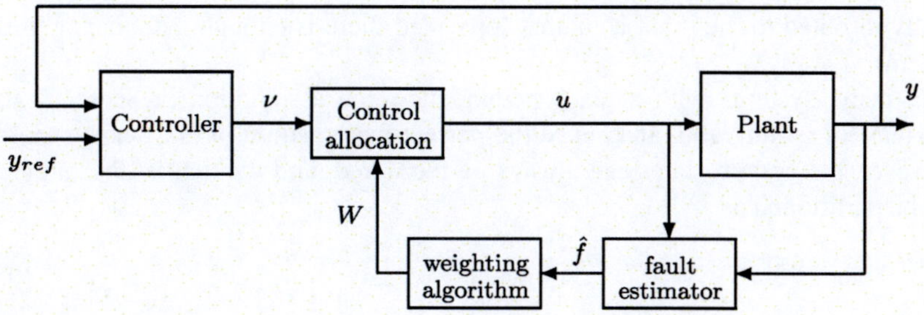

Fig. 9.1 Control allocation strategy

9.2.1 Problem Formulation

This chapter considers a situation where a fault develops in system (9.1) associated with the actuators. It will be assumed that in the event of actuator faults or failures, (9.1) can be rewritten as

$$\dot{x}(t) = Ax(t) + Bu(t) - BK(t)u(t) \tag{9.7}$$

with $K(t) = \text{diag}(k_1(t), \ldots, k_m(t))$ where the $k_i(t)$ are scalars satisfying $0 \leq k_i(t) \leq 1$. These scalars model a decrease in effectiveness of a particular actuator. If $k_i(t) = 0$, the ith actuator is working perfectly whereas if $k_i(t) > 0$, a fault is present, and if $k_i(t) = 1$, the actuator has failed completely.

In this chapter, a novel choice of weighting matrix W will be considered. If information about the actuator faults is available from a fault detection and isolation (FDI) scheme so that the actuator effectiveness values k_i are known, the control signal from the 'virtual control' $\nu(t)$ can be redistributed to the remaining working actuators using W in (9.6). Here, the weight W has been chosen as

$$W = I - K \tag{9.8}$$

As a direct consequence, $W = \text{diag}(w_1, \ldots, w_m)$ and the diagonal elements $w_i = 1 - k_i$. As $k_i \to 1$, $w_i \to 0$ and so the associated control component u_i in (9.5) is weighted heavily since $1/w_i$ becomes large. Note that in a fault-free situation $W = I$ (a common choice in the CA literature [222, 226, 269]).

Figure 9.1 illustrates the proposed FTC control allocation strategy. The control allocation will depend on the effectiveness of the actuators. The information necessary to compute W online can be supplied by a fault reconstruction scheme as described in Chap. 4 for example, or by using a measurement of the actual actuator deflection compared to the demand which is available in many systems, e.g., passenger aircraft [35]. Alternatively other fault reconstruction schemes based on Kalman filters [296] can be used. From (9.8) if an actuator fault occurs, the weighting W will be changed online and the control input $u(t)$ is reallocated to minimise the use of the faulty control surface. In the event of total failure of the ith control surface, $k_i \to 1$ and therefore the ith component of W^{-1} becomes large. Hence, $u_i(t)$ is totally rerouted to the other actuators (provided there is enough redundancy in the system).

In many systems with actuator redundancy, the assumption in Sect. 9.2 that $\text{rank}(B) = l$ is not valid, and hence the perfect factorisation in (9.2) cannot hold.[2] However, the system states can always be reordered, and the matrix B from (9.2) can be partitioned as

$$B = \begin{bmatrix} B_1 \\ B_2 \end{bmatrix} \tag{9.9}$$

where $B_1 \in \mathbb{R}^{(n-l)\times m}$ and $B_2 \in \mathbb{R}^{l\times m}$ has rank l. The separating of the control law from the control allocation task comes naturally with design methods like feedback linearisation and backstepping, which employ intermediate 'virtual control' signals [20, 121]. In most aircraft and submersible systems the control objectives can be achieved by commanding some desired moment to be generated by the control surfaces [20, 121]: in aircraft systems for example, the channels associated with B_2 are the equations of angular acceleration in roll, pitch and yaw [122]. *Here it is assumed that the matrix B_2 will represent the dominant contribution of the control action on the system.* This will be discussed formally later in the chapter. In [222], it is assumed that $B_1 = 0$. This represents the extreme situation where the total effect of the control is through B_2 only. Here $B_1 \neq 0$ will be considered explicitly in the controller design and in the stability analysis. It will be assumed without loss of generality that the states of the system in (9.1) have been transformed so that $B_2 B_2^{\mathrm{T}} = I_l$ and therefore $\|B_2\| = 1$. This is always possible since $\text{rank}(B_2) = l$. Let the 'virtual control'

$$\nu(t) = B_2 u(t) \tag{9.10}$$

so that

$$u(t) = B_2^{\dagger}\nu(t) \tag{9.11}$$

where the right pseudo-inverse is

$$B_2^{\dagger} := W B_2^{\mathrm{T}} \left(B_2 W B_2^{\mathrm{T}}\right)^{-1} \tag{9.12}$$

The fault term from (9.7), $BKu(t) = BKB_2^{\dagger}\nu(t)$, and therefore (9.7) becomes

$$\dot{x}(t) = Ax(t) + \underbrace{\begin{bmatrix} B_1 B_2^{\dagger} \\ I_l \end{bmatrix}}_{B_\nu} \nu(t) - \begin{bmatrix} B_1 K B_2^{\dagger} \\ B_2 K B_2^{\dagger} \end{bmatrix} \nu(t) \tag{9.13}$$

The objective is to use SMC techniques, to synthesise the 'virtual control' $\nu(t)$. Define a switching function $s(t) : \mathbb{R}^n \to \mathbb{R}^l$ to be

$$s(t) = Sx(t) \tag{9.14}$$

[2] See for example (9.70), where $\text{rank}(B) = 4$ and l, which relates to the choice of B_2 in (9.9), is $l = \text{rank}(B_2) = 3$. Similarly in (11.2) and (11.4), $\text{rank}(B) = 3$ whilst $l = \text{rank}(B_2) = 2$.

where $S \in \mathbb{R}^{l \times n}$ and $\det(SB_\nu) \neq 0$. Let $\mathcal{S}$ be the hyperplane defined by $\mathcal{S} = \{x(t) \in \mathbb{R}^n : Sx(t) = 0\}$. If a control law can be developed which forces the closed-loop trajectories onto the surface $\mathcal{S}$ in finite time and constrains the states to remain there, then an ideal sliding motion is said to have been attained. The selection of the sliding surface is the first part of any design and defines the system's closed-loop performance. The sliding surface will be designed based on the nominal no fault condition ($K = 0$). The second aspect of the control design is the synthesis of a control law to guarantee that the surface is reached in finite time and a sliding mode is subsequently maintained.

Using the fact that $K = I - W$ and $B_2 B_2^\dagger = I_l$, it follows that (9.13) can be written as

$$\dot{x}(t) = Ax(t) + \begin{bmatrix} B_1 W^2 B_2^{\mathrm{T}} (B_2 W B_2^{\mathrm{T}})^{-1} \\ B_2 W^2 B_2^{\mathrm{T}} (B_2 W B_2^{\mathrm{T}})^{-1} \end{bmatrix} \nu(t) \tag{9.15}$$

If

$$\bar{\nu}(t) := \left(B_2 W B_2{}^{\mathrm{T}}\right)^{-1} \nu(t) \tag{9.16}$$

then it is easy to see that (9.15) can be written as

$$\dot{x}(t) = Ax(t) + \begin{bmatrix} B_1 B_2^{\mathrm{T}} \\ I \end{bmatrix} \bar{\nu}(t) - \begin{bmatrix} B_1 (I - W^2) B_2^{\mathrm{T}} \\ B_2 (I - W^2) B_2^{\mathrm{T}} \end{bmatrix} \bar{\nu}(t) \tag{9.17}$$

In the fault-free case $W = I$ and the nominal system is

$$\dot{x}(t) = Ax(t) + \begin{bmatrix} B_1 B_2^{\mathrm{T}} \\ I \end{bmatrix} \bar{\nu}(t) \tag{9.18}$$

Notice that the virtual input distribution matrix in (9.17) and (9.18) is independent of W. From the representation in (9.18), a coordinate transformation $x \mapsto T_r x(t) = \hat{x}(t)$ will be introduced to obtain the so-called 'regular form' as discussed in Sect. 3.1.1, which is a convenient representation from which to design the hyperplane. If

$$T_r := \begin{bmatrix} I & -B_1 B_2^T \\ 0 & I \end{bmatrix} \tag{9.19}$$

then it is easy to check that (9.17) becomes

$$\dot{\hat{x}}(t) = \hat{A}\hat{x}(t) + \underbrace{\begin{bmatrix} 0 \\ I \end{bmatrix}}_{\hat{B}_\nu} \bar{\nu}(t) - \begin{bmatrix} B_1 B_2^N (I - W^2) B_2^{\mathrm{T}} \\ B_2 (I - W^2) B_2^{\mathrm{T}} \end{bmatrix} \bar{\nu}(t) \tag{9.20}$$

where $\hat{A} := T_r A T_r^{-1}$ and

$$B_2^N := \left(I - B_2^{\mathrm{T}} B_2\right) \tag{9.21}$$

Because by construction the matrix $B_2 B_2^{\mathrm{T}} = I_l$, it follows that $B_2^N B_2^{\mathrm{T}} = (I - B_2^{\mathrm{T}} B_2) B_2^{\mathrm{T}} = 0$, and

$$B_1 B_2^N (I - W^2) B_2^{\mathrm{T}} = -B_1 B_2^N W^2 B_2^{\mathrm{T}} \tag{9.22}$$

Therefore (9.20) becomes

$$\dot{\hat{x}}(t) = \hat{A}\hat{x}(t) + \begin{bmatrix} 0 \\ B_2 W^2 B_2^{\mathrm{T}} \end{bmatrix} \bar{\nu}(t) + \begin{bmatrix} B_1 B_2^N W^2 B_2^{\mathrm{T}} \\ 0 \end{bmatrix} \bar{\nu}(t) \tag{9.23}$$

The last term in (9.23) is zero in the fault-free case, but is treated as (unmatched) uncertainty when $W \neq I$. Define another nonsingular scaling of the virtual control signal as

$$\hat{\nu}(t) := \left(B_2 W^2 B_2^{\mathrm{T}}\right) \bar{\nu}(t) \tag{9.24}$$

therefore, (9.23) becomes

$$\begin{bmatrix} \dot{\hat{x}}_1(t) \\ \dot{\hat{x}}_2(t) \end{bmatrix} = \begin{bmatrix} \hat{A}_{11} & \hat{A}_{12} \\ \hat{A}_{21} & \hat{A}_{22} \end{bmatrix} \begin{bmatrix} \hat{x}_1(t) \\ \hat{x}_2(t) \end{bmatrix} + \begin{bmatrix} 0 \\ I \end{bmatrix} \hat{\nu}(t) + \begin{bmatrix} B_1 B_2^N B_2^{+} \\ 0 \end{bmatrix} \hat{\nu}(t) \tag{9.25}$$

where

$$B_2^{+} := W^2 B_2^{\mathrm{T}} \left(B_2 W^2 B_2^{\mathrm{T}}\right)^{-1} \tag{9.26}$$

It is important to point out that there is an upper bound on the norm of the pseudo-inverse B_2^{+} in (9.26) which is independent of W. Formally:

Proposition 9.1 *There exists a scalar γ_0 which is finite and independent of W such that*

$$\|B_2^{+}\| = \left\| W^2 B_2^{\mathrm{T}} \left(B_2 W^2 B_2^{\mathrm{T}}\right)^{-1} \right\| < \gamma_0 \tag{9.27}$$

for all $W = \mathrm{diag}(w_1, \ldots, w_m)$ such that $0 < w_i \leq 1$.

Proof This follows from a modification of the proof of Theorem 1 in [241]. The work in [241] considers left pseudo-inverses but since $(B_2^{+})^{\mathrm{T}} = (B_2^{\mathrm{T}})^{+}$, the result follows. □

Remark 9.1 As shown in [241], if W is not diagonal, $\|B_2^{+}\|$ is no longer necessarily bounded.

The virtual control law will now be designed based on the nominal fault-free system in which the top partition of the last term in (9.25) is zero since $B_1 B_2^N B_2^{+}|_{W=I} = 0$. In the $\hat{x}(t)$ coordinates in (9.25), a suitable choice for the sliding surface is

$$\hat{S} = S T_r^{-1} = [\mathcal{K} \quad I_l] \tag{9.28}$$

where $\mathcal{K} \in \mathbb{R}^{l \times (n-l)}$ represents design freedom. Introduce another transformation so that $(\hat{x}_1, \hat{x}_2) \mapsto (\hat{x}_1, s(t))$, associated with the nonsingular matrix

$$T_s = \begin{bmatrix} I & 0 \\ \mathcal{K} & I \end{bmatrix} \tag{9.29}$$

Equation (9.25) then becomes

$$\begin{bmatrix} \dot{\hat{x}}_1(t) \\ \dot{s}(t) \end{bmatrix} = \begin{bmatrix} \tilde{A}_{11} & \tilde{A}_{12} \\ \tilde{A}_{21} & \tilde{A}_{22} \end{bmatrix} \begin{bmatrix} \hat{x}_1(t) \\ s(t) \end{bmatrix} + \begin{bmatrix} B_1 B_2^N B_2^+ \\ I + \mathcal{K} B_1 B_2^N B_2^+ \end{bmatrix} \hat{\nu}(t) \tag{9.30}$$

where B_2^+ is from (9.26) and $\tilde{A}_{11} := \hat{A}_{11} - \hat{A}_{12}\mathcal{K}$, $\tilde{A}_{21} := \mathcal{K}\tilde{A}_{11} + \hat{A}_{21} - \hat{A}_{22}\mathcal{K}$. If a control law can be designed to induce sliding, then during sliding $\dot{s}(t) = s(t) = 0$ and so the equivalent control necessary to maintain sliding is obtained from solving for $\hat{\nu}_{eq}(t)$ from the lower equations of (9.30) to give

$$\hat{\nu}_{eq}(t) = -\left(I + \mathcal{K} B_1 B_2^N B_2^+\right)^{-1} \tilde{A}_{21} \hat{x}_1(t) \tag{9.31}$$

where B_2^N is defined in (9.21).

Define

$$\gamma_1 := \left\| \mathcal{K} B_1 B_2^N \right\| \tag{9.32}$$

It follows that $\|\mathcal{K} B_1 B_2^N B_2^+\| < \|\mathcal{K} B_1 B_2^N\| \|B_2^+\| < \gamma_0 \gamma_1$. Since B_2^+ is independent of $\mathcal{K}$, the term γ_0 can be calculated a priori using the boundedness result from Proposition 9.1. If (A, B_ν) is controllable, then as argued in Proposition 3.1, $(\hat{A}_{11}, \hat{A}_{12})$ is controllable and so $\mathcal{K}$ can always be chosen to make $\hat{A}_{11} - \hat{A}_{12}\mathcal{K}$ stable. If the design matrix $\mathcal{K}$ can also be chosen so that γ_1 from (9.32) satisfies $\gamma_0 \gamma_1 < 1$, this guarantees that the inverse in (9.31) exists for all W.

Substituting (9.31) into the top partition of (9.30), yields the following reduced order system which governs the sliding motion:

$$\dot{\hat{x}}_1(t) = \tilde{A}_{11} \hat{x}_1(t) - B_1 B_2^N B_2^+ \left(I + \mathcal{K} B_1 B_2^N B_2^+\right)^{-1} \tilde{A}_{21} \hat{x}_1(t) \tag{9.33}$$

Remark 9.2 In a fault-free condition, $W = I$ and therefore $B_2^+|_{W=I} = B_2^{\mathrm{T}}$ since $B_2 B_2^{\mathrm{T}} = I$. Also

$$B_2^N B_2^+ = \left(I - B_2^{\mathrm{T}} B_2\right) B_2^+ = \left(I - B_2^{\mathrm{T}} B_2\right) B_2^{\mathrm{T}} = 0$$

and the system in (9.33) 'collapses' to $\dot{\hat{x}}_1(t) = \tilde{A}_{11} \hat{x}_1(t)$ which is the nominal sliding mode reduced order system for which $\mathcal{K}$ has been designed to guarantee stability. The system in (9.33) depends on W and so stability needs to be established. The stability analysis which follows examines what will happen to the reduced order sliding motion when the system is subjected to faults and failures. The idea is to use available design tools from the literature e.g., [85, 256], to design the sliding surface for the fault-free condition, and then extend the stability analysis to the faulty situation.

9.2.2 Stability Analysis

The stability of the sliding mode is dependent on the reduced order system (9.33). Typically in SMC the stability of the system only depends on $\tilde{A}_{11}$ which is guaranteed to be stable by choice of $\mathcal{K}$. Note that $\mathcal{K}$ can be designed using standard sliding hyperplane design methods (such as Sect. 3.4) assuming a nominal no fault condition—i.e., $W = I$ in (9.25). To facilitate the subsequent analysis, define

$$\tilde{G}(\mathrm{s}) := \tilde{A}_{21}(\mathrm{s}I - \tilde{A}_{11})^{-1}B_1B_2^N \tag{9.34}$$

where s represents the Laplace variable. By construction the transfer function matrix $\tilde{G}(\mathrm{s})$ is stable. Suppose

$$\gamma_2 = \|\tilde{G}(\mathrm{s})\|_\infty \tag{9.35}$$

Proposition 9.2 *During a fault or failure condition, for any combination of* $0 < w_i \le 1$, *the closed-loop system will be stable if*

$$0 \le \frac{\gamma_2\gamma_0}{1-\gamma_1\gamma_0} < 1 \tag{9.36}$$

where the positive scalar γ_0 *is defined in Proposition* 9.1, *the positive scalar* γ_1 *is defined in* (9.32) *and* $\gamma_2 = \|\tilde{G}(\mathrm{s})\|_\infty$.

Proof Consider the reduced order system from (9.33) rewritten as follows:

$$\dot{\hat{x}}_1(t) = \tilde{A}_{11}\hat{x}_1(t) - B_1B_2^N\tilde{u}(t) \tag{9.37}$$

$$\tilde{y}(t) = \tilde{A}_{21}\hat{x}_1(t) \tag{9.38}$$

where

$$\tilde{u}(t) := B_2^+\big(I + \mathcal{K}B_1B_2^NB_2^+\big)^{-1}\tilde{y}(t) \tag{9.39}$$

Let $\tilde{G}(\mathrm{s})$ be defined as (9.34). Consequently (9.33) may be viewed as the closed-loop dynamics of the negative feedback interconnection of $\tilde{G}(\mathrm{s})$ and the varying (with respect to W) 'feedback gain' associated with (9.39). Since (9.36) is assumed to hold, $\gamma_0\gamma_1 < 1$ and it follows that $\|\mathcal{K}B_1B_2^NB_2^+\| < \|\mathcal{K}B_1B_2^N\|\|B_2^+\| < \gamma_0\gamma_1 < 1$. Consequently,[3] $\det(I + \mathcal{K}B_1B_2^NB_2^+) \ne 0$. Furthermore, using the fact that in general $\|(I+X)^{-1}\| \le (I - \|X\|)^{-1}$ if $\|X\| < 1$ (page 301 [129]), then

$$\big\|B_2^+\big(I + \mathcal{K}B_1B_2^NB_2^+\big)^{-1}\big\| < \|B_2^+\|\big\|\big(I + \mathcal{K}B_1B_2^NB_2^+\big)^{-1}\big\| < \frac{\gamma_0}{1-\gamma_1\gamma_0} \tag{9.40}$$

[3] Suppose $\det(I + X) = 0$, then $\lambda = -1$ is an eigenvalue of X so there exists a $v \in \mathbb{R}^m$ s.t. $Xv = -v$. Therefore,

$$v^\mathrm{T}X^\mathrm{T}Xv = v^\mathrm{T}v \Rightarrow \sup_{\|v\|}\frac{v^\mathrm{T}X^\mathrm{T}Xv}{\|v\|^2} \ge 1 \Rightarrow \|X\| \ge 1$$

which is a contradiction since by assumption $\|X\| < 1$.

From the Small Gain Theorem [152], if

$$\|\tilde{G}(s)\|_{\infty}\|B_2^+(I+\mathcal{K}B_1B_2^N B_2^+)^{-1}\| < 1 \tag{9.41}$$

then (9.33) is stable. Using (9.40) and the fact that $\|\tilde{G}(s)\|_{\infty} < \gamma_2$, inequality (9.36) implies (9.41) holds and so (9.33) is stable. □

Remark 9.3 Both γ_1 and γ_2 depend on the design of the sliding surface since they depend on $\mathcal{K}$, however, they are independent of W. The scalar γ_0 depends on W but is independent of $\mathcal{K}$.

Remark 9.4 If $B_1 = 0$ (which is an assumption in many schemes: for example [222]), then $\gamma_1 = 0$ and $\gamma_2 = 0$ and Proposition 9.2 is trivially satisfied. Furthermore, as $\|B_1\| \to 0$, the scalar $\frac{\gamma_2\gamma_0}{1-\gamma_1\gamma_0} \to 0$ and so the requirements of Proposition 9.2 are satisfied. This means that, loosely speaking, for weakly coupled systems in which $\|B_1\|$ is small, the approach will be feasible. The situation where $B_1 = 0$ can be regarded as the special extreme case as $\|B_1\| \to 0$.

Equation (9.36) represents a test to guarantee the stability of the closed-loop system when faults occur (i.e., when the w_i vary). One important feature is that in order for (9.33) to hold, the norm of the pseudo-inverse B_2^+ which depends on W must be bounded for all $0 < w_i \le 1$ (which was proved in Proposition 9.1).

9.2.3 Sliding Mode Control Laws

Next, a sliding mode controller is designed based on the system in (9.30) with respect to $\hat{\nu}$. As in Sect. 3.3, the proposed control law has a structure given by $\hat{\nu}(t) = \hat{\nu}_l(t) + \hat{\nu}_n(t)$ where

$$\hat{\nu}_l(t) := -\tilde{A}_{21}\hat{x}_1(t) - \tilde{A}_{22}s(t) \tag{9.42}$$

and the nonlinear component is defined to be

$$\hat{\nu}_n(t) := -\rho(t,x)\frac{s(t)}{\|s(t)\|} \quad \text{if } s(t) \neq 0 \tag{9.43}$$

where $s(t) = \hat{S}\hat{x}(t)$.

Proposition 9.3 *Suppose the hyperplane matrix $\mathcal{K}$ has been chosen so that $\tilde{A}_{11} = \hat{A}_{11} - \hat{A}_{12}\mathcal{K}$ is stable and condition (9.36) from Proposition 9.2 holds, then choosing*

$$\rho(t,x) := \frac{\gamma_1\gamma_0\|\hat{\nu}_l(t)\| + \eta}{1-\gamma_1\gamma_0} \tag{9.44}$$

ensures a sliding motion takes place on $\mathcal{S}$ in finite time.

Proof Notice if (9.36) holds, then $\gamma_0\gamma_1 < 1$ and so the modulation function defined in (9.44) is well defined. From (9.30),

$$\begin{aligned}\dot{s}(t) &= \tilde{A}_{21}\hat{x}_1(t) + \tilde{A}_{22}s(t) + \big(I + \mathcal{K}B_1B_2^N B_2^+\big)\hat{\nu}(t)\\ &= \big(I + \mathcal{K}B_1B_2^N B_2^+\big)\hat{\nu}_n(t) + \big(\mathcal{K}B_1B_2^N B_2^+\big)\hat{\nu}_l(t)\end{aligned}$$

and so

$$\begin{aligned}s^{\mathrm{T}}\dot{s} &= -\rho\|s\| + s^{\mathrm{T}}\mathcal{K}B_1B_2^N B_2^+\hat{\nu}_n(t) + s^{\mathrm{T}}\big(\mathcal{K}B_1B_2^N B_2^+\big)\hat{\nu}_l(t)\\ &\le \|s\|\big(\rho\big\|\mathcal{K}B_1B_2^N B_2^+\big\| + \big\|\mathcal{K}B_1B_2^N B_2^+\big\|\big\|\hat{\nu}_l(t)\big\| - \rho\big)\\ &\le \|s\|\big(\rho\gamma_1\gamma_0 + \gamma_1\gamma_0\big\|\hat{\nu}_l(t)\big\| - \rho\big)\end{aligned} \tag{9.45}$$

So choosing $\rho(.)$ as described in (9.44) and substituting into (9.45) implies

$$s(t)^{\mathrm{T}}\dot{s}(t) \le -\eta\big\|s(t)\big\| \tag{9.46}$$

The differential inequality (9.46) is a standard 'reachability condition' and implies $s(t) = 0$ in finite time and a sliding motion is maintained for all subsequent time. □

Remark 9.5 It can be shown that $\hat{\nu}_l(t)$ as defined in (9.42) can be written as $\hat{\nu}_l(t) = -(\hat{S}\hat{B})^{-1}\hat{S}\hat{A}\hat{x}(t)$ which is more in keeping with the notation in [85]. Note that here $\hat{S}\hat{B} = I_l$ and so this simplifies to $\hat{\nu}_l(t) = -\hat{S}\hat{A}x(t)$.

It follows that the actual control which is sent to the actuators is resolved from the 'virtual control law' $\nu(t)$ (from (9.42)–(9.43)), using (9.11), (9.12), (9.16) and (9.24). Some straightforward calculations shown

$$u(t) = WB_2^{\mathrm{T}}\big(B_2W^2B_2^{\mathrm{T}}\big)^{-1}\hat{\nu}(t) \tag{9.47}$$

i.e., the control which is sent to the actuators depends on the effectiveness gains k_i (through the matrix W).

Note that in most of the literature, whilst SMC has been successfully tested on systems with faulty actuators, it was claimed that SMC cannot deal directly with total failures [143]. However, provided the matrix $\mathcal{K}$ satisfies the stability condition (9.36), the sliding mode controller for the 'virtual' system proposed above, can handle total actuator failures in the original system provided that $\det(B_2WB_2^{\mathrm{T}}) \neq 0$.

Remark 9.6 Formally, the effect of position and rate limits on the actuators has not been considered. However, if a rate limit or position limit is exceeded, it would be interpreted by the estimation mechanism as a fault, because the actual actuator position would be different from the expected one based on the commanded control signal. This would result in $k_i > 0$ in the channel in which the saturation or rate limit is reached. The proposed scheme would then inherently attempt to reduce the 'burden' in this channel and redistribute the control effort to other actuators, which would mitigate the effect of the saturation.

So far, it has been assumed that the effectiveness gains $k_i(t)$ that make up K and hence W are known perfectly. In real engineering systems, there will always be some error in the computation/measurements of the $k_i(t)$. The next section considers the impact of this on the proposed scheme.

9.3 The Effect of Non-perfect Fault Reconstruction

Consider a faulty system represented by (9.7). Let $\bar{K}$ correspond to the estimated actuator efficiency based on the information provided by the FDI scheme. Define

$$\bar{W} = I - \bar{K} \tag{9.48}$$

and suppose $\bar{K} \neq K$, where, as described earlier, K represents the actual reduction in actuator efficiency. Suppose

$$W = (I - \Delta)\bar{W} \tag{9.49}$$

were $\Delta = \text{diag}(\delta_1, \ldots, \delta_m)$ and (the unknown) δ_i are elements which represent the level of imperfection in the fault reconstructions. Since $(I - K) = W$, from (9.7)

$$\dot{x}(t) = Ax(t) + B(I - K)u(t) = Ax(t) + B(I - \Delta)\bar{W}u(t) \tag{9.50}$$

Now suppose $u(t) = \bar{B}_2^{\dagger}\nu(t)$ where

$$\bar{B}_2^{\dagger} := \bar{W}B_2^{\mathrm{T}}\big(B_2\bar{W}B_2^{\mathrm{T}}\big)^{-1} \tag{9.51}$$

This represents the fact that $\bar{W}$ (i.e., the estimate rather than the true value of W) will be used to compute the control signal. Then (9.50) becomes

$$\dot{x}(t) = Ax(t) + B(I - \Delta)\bar{W}\bar{B}_2^{\dagger}\nu(t) \tag{9.52}$$

Also define $\bar{\nu}(t) = (B_2\bar{W}B_2^{\mathrm{T}})^{-1}\nu(t)$, then it follows from (9.52) that

$$\begin{aligned}
\dot{x}(t) &= Ax(t) + \begin{bmatrix} B_1\bar{W}^2B_2^{\mathrm{T}} \\ B_2\bar{W}^2B_2^{\mathrm{T}} \end{bmatrix}\bar{\nu}(t) - \begin{bmatrix} B_1\Delta\bar{W}^2B_2^{\mathrm{T}} \\ B_2\Delta\bar{W}^2B_2^{\mathrm{T}} \end{bmatrix}\bar{\nu}(t) \\
&= Ax(t) + \begin{bmatrix} B_1B_2^{\mathrm{T}} \\ I \end{bmatrix}\bar{\nu}(t) - \begin{bmatrix} B_1(I - \bar{W}^2)B_2^{\mathrm{T}} \\ B_2(I - \bar{W}^2)B_2^{\mathrm{T}} \end{bmatrix}\bar{\nu}(t) \\
&\quad - \begin{bmatrix} B_1\Delta\bar{W}^2B_2^{\mathrm{T}} \\ B_2\Delta\bar{W}^2B_2^{\mathrm{T}} \end{bmatrix}\bar{\nu}(t)
\end{aligned} \tag{9.53}$$

Notice that compared with (9.17), (9.53) has an additional term dependent on both the faults and the error in fault reconstruction. Again consider a transformation to regular form using the transformation matrix T_r defined in (9.19). Equation (9.53) becomes

$$\begin{aligned}\dot{\hat{x}} &= \hat{A}\hat{x}(t) + \begin{bmatrix}0\\ I\end{bmatrix}\bar{\nu}(t) - \begin{bmatrix}-B_1 B_2^N \bar{W}^2 B_2^{\mathrm{T}}\\ I - B_2\bar{W}^2 B_2^{\mathrm{T}}\end{bmatrix}\bar{\nu}(t) - \begin{bmatrix}B_1 B_2^N \Delta\bar{W}^2 B_2^{\mathrm{T}}\\ B_2\Delta\bar{W}^2 B_2^{\mathrm{T}}\end{bmatrix}\bar{\nu}(t)\\ &= \hat{A}\hat{x}(t) + \begin{bmatrix}0\\ B_2\bar{W}^2 B_2^{\mathrm{T}}\end{bmatrix}\bar{\nu}(t) + \begin{bmatrix}B_1 B_2^N \bar{W}^2 B_2^{\mathrm{T}}\\ 0\end{bmatrix}\bar{\nu}(t)\\ &\quad - \begin{bmatrix}B_1 B_2^N \Delta\bar{W}^2 B_2^{\mathrm{T}}\\ B_2\Delta\bar{W}^2 B_2^{\mathrm{T}}\end{bmatrix}\bar{\nu}(t)\end{aligned} \tag{9.54}$$

where B_2^N is given originally in (9.21). Define $\hat{\nu}(t) := B_2\bar{W}^2 B_2^{\mathrm{T}}\bar{\nu}(t)$ and $\bar{B}_2^+ := \bar{W}^2 B_2^{\mathrm{T}}(B_2\bar{W}^2 B_2^{\mathrm{T}})^{-1}$; then (9.54) becomes

$$\dot{\hat{x}} = \hat{A}\hat{x}(t) + \begin{bmatrix}0\\ I\end{bmatrix}\hat{\nu}(t) + \begin{bmatrix}B_1 B_2^N (I-\Delta)\bar{B}_2^+\\ -B_2\Delta\bar{B}_2^+\end{bmatrix}\hat{\nu}(t) \tag{9.55}$$

Consider another coordinate transformation T_s defined in (9.29); then (9.55) becomes

$$\begin{aligned}\begin{bmatrix}\dot{\hat{x}}_1(t)\\ \dot{s}(t)\end{bmatrix} &= \begin{bmatrix}\tilde{A}_{11} & \tilde{A}_{12}\\ \tilde{A}_{21} & \tilde{A}_{22}\end{bmatrix}\begin{bmatrix}\hat{x}_1(t)\\ s(t)\end{bmatrix} + \begin{bmatrix}0\\ I\end{bmatrix}\hat{\nu}(t)\\ &\quad + \begin{bmatrix}B_1 B_2^N (I-\Delta)\bar{B}_2^+\\ \mathcal{K}B_1 B_2^N (I-\Delta)\bar{B}_2^+ - B_2\Delta\bar{B}_2^+\end{bmatrix}\hat{\nu}(t)\end{aligned} \tag{9.56}$$

where $\tilde{A}_{11} := \hat{A}_{11} - \hat{A}_{12}\mathcal{K}$ and $\tilde{A}_{21} := \mathcal{K}\tilde{A}_{11} + \hat{A}_{21} - \hat{A}_{22}\mathcal{K}$.

If a control law can be designed to induce sliding, then during the sliding motion $\dot{s}(t) = s(t) = 0$ and the equivalent control necessary to maintain sliding is obtained from solving for $\hat{\nu}_{eq}(t)$ from the lower of (9.56) to give

$$\hat{\nu}_{eq}(t) = -\big(I + \mathcal{K}B_1 B_2^N (I-\Delta)\bar{B}_2^+ - B_2\Delta\bar{B}_2^+\big)^{-1}\tilde{A}_{21}\hat{x}_1(t) \tag{9.57}$$

Substituting into the first equation of (9.56) gives the following reduced order system:

$$\begin{aligned}\dot{\hat{x}}_1(t) &= \tilde{A}_{11}\hat{x}_1(t) - B_1 B_2^N (I-\Delta)\bar{B}_2^+\big(I + \mathcal{K}B_1 B_2^N (I-\Delta)\bar{B}_2^+\\ &\quad - B_2\Delta\bar{B}_2^+\big)^{-1}\tilde{A}_{21}\hat{x}_1(t)\end{aligned} \tag{9.58}$$

Remark 9.7 If the information about the actual degradation of the control surface efficiency is 'perfect', then $\Delta = 0$, and (9.58) reduces to (9.33) in the stability analysis that follows. However, in the event of non-perfect fault reconstruction, there is a bound on Δ for which stability is still guaranteed.

Proposition 9.4 *Assume* (*as in Proposition* 9.2), *that* (9.36) *holds. During a fault or failure condition, for any combinations of* $0 < w_i \leq 1$, *the closed-loop system will* *be stable if the mismatch between the actual and reconstructed fault Δ satisfies:*

$$\|\Delta\| < \frac{1 - \gamma_1\gamma_0 - \gamma_2\gamma_0}{\gamma_0(\gamma_1 + \gamma_2 + 1)} \tag{9.59}$$

where γ_0 and γ_1 are defined in Propositions 9.1 and 9.2, respectively, and γ_2 is defined in (9.35).

Proof Consider the reduced order system from (9.58) which can be rewritten as (9.37)–(9.39), where now

$$\tilde{u}(t) := (I - \Delta)\bar{B}_2^+\big(I + \mathcal{K}B_1B_2^N(I - \Delta)\bar{B}_2^+ - B_2\Delta\bar{B}_2^+\big)^{-1}\tilde{y}(t) \tag{9.60}$$

Assume that (9.36) and (9.59) hold. Inequality (9.59) implies

$$\|\Delta\| < \frac{1 - \gamma_1\gamma_0}{\gamma_0(\gamma_1 + 1)} \tag{9.61}$$

because

$$\frac{1 - \gamma_1\gamma_0}{\gamma_0(\gamma_1 + 1)} > \frac{1 - \gamma_1\gamma_0 - \gamma_2\gamma_0}{\gamma_0(\gamma_1 + \gamma_2 + 1)} > 0 \tag{9.62}$$

Since

$$\big\|\mathcal{K}B_1B_2^N(I - \Delta)\bar{B}_2^+ - B_2\Delta\bar{B}_2^+\big\| \le \big\|\mathcal{K}B_1B_2^N\big\|\big(1 + \|\Delta\|\big)\|\bar{B}_2^+\| + \|B_2\|\|\Delta\|\|\bar{B}_2^+\|$$

and in addition $\|B_2\| = 1$ and $\|\bar{B}_2^+\| \le \gamma_0$ (since $\|B_2^+\| \le \gamma_0 \Rightarrow \|\bar{B}_2^+\| \le \gamma_0$), inequality (9.61) implies

$$\big\|\mathcal{K}B_1B_2^N(I - \Delta)\bar{B}_2^+ - B_2\Delta\bar{B}_2^+\big\| \le \gamma_1\gamma_0 + \|\Delta\|\gamma_0(\gamma_1 + 1) < 1 \tag{9.63}$$

Therefore $(I + \mathcal{K}B_1B_2^N(I - \Delta)\bar{B}_2^+ - B_2\Delta\bar{B}_2^+)^{-1}$ exists for all $0 < w_i \le 1$. Furthermore, using arguments similar to those employed in the proof of Proposition 9.2

$$\begin{aligned} &\big\|(I - \Delta)\bar{B}_2^+(I + \mathcal{K}B_1B_2^N(I - \Delta)\bar{B}_2^+ - B_2\Delta\bar{B}_2^+)^{-1}\big\| \\ &\quad < \frac{(1 + \|\Delta\|)\gamma_0}{(1 - \gamma_1(1 + \|\Delta\|)\gamma_0 - \|\Delta\|\gamma_0)} \end{aligned} \tag{9.64}$$

From the Small Gain Theorem, since (9.58) is the closed-loop system obtained from $\tilde{G}(s)$ interconnected with (9.60), if

$$\frac{(1 + \|\Delta\|)\gamma_0}{(1 - \gamma_1(1 + \|\Delta\|)\gamma_0 - \|\Delta\|\gamma_0)} < \frac{1}{\gamma_2} \tag{9.65}$$

holds, then (9.58) is stable. By direct manipulation, (9.65) holds if (9.59) holds, and the proof is complete. □

A sliding mode controller will now be designed based on the 'virtual' system in (9.56) with respect to $\hat{\nu}$, as defined in (9.42) and (9.43).

Proposition 9.5 *Suppose that the hyperplane matrix $\mathcal{K}$ has been chosen so that $\tilde{A}_{11} = \hat{A}_{11} - \hat{A}_{12}\mathcal{K}$ is stable and*

$$\|\Delta\| \le \Delta_{\max} < \frac{1 - \gamma_1\gamma_0}{\gamma_0(1+\gamma_1)} \tag{9.66}$$

where γ_0, γ_1 and γ_2 are defined in (9.27), (9.32) *and* (9.35)*, respectively, and $\Delta_{\max}$ is a fixed positive scalar. Then choosing*

$$\rho(t,x) = \frac{(\gamma_1(1+\Delta_{\max})\gamma_0 + \Delta_{\max}\gamma_0)\|\hat{\nu}_l(t)\| + \eta}{1 - \gamma_1(1+\Delta_{\max})\gamma_0 - \Delta_{\max}\gamma_0} \tag{9.67}$$

ensures a sliding motion takes place on $\mathcal{S}$ in finite time.

Proof The assumption on Δ in (9.66) implies $\gamma_0\gamma_1 + (1+\gamma_1)\Delta_{\max}\gamma_0 < 1$ and so the gain $\rho(t,x)$ in (9.67) is well defined. From (9.56),

$$\begin{aligned}
\dot{s}(t) &= \tilde{A}_{21}\hat{x}_1(t) + \tilde{A}_{22}s(t) + \big(I + \mathcal{K}B_1B_2^N(I-\Delta)\bar{B}_2^+ - B_2\Delta\bar{B}_2^+\big)\hat{\nu}(t) \\
&= \big(I + \mathcal{K}B_1B_2^N(I-\Delta)\bar{B}_2^+ - B_2\Delta\bar{B}_2^+\big)\hat{\nu}_n(t) \\
&\quad + \big(\mathcal{K}B_1B_2^N(I-\Delta)\bar{B}_2^+ - B_2\Delta\bar{B}_2^+\big)\hat{\nu}_l(t)
\end{aligned}$$

after substituting for $\nu_l(t)$ from (9.42). Consequently substituting for $\nu_n(t)$ from (9.43) into the above yields

$$\begin{aligned}
s^{\mathrm{T}}\dot{s} &= -\rho\|s\| + s^{\mathrm{T}}\big(\mathcal{K}B_1B_2^N(I-\Delta)\bar{B}_2^+ - B_2\Delta\bar{B}_2^+\big)\hat{\nu}_n(t) \\
&\quad + s^{\mathrm{T}}\big(\mathcal{K}B_1B_2^N(I-\Delta)\bar{B}_2^+ - B_2\Delta\bar{B}_2^+\big)\hat{\nu}_l(t) \\
&\le \|s\|\big(\rho\big\|\mathcal{K}B_1B_2^N(I-\Delta)\bar{B}_2^+ - B_2\Delta\bar{B}_2^+\big\| \\
&\quad + \big\|\mathcal{K}B_1B_2^N(I-\Delta)\bar{B}_2^+ - B_2\Delta\bar{B}_2^+\big\|\big\|\hat{\nu}_l(t)\big\| - \rho\big) \\
&\le \|s\|\big(-\rho\big(1 - \gamma_1(1+\Delta_{\max})\gamma_0 - \Delta_{\max}\gamma_0\big) \\
&\quad + \big(\gamma_1(1+\Delta_{\max})\gamma_0 + \Delta_{\max}\gamma_0\big)\big\|\hat{\nu}_l(t)\big\|\big)
\end{aligned} \tag{9.68}$$

So choosing $\rho(.)$ as described in (9.67) and substituting into (9.68) implies

$$s(t)^{\mathrm{T}}\dot{s}(t) \le -\eta\big\|s(t)\big\| \tag{9.69}$$

Again, as in Sect. 9.2.3, the differential inequality (9.69) implies that the 'reachability condition' is achieved. Therefore $s(t) = 0$ in finite time and a sliding motion is maintained for all subsequent time. □

9.4 Sliding Mode Design Issues

Based on the stability analysis above, the sliding mode design problem can be summarised as follows:

1. Pre-design calculations:
   a. Make an appropriate re-ordering of the states in (9.7) so that the input distribution matrix B is partitioned to identify B_1 and B_2.
   b. Scale the states so that $B_2 B_2^{\mathrm{T}} = I$.
   c. Change coordinates using the linear transformation $x(t) \mapsto \hat{x}(t) = T_r x(t)$, where T_r is given in (9.19), to achieve the canonical form in (9.25) and isolate the matrices $\hat{A}_{11}$, $\hat{A}_{12}$, $\hat{A}_{21}$ and $\hat{A}_{22}$.
   d. Compute the smallest scalar γ_0 so that $\| W^2 B_2^{\mathrm{T}} (B_2 W^2 B_2^{\mathrm{T}})^{-1} \| < \gamma_0$, for all $0 < W \leq I$. This value is an a priori calculation and is independent of the choice of sliding surface and control law.
2. Design of matrix $\mathcal{K}$:
   a. The design objective is to compute $\mathcal{K}$ from (9.28) so that the matrix $\tilde{A}_{11} := \hat{A}_{11} - \hat{A}_{12}\mathcal{K}$ is stable. This is always possible if (A, B_ν) is controllable.
3. Stability analysis:
   a. Compute and check $\gamma_1 := \|\mathcal{K} B_1 B_2^N\| < 1/\gamma_0$ is satisfied. Otherwise re-design $\mathcal{K}$.
   b. Calculate $\tilde{G}(\mathrm{s}) := \tilde{A}_{21}(\mathrm{s}I - \tilde{A}_{11})^{-1} B_1 B_2^N$. If $\|\tilde{G}(\mathrm{s})\|_\infty := \gamma_2 < 1/\gamma_0 - \gamma_1$, the closed-loop is guaranteed to be stable $\forall 0 < W \leq I$. (Since $\gamma_2 < 1/\gamma_0 - \gamma_1$ ensures inequality (9.36) in Proposition 9.2 holds.) Otherwise consider re-designing the matrix $\mathcal{K}$.
   c. Calculate $\|\Delta\|$ from Proposition 9.5. This is the maximum tolerable mismatch between the actual and the estimated faults that guarantees the closed-loop system is stable $\forall 0 < W \leq I$. This might dictate the choice of the fault estimation scheme.
4. Obtain the virtual control law using (9.42), (9.43) and the actual control law using (9.47).

9.5 ADMIRE Simulations

9.5.1 Controller Design

The small, rigid, delta-canard ADMIRE aircraft model from Sect. 6.4.2.1 is considered here. The states are $x = [\alpha\ \beta\ p\ q\ r]^{\mathrm{T}}$ with controlled outputs $y_c = [\alpha, \beta, p]^{\mathrm{T}}$. The control surfaces are $\delta = [\delta_c\ \delta_{re}\ \delta_{le}\ \delta_r]^{\mathrm{T}}$. The linearised (A, B) matrices are given in (6.135) and (6.136) where the partition of the B matrix is shown below:

$$B = \left[\begin{array}{cccc} 0.0069 & -0.0866 & -0.0866 & 0.0004 \\ 0 & 0.0119 & -0.0119 & 0.0287 \\ \hline 0 & -4.2423 & 4.2423 & 1.4871 \\ 1.6532 & -1.2735 & -1.2735 & 0.0024 \\ 0 & -0.2805 & 0.2805 & -0.8823 \end{array}\right] \begin{array}{l} \big\} B_1 \\ \\ \Bigg\} B_2 \end{array} \tag{9.70}$$

The partition of B in (9.70) shows the terms B_1 and B_2 (although a further change of coordinate is necessary to scale B_2 to ensure $B_2 B_2^{\mathrm{T}} = I$). It can be shown that in the coordinates in which $\|B_2\| = 1$, $\|B_1\| = 0.1227$ and so the dominant effect of the control signal is through the B_2 channels. To include a tracking facility, integral action (as discussed in Sect. 3.5.1) has been included. Let $x_r(t)$ represent integral action states. Specifically define

$$\dot{x}_r(t) = r(t) - C_c x(t) \tag{9.71}$$

where

$$C_c = [I_3 \quad 0_{3\times 2}] \tag{9.72}$$

is the distribution matrix associated with the controlled outputs, and the differentiable (filtered reference) signal $r(t)$ satisfies

$$\dot{r}(t) = \Gamma\big(r(t) - r_c\big) \tag{9.73}$$

with $\Gamma \in \mathbb{R}^{3\times 3}$ a stable design matrix and r_c a piecewise constant demand vector. Augmenting the states from (9.70) with the integral action states and defining $x_a(t) = \mathrm{col}(x_r(t), x(t))$ (as discussed in Sect. 3.5.1) it follows that

$$\dot{x}_a(t) = A_a x_a(t) + B_a u(t) + B_r r(t) \tag{9.74}$$

where

$$A_a = \begin{bmatrix} 0 & -C_c \\ 0 & A \end{bmatrix}, \qquad B_a = \begin{bmatrix} 0 \\ B \end{bmatrix}, \qquad B_r = \begin{bmatrix} I_3 \\ 0 \end{bmatrix} \tag{9.75}$$

If (A, B) is controllable and (A, B, C_c) does not have any zeros at the origin, then as argued in Proposition 3.1 and Lemma 3.1 the matrix pair (A_a, B_a) is controllable. Define a switching function $s_a(t) : \mathbb{R}^{(n+l)} \to \mathbb{R}^l$ of the form

$$s_a(t) = S_a x_a(t) \tag{9.76}$$

where $S_a \in \mathbb{R}^{l\times(n+l)}$. As in (9.42)–(9.43), the proposed 'virtual control' law comprises two components $\hat{\nu}(t) = \hat{\nu}_l(t) + \hat{\nu}_n(t)$. Now because of the reference signal $r(t)$, the linear component has a feed-forward reference term and so $\hat{\nu}_l(t) = L_{x_a} x_a(t) + L_r r(t)$ where $L_{x_a} = -\hat{S}_a \hat{A}_a$ and $L_r = -\hat{S}_a \hat{B}_r$. Here $\hat{A}$, $\hat{B}_r$ and $\hat{S}$ are the matrices from (9.75) and (9.76) after a transformation to achieve regular form (analogous to (9.19)) has been performed. Note that an extract term L_r has appeared in this tracking formulation compared to the one in Sect. 9.2.3. The nonlinear component is defined as

$$\hat{\nu}_n(t) = -\rho(t, x_a)\frac{s_a(t)}{\|s_a(t)\|} \quad \text{if } s_a(t) \neq 0 \tag{9.77}$$

The actual control sent to the actuator is given in (9.47). A quadratic optimal design has been used to obtain the sliding surface matrix S_a (see for example

Sect. 3.4.1). The symmetric positive definite weighting matrix has been chosen as $\mathbf{Q} = \text{diag}(20, 20, 20, 7, 10, 10, 1, 1)$. This results in

$$S_a = \begin{bmatrix} -0.0002 & -0.0844 & -1.4117 & 0.0001 & 0.0701 & 1.0000 & 0.0000 & 0.0000 \\ -4.4721 & 0.0074 & 0.0001 & 3.5011 & 0.0058 & 0.0000 & 1.0000 & 0.0000 \\ 0.0074 & 4.4642 & -0.2668 & -0.0058 & -4.2606 & 0.0000 & 0.0000 & 1.0000 \end{bmatrix}$$

and linear gains

$$L_{x_a} = \begin{bmatrix} 0.0015 & 0.7592 & 12.7053 & -0.0012 & 9.8061 & -9.4305 & -0.0001 & -0.5499 \\ 40.2492 & -0.0665 & -0.0006 & -36.7020 & -0.0895 & -0.0012 & -11.9176 & 0.0056 \\ -0.0664 & -40.1775 & 2.4009 & 0.0562 & 41.5999 & 0.7709 & 0.0056 & -12.9035 \end{bmatrix}$$

and

$$L_r = \begin{bmatrix} 0.0002 & 0.0844 & 1.4117 \\ 4.4721 & -0.0074 & -0.0001 \\ -0.0074 & -4.4642 & 0.2668 \end{bmatrix}$$

from (9.5.1). The pre-filter from (9.73) has been chosen as $\Gamma = -20I_3$. In the simulations the discontinuity in the nonlinear control term in (9.77) has been smoothed (as in Sect. 3.2.2) by using a sigmoidal approximation $\frac{s_a}{\|s_a\|+\delta_a}$, where the scalar δ_a has been chosen as $\delta_a = 0.001$. This removes the discontinuity and introduces a further degree of tuning to accommodate the actuator limits, especially during actuator fault or failure conditions.

In normal flight, either the canard or elevons (left and right) are sufficient to provide the pitch moment and therefore redundancy is available. In the event of faults or failures, elevons can replace the canard to obtain a pitch moment. However, for roll, the elevons will become the only active control surface (the rudder is used for yaw). During the design stage, and based on analysis from (9.44), it was found that $\text{rank}(B_2 W B_2^{\mathrm{T}}) < 3$ when the rudder completely fails or any two surfaces from the set consisting of the canard and the left and right elevons completely fail. This is an expected result since there is no redundancy for the rudder to provide yaw; and when two actuators fail from either the canard or elevons, it means that there is no redundancy left in the system and all possible actuators to provide pitch or roll have failed. Based on this assumption, it can be verified from a numerical search that $\gamma_0 = 2.0913$. Simple calculations show that $\gamma_1 = 0.0980$, therefore $\gamma_1\gamma_0 = 0.2050 < 1$ and the requirement of Proposition 9.2 is satisfied. Also for this particular choice of sliding surface $\|\tilde{G}(s)\|_\infty < \gamma_2 = 0.0819$. Therefore from Proposition 9.2,

$$\frac{\gamma_2\gamma_0}{1-\gamma_1\gamma_0} = 0.2154 < 1$$

which shows that the closed-loop system is stable for all choices of $0 < w_i \leq 1$. From Proposition 9.4, the limits of the tolerable mismatch between the actual and estimated fault signal (for guaranteed stability) is $\Delta_{\max} = 0.3519$.

9.5.2 Actuator Fault Estimation Using an Observer

In many systems e.g., passenger aircraft [35], the information necessary to compute W online can be obtained by using a measurement of the actual actuator deflection compared to the demand signals. If measurements of the actual actuator deflections are not available, a sliding mode fault reconstruction scheme, as described in Chap. 4 for example, can be employed. Alternatively other fault reconstruction schemes based on Kalman filters [296] can also be used. Here it is assumed that direct measurements of the actuator deflections are not available and a sliding mode reconstruction approach similar to the one in Sect. 4.3 will be used. Consider the system affected by actuator faults described by (9.7). The objective is to design a sliding mode observer in order to reconstruct $Ku(t)$.

Suppose all the states are available as measured output information, therefore the proposed sliding mode observer has the form:

$$\dot{z}(t) = Az(t) + Bu(t) + \vartheta(t) \tag{9.78}$$

where $u(t)$ is the actual control signal sent to each actuator. The discontinuous injection term is defined as

$$\vartheta(t) := -\rho_o(t, y, u)\frac{e(t)}{\|e(t)\|} \quad \text{if } e(t) \neq 0 \tag{9.79}$$

where $e(t) := z(t) - x(t)$ is the (state) estimation error. As argued in Chap. 4, a sliding mode observer of the form (9.78)–(9.79) can be designed to be completely insensitive to the faults. For an appropriate choice of $\rho_o(t, y, u)$ in (9.79), which must bound the fault signal $\|Ku(t)\|$, it can be shown that an ideal sliding motion takes place on $\mathcal{S}_o = \{e(t) : e(t) = 0\}$ in finite time. During the ideal sliding motion, $e(t) = \dot{e}(t) = 0$ and the discontinuous signal $\vartheta(t)$ must take an average value to compensate for $Ku(t)$ to maintain sliding. This equivalent error injection term, can be approximated to any degree of accuracy, and is computable online as

$$\vartheta_\delta(t) = -\rho_o(t, y, u)\frac{e(t)}{\|e(t)\| + \delta_o} \tag{9.80}$$

where $\delta_o > 0$ is a (small) design parameter. The observer state estimation error system is given by

$$\dot{e}(t) = Ae(t) + \vartheta(t) + BKu(t) \tag{9.81}$$

During ideal sliding, $\dot{e}(t) = e(t) = 0$, and therefore (9.81) reduces to

$$-BKu(t) = \vartheta_{eq}(t)$$

Using the online computed approximation $\vartheta_\delta(t)$ from (9.80), the fault reconstruction can be obtained as

$$-Ku(t) \approx \left(B^{\mathrm{T}}B\right)^{-1}B^{\mathrm{T}}\vartheta_\delta(t) \tag{9.82}$$

Then, provided that $u_i(t) \neq 0$, the effectiveness gains k_i can be computed from (9.82).

For this example, the observer gains $\rho_o = 30$ and $\delta_o = 0.0001$ from (9.78) have been chosen. A saturation $(0, 1]$ block has been included to 'clip' the w_i before they are used for the online control allocation. This ensures that the weight W stays within the theoretical limits. During implementation, as $u_i(t) \to 0$, the estimation of the actuator effectiveness K from (9.7), becomes unreliable. A small threshold has been introduced so that if t_ϵ is the last time when $|u_i(t)| \leq \epsilon$, then

$$k_i(t) = \begin{cases} ((B^{\mathrm{T}}B)^{-1}B^{\mathrm{T}}v_\delta)_i / u_i(t) & \text{if } |u_i(t)| > \epsilon \\ k_i(t_\epsilon) & \text{otherwise} \end{cases} \tag{9.83}$$

The idea is to hold the component of the weighting matrix $w_i := 1 - k_i$ constant if $|u_i(t)| \leq \epsilon$, otherwise $k_i(t)$ is provided by the fault estimator. Here, the threshold defined in (9.83) is set as $\epsilon = 1 \times 10^{-3}$(rad).

9.5.3 ADMIRE: Simulation Results

In the following simulations (which assume that there is no saturation or rate limits on the actuators), the linear aircraft model from [122] undertakes a manoeuvre called 'α roll' [122], where a step demand of magnitude 10 deg is applied to α during 1–5 s and a step of 150 deg for p is applied during 3–7 s. (There is no reference command for β—see Fig. 9.2(a).) Figures 9.2(a), 9.2(b), 9.2(c) and 9.2(d) show the responses of the closed-loop system under eleven different canard fault conditions ranging from 0% → 100% (including total failure). It can be seen that the control signal is systematically rerouted to the right and left elevon (Fig. 9.2(b)). The tracking responses (Fig. 9.2(a)) show no degradation in performance. The control allocation redistributes the control signal to obtain the required performance. Figure 9.2(a) shows that the observer designed for fault reconstruction, tracks the plant output 'perfectly'. Figure 9.2(c) shows the evolution of the fault reconstruction signal from the observer. These signals are used for the online control allocation through the term W as shown in Fig. 9.2(d).

Figures 9.3(a), 9.3(b), 9.3(c) and 9.3(d) show that in the event the left elevon fails, the control signal is redistributed to the remaining actuators without reconfiguring the structure of the controller. The control signals are rerouted to other control surfaces when the fault is detected and estimated (Fig. 9.3(c)). Initially in Fig. 9.3(b), a control signal is sent to the failed actuator. After the failure has been detected by the observer (Fig. 9.3(c)), the weight w_i in the control allocation is changed (Fig. 9.3(d)) and the control signal sent to the left elevon is 'switched off' and redistributed to the canard and right elevon.

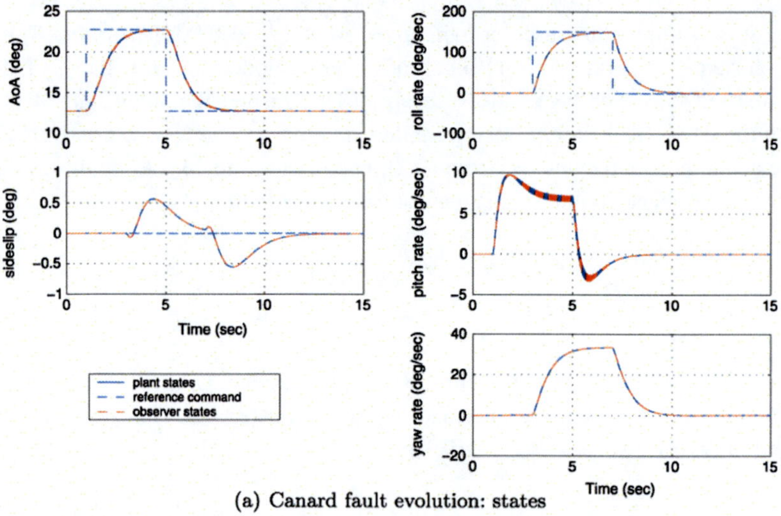

(a) Canard fault evolution: states

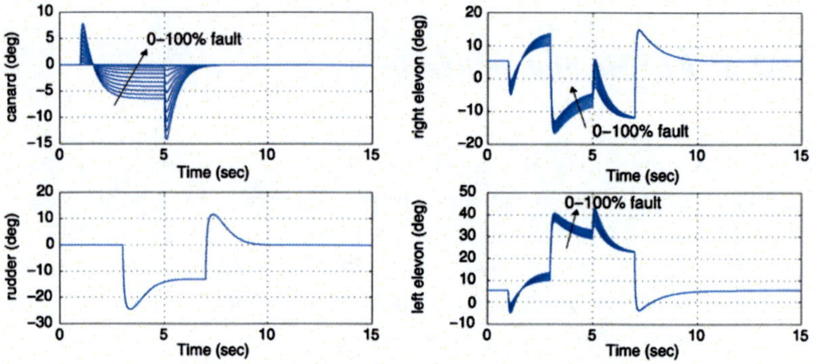

(b) Canard fault evolution: actuator deflection

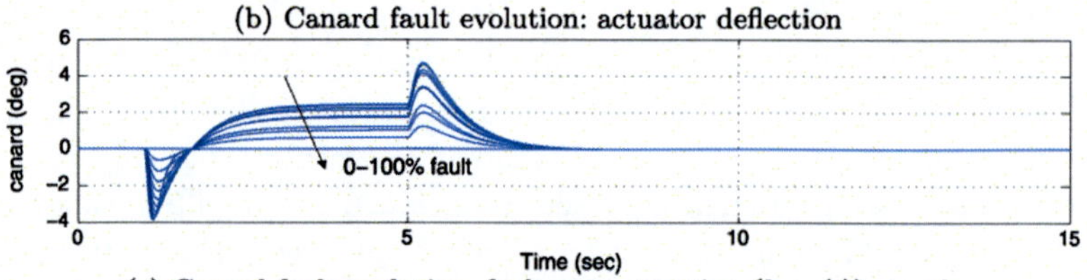

(c) Canard fault evolution: fault reconstruction $(k_i u_i(t))$ signals

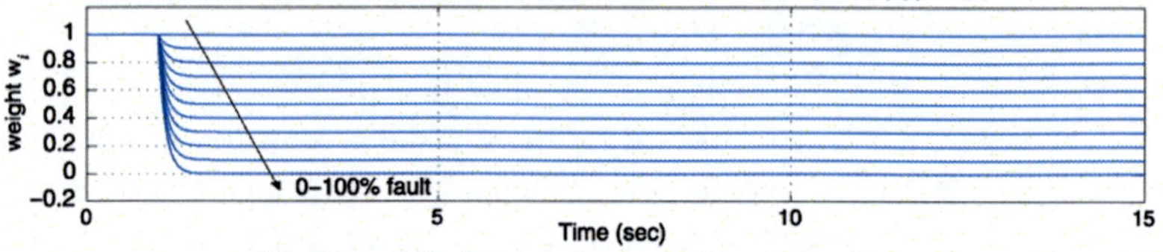

(d) Canard fault evolution: weight w_i signals

Fig. 9.2 Responses to a fault and failure on canard

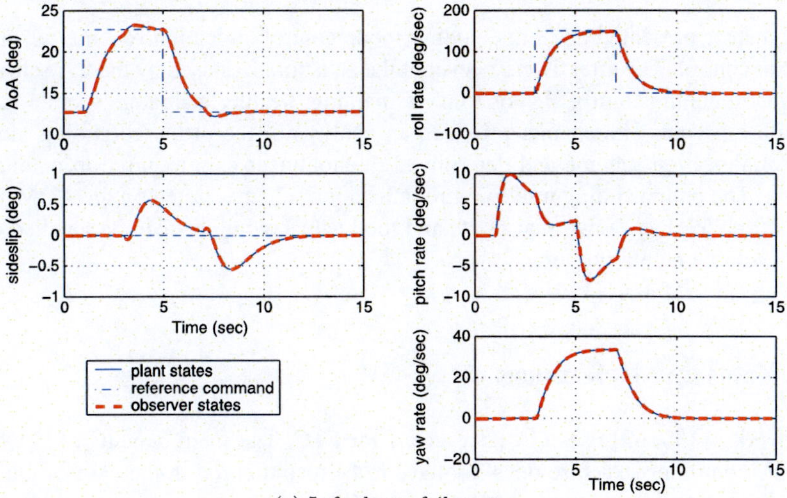

(a) Left elevon fail: states

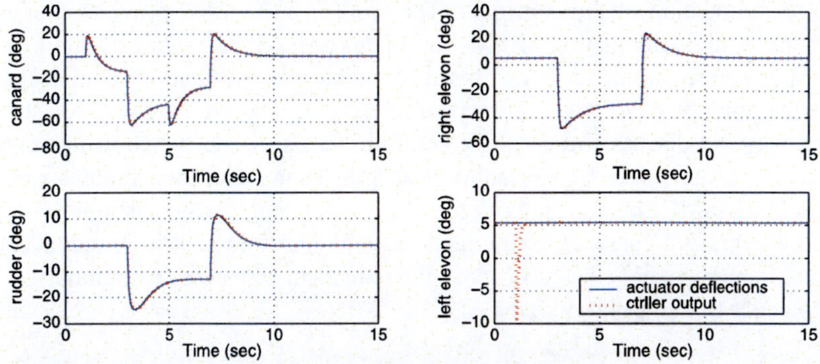

(b) Left elevon fail: actuator deflection

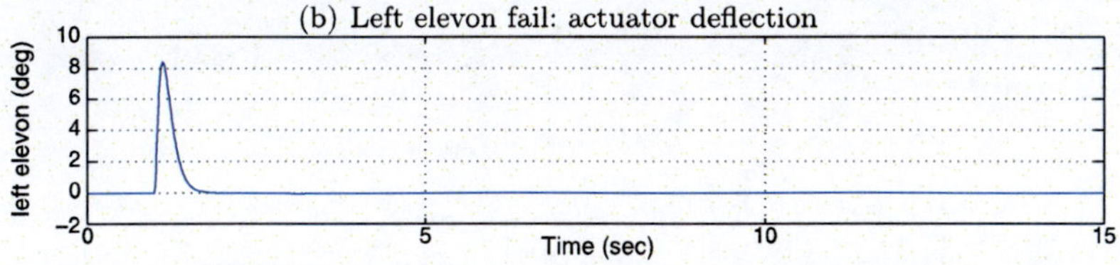

(c) Left elevon fail: fault reconstruction $(k_i\, u_i\,(t))$ signals

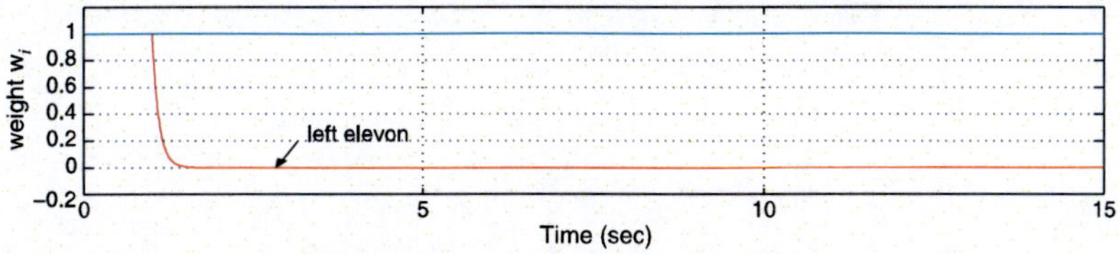

(d) Left elevon fail: weight w_i signals

Fig. 9.3 Responses to a failure on left elevon

9.6 Summary

This chapter presented an online sliding mode control allocation scheme for fault tolerant control. The effectiveness level of the actuators was used by the CA scheme to redistribute the control signals to the remaining 'healthy' actuators when a fault or failure occurs. The chapter provided an analysis of the proposed sliding mode control allocation scheme and determined the nonlinear gain required to maintain sliding. The online sliding mode control allocation scheme, tested on the ADMIRE aircraft model, has shown that faults and total actuator failures can be handled directly without reconfiguration.

9.7 Notes and References

The work in [38, 63] uses CA as a means for FTC. The main benefits of CA are that the controller structure does not have to be reconfigured in the case of faults, and it can deal directly with total actuator failures without requiring reconfiguration/accommodation of the controller. The work in [226, 269] provides practical examples of the combination of SMC and CA for FTC. The work by Shin et al. [222] uses control allocation ideas, but it formulates the problem from an adaptive controller point of view. However, neither of these papers provide a detailed stability analysis and discuss sliding mode controller design issues when using control allocation. Recent work by Corradini et al. [58] shows that total failures can be dealt with by SMC schemes provided that there is enough redundancy in the system. However, [58] considers exact duplication of actuators to achieve redundancy, whereas in many over actuated systems, the redundant actuators do not have identical dynamics to the 'primary' ones.

Chapter 10
Model-Reference Sliding Mode FTC

In the last chapter, a combination of CA and SMC was introduced in an integral action framework for tracking purposes. In this chapter, two different CA strategies will be considered. The first CA strategy is based on the effectiveness of the actuators, whereas in the second strategy, the control signal will be distributed equally among all actuators in a fixed fashion. The difference between this chapter and the previous one is the use of a model-reference framework for tracking purposes. The use of a model-reference tracking strategy is well known in the literature to have the benefit of avoiding the problem of 'windup' which can blight integral action tracking methods—especially when significant faults/failures occur. This chapter will highlight the benefit of combining SMC, CA and a model-reference framework for achieving FTC. An adaptive gain and an adaptive reference model are used to increase the flexibility of the design and to provide further tuning for the controller.

10.1 Introduction

The so-called model-reference framework is one of the many ways of achieving control reconfiguration or adaptation [164]. Therefore it is not surprising that model-reference control is quite synonymous with FTC. One advantage of the approach in this chapter is the absence of integrators within the control structure; this eliminates the dangers of windup in the face of saturation and rate limits being exceeded (because of the increased burden imposed on the remaining working actuators as a result of faults). As in Chap. 8 the introduction of the adaptive gain in the SMC controller obviates the need for unnecessarily large gains in the nonlinear control terms in the fault-free case. An adaptive reference model is also discussed to provide a safe level of degraded performance.

As in Chap. 9, this chapter considers a situation where a fault associated with the actuators develops in a system. It will be assumed that the system subject to actuator faults or failures, can be written as

$$\dot{x}(t) = Ax(t) + Bu(t) - BK(t)u(t) \tag{10.1}$$

H. Alwi et al., *Fault Detection and Fault-Tolerant Control Using Sliding Modes*, Advances in Industrial Control,
DOI 10.1007/978-0-85729-650-4_10,

where $A \in \mathbb{R}^{n\times n}$ and $B \in \mathbb{R}^{n\times m}$. As in earlier chapters, the effectiveness gain $K(t) := \mathrm{diag}(k_1(t), \ldots, k_m(t))$, where the $k_i(t)$ are scalars satisfying $0 \le k_i(t) \le 1$. These scalars model a decrease in effectiveness of a particular actuator. In most CA strategies in the literature, the control signal is distributed equally among all the actuators [222, 226, 269] or distributed based on the limits (position and rate) of the actuators [122]. In this chapter, two different CA strategies will be considered. First (as in Chap. 9), information about $K(t)$ will be incorporated into the allocation algorithm through a weighting matrix $W(t)$, so that the control is redistributed to the remaining actuators when faults/failures occur. The idea is that if an actuator fault occurs, the control input $u(t)$ is reallocated to minimise the use of the faulty control surface. The second strategy is based on a widely used CA approaches from the literature; i.e., fixed and equal distribution of the control signals. This is motivated by the fact that the information about $K(t)$ in (10.1) is not always available. Here, the CA is set to be fixed and the control signals are distributed equally to all actuators and is therefore independent of the fault information.

Again, as in Chap. 9, assume that the system states can be reordered, and the input distribution matrix B from (10.1) can be partitioned as

$$B = \begin{bmatrix} B_1 \\ B_2 \end{bmatrix} \tag{10.2}$$

where $B_1 \in \mathbb{R}^{(n-l)\times m}$ and $B_2 \in \mathbb{R}^{l\times m}$ has rank l. It will be assumed without loss of generality that the states of the system in (10.1) have been transformed so that $B_2 B_2^{\mathrm{T}} = I_l$ and therefore $\|B_2\| = 1$. As in Chap. 9, let the 'virtual control' be

$$\nu(t) := B_2 u(t) \tag{10.3}$$

so that

$$u(t) = B_2^{\dagger}\nu(t) \tag{10.4}$$

where the right pseudo-inverse is chosen as

$$B_2^{\dagger} := \Omega B_2^{\mathrm{T}}\big(B_2 \Omega B_2^{\mathrm{T}}\big)^{-1} \tag{10.5}$$

and $\Omega \in \mathbb{R}^{m\times m}$ is a symmetric positive definite diagonal weighting matrix. In the subsequent sections different choices for Ω will be considered.

10.2 Online Control Allocation

In this section, the idea is to use the information about $K(t)$ and incorporate this into the allocation algorithm via a weighting matrix $W(t) = I - K(t)$. As in the previous chapter the idea is that the control is redistributed to the remaining actuators when faults/failures occur to minimise the use of the faulty control surface (see Fig. 10.1).

Equation (10.1) can be written as

$$\dot{x}(t) = Ax(t) + B(I - K)u(t) \tag{10.6}$$

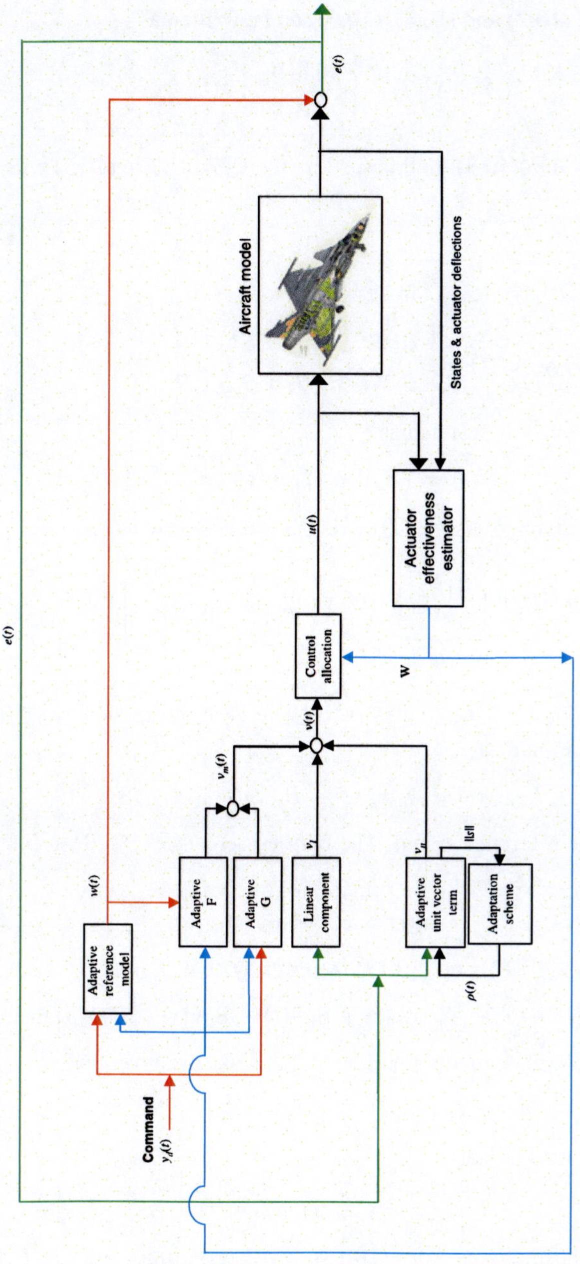

Fig. 10.1 Overall structure of proposed FTC

Define

$$W := I - K \tag{10.7}$$

then using (10.7), (10.2) and (10.4), (10.6) can be written as

$$\dot{x}(t) = Ax(t) + \begin{bmatrix} B_1 W B_2^{\dagger} \\ B_2 W B_2^{\dagger} \end{bmatrix} \nu(t) \tag{10.8}$$

where $B_2^{\dagger}$ is defined in (10.5). In this section, the weight Ω in (10.5) will be chosen as

$$\Omega = W \tag{10.9}$$

Using (10.9) and (10.5), (10.8) can be written as

$$\dot{x}(t) = Ax(t) + \begin{bmatrix} B_1 W^2 B_2^{\mathrm{T}} (B_2 W B_2^{\mathrm{T}})^{-1} \\ B_2 W^2 B_2^{\mathrm{T}} (B_2 W B_2^{\mathrm{T}})^{-1} \end{bmatrix} \nu(t) \tag{10.10}$$

If a new virtual control is selected as

$$\bar{\nu}(t) := \left(B_2 W B_2^{\mathrm{T}}\right)^{-1} \nu(t) \tag{10.11}$$

then as shown in Chap. 9 (see (9.17)), (10.10) can be written as

$$\dot{x}(t) = Ax(t) + \underbrace{\begin{bmatrix} B_1 B_2^{\mathrm{T}} \\ I \end{bmatrix}}_{B_\nu} \bar{\nu}(t) - \underbrace{\begin{bmatrix} B_1 (I - W^2) B_2^{\mathrm{T}} \\ B_2 (I - W^2) B_2^{\mathrm{T}} \end{bmatrix}}_{\bar{B}_\nu} \bar{\nu}(t) \tag{10.12}$$

In the fault-free case $W = I$ and $\bar{B}_\nu$ in (10.12) is zero. As in Sect. 3.5.2, consider a reference model defined as

$$\dot{x}_m(t) = A_m x_m(t) + B_m y_d(t) \tag{10.13}$$

where $y_d(t)$ is the reference signal. The matrices $A_m \in \mathbb{R}^{n \times n}$, $B_m \in \mathbb{R}^{n \times l}$ and A_m is stable. Define

$$e(t) = x(t) - x_m(t) \tag{10.14}$$

and therefore from (10.12) and (10.13) the error system

$$\dot{e}(t) = Ae(t) + (A - A_m)x_m(t) + B_\nu \bar{\nu}(t) - \bar{B}_\nu \bar{\nu}(t) - B_m y_d(t) \tag{10.15}$$

Suppose the reference model matrices A_m and B_m are given by

$$A_m = A + B_\nu F, \qquad B_m = B_\nu G \tag{10.16}$$

Define a feed-forward signal

$$\nu_m(t) = F x_m(t) + G y_d(t) \tag{10.17}$$

associated with the matrices from (10.16). SMC techniques, will now be used to synthesise $\bar{\nu}(t)$ from (10.12). Define a switching function $s : \mathbb{R}^n \to \mathbb{R}^l$ to be

$$s(t) = Se(t) \tag{10.18}$$

where $S \in \mathbb{R}^{l\times n}$ and $\det(SB_\nu) \neq 0$. Let $\mathcal{S}$ be the hyperplane defined by

$$\mathcal{S} = \{e(t) \in \mathbb{R}^n : Se(t) = 0\}$$

If a control law can be developed which forces the closed-loop trajectories onto the surface $\mathcal{S}$ in finite time and constrains the states to remain there, then an ideal sliding motion is said to have been attained. The sliding surface will be designed based on the nominal no fault condition ($K = 0$). Using (10.16), (10.15) can be rewritten as

$$\dot{e}(t) = Ae(t) - \bar{B}_\nu \bar{\nu}(t) + B_\nu\big(\bar{\nu}(t) \underbrace{-Fx_m(t) - Gy_d(t)}_{-\nu_m(t)}\big) \tag{10.19}$$

where B_ν and $\bar{B}_\nu$ are defined in (10.12). As in Chap. 9, a coordinate transformation $e \mapsto T_r e(t) = \hat{e}(t)$ is introduced to obtain 'regular form'. If

$$T_r := \begin{bmatrix} I & -B_1 B_2^{\mathrm{T}} \\ 0 & I \end{bmatrix} \tag{10.20}$$

then it is easy to check that (10.19) becomes

$$\dot{\hat{e}}(t) = \hat{A}\hat{e}(t) + \underbrace{\begin{bmatrix} 0 \\ I \end{bmatrix}}_{\hat{B}_\nu}\big(\bar{\nu}(t) - \nu_m(t)\big) - \begin{bmatrix} B_1 B_2^N (I - W^2) B_2^{\mathrm{T}} \\ B_2 (I - W^2) B_2^{\mathrm{T}} \end{bmatrix} \bar{\nu}(t) \tag{10.21}$$

where $\hat{A} := T_r A T_r^{-1}$ and

$$B_2^N := \big(I - B_2^{\mathrm{T}} B_2\big) \tag{10.22}$$

Because $B_2 B_2^{\mathrm{T}} = I_l$, it follows that $B_2^N B_2^{\mathrm{T}} = (I - B_2^{\mathrm{T}} B_2) B_2^{\mathrm{T}} = 0$, and

$$B_1 B_2^N \big(I - W^2\big) B_2^{\mathrm{T}} = -B_1 B_2^N W^2 B_2^{\mathrm{T}} \tag{10.23}$$

Define another scaling of the virtual control signal as

$$\hat{\nu}(t) := \big(B_2 W^2 B_2^{\mathrm{T}}\big) \bar{\nu}(t) \tag{10.24}$$

Then using similar arguments to those in Chap. 9, (10.21) becomes

$$\begin{bmatrix} \dot{\hat{e}}_1(t) \\ \dot{\hat{e}}_2(t) \end{bmatrix} = \begin{bmatrix} \hat{A}_{11} & \hat{A}_{12} \\ \hat{A}_{21} & \hat{A}_{22} \end{bmatrix} \begin{bmatrix} \hat{e}_1(t) \\ \hat{e}_2(t) \end{bmatrix} + \begin{bmatrix} 0 \\ I \end{bmatrix} \big(\hat{\nu}(t) - \nu_m(t)\big) + \begin{bmatrix} B_1 B_2^N B_2^+ \\ 0 \end{bmatrix} \hat{\nu}(t) \tag{10.25}$$

where

$$B_2^+ := W^2 B_2^{\mathrm{T}} \big(B_2 W^2 B_2^{\mathrm{T}}\big)^{-1} \tag{10.26}$$

As shown in Proposition 9.1 in Chap. 9, there exists a scalar γ_0 which is finite and independent of W such that

$$\|B_2^+\| = \big\|W^2 B_2^{\mathrm{T}} \big(B_2 W^2 B_2^{\mathrm{T}}\big)^{-1}\big\| < \gamma_0 \tag{10.27}$$

for all $W = \mathrm{diag}(w_1, \ldots, w_m)$ such that $0 < w_i \leq 1$.

The virtual control law will now be designed based on the nominal fault-free system in which the last term in (10.25) is zero since $B_1 B_2^N B_2^+|_{W=I} = 0$. In the $\hat{e}(t)$ coordinates, a suitable choice for the sliding surface matrix is

$$\hat{S} = ST_r^{-1} = [\mathcal{K} \quad I] \tag{10.28}$$

where $\mathcal{K} \in \mathbb{R}^{l \times (n-l)}$ represents design freedom. Introduce another transformation $(\hat{e}_1, \hat{e}_2) \mapsto (\hat{e}_1, s)$, associated with

$$T_s = \begin{bmatrix} I & 0 \\ \mathcal{K} & I \end{bmatrix} \tag{10.29}$$

Then (10.25) becomes

$$\begin{bmatrix} \dot{\hat{e}}_1(t) \\ \dot{s}(t) \end{bmatrix} = \begin{bmatrix} \tilde{A}_{11} & \tilde{A}_{12} \\ \tilde{A}_{21} & \tilde{A}_{22} \end{bmatrix} \begin{bmatrix} \hat{e}_1(t) \\ s(t) \end{bmatrix} + \begin{bmatrix} 0 \\ I \end{bmatrix} \big(\hat{\nu}(t) - \nu_m(t)\big) + \begin{bmatrix} B_1 B_2^N B_2^+ \\ \mathcal{K} B_1 B_2^N B_2^+ \end{bmatrix} \hat{\nu}(t) \tag{10.30}$$

where

$$\tilde{A}_{11} := \hat{A}_{11} - \hat{A}_{12}\mathcal{K} \tag{10.31}$$

and

$$\tilde{A}_{21} := \mathcal{K}\tilde{A}_{11} + \hat{A}_{21} - \hat{A}_{22}\mathcal{K} \tag{10.32}$$

If a control law can be designed to induce a sliding motion, then during sliding $\dot{s}(t) = s(t) = 0$ and the equivalent control necessary to maintain sliding is obtained from solving for $\hat{\nu}_{\text{eq}}(t)$ from the lower equations of (10.30) to give

$$\hat{\nu}_{\text{eq}}(t) = -\big(I + \mathcal{K} B_1 B_2^N B_2^+\big)^{-1} \big(\tilde{A}_{21}\hat{e}_1(t) - \nu_m(t)\big) \tag{10.33}$$

where B_2^N is defined in (10.22).

Assume the sliding surface matrix $\mathcal{K}$ has been designed, so that $\tilde{A}_{11} := \hat{A}_{11} - \hat{A}_{12}\mathcal{K}$ is stable and $\|\mathcal{K} B_1 B_2^N B_2^+\| < 1$ for all $0 < W \leq I$.

Note: $\|\mathcal{K} B_1 B_2^N B_2^+\| < 1$ guarantees the inverse in (10.33) exists and uses the boundedness result from Proposition 9.1 in Chap. 9. If (A, B_ν) is controllable, then $(\hat{A}_{11}, \hat{A}_{12})$ is controllable and so $\mathcal{K}$ can be chosen to make $\hat{A}_{11} - \hat{A}_{12}\mathcal{K}$ stable. Substituting (10.33) into the top partition of (10.30) yields the following reduced order system which governs the sliding motion:

$$\begin{aligned} \dot{\hat{e}}_1(t) &= \big(\tilde{A}_{11} - B_1 B_2^N B_2^+ \big(I + \mathcal{K} B_1 B_2^N B_2^+\big)^{-1} \tilde{A}_{21}\big)\hat{e}_1(t) \\ &\quad + B_1 B_2^N B_2^+ \big(I + \mathcal{K} B_1 B_2^N B_2^+\big)^{-1} \nu_m(t) \end{aligned} \tag{10.34}$$

As shown in (9.33), when $W = I$ (fault-free situation), $B_2^+|_{W=I} = B_2^{\mathrm{T}}$ and the system in (10.34) 'collapses' to $\dot{\hat{e}}_1(t) = \tilde{A}_{11}\hat{e}_1(t)$ which is the nominal sliding mode reduced order system for which $\mathcal{K}$ has been designed to guarantee stability. However, during fault/failure conditions, stability of the system in (10.34) (which depends on W through B_2^+) needs to be established.

10.2.1 Stability Analysis

The stability of the sliding mode is dependent on the reduced order system (10.34). Since by construction, the reference model is stable, for a bounded signal $y_d(t)$, the signal $x_m(t)$ is bounded and hence ν_m is bounded. Therefore, the stability of the reduced order system which governs the sliding motion depends on

$$\dot{\hat{e}}_1(t) = \big(\tilde{A}_{11} - B_1 B_2^N B_2^+ \big(I + \mathcal{K} B_1 B_2^N B_2^+\big)^{-1} \tilde{A}_{21}\big)\hat{e}_1(t) \tag{10.35}$$

Define

$$\tilde{G}(\mathrm{s}) := -\tilde{A}_{21}(\mathrm{s}I - \tilde{A}_{11})^{-1} B_1 B_2^N \tag{10.36}$$

where s represents the Laplace variable. By construction $\tilde{G}(\mathrm{s})$ is stable. Define scalars γ_1 and γ_2 according to

$$\gamma_2 = \big\|\tilde{G}(\mathrm{s})\big\|_\infty \tag{10.37}$$

and

$$\gamma_1 := \big\|\mathcal{K} B_1 B_2^N\big\| \tag{10.38}$$

As proven in Proposition 9.2 in Chap. 9, during a fault or failure condition, for any combination of $0 < w_i \le 1$, the closed-loop system will be stable if

$$0 \le \frac{\gamma_2 \gamma_0}{1 - \gamma_1 \gamma_0} < 1 \tag{10.39}$$

where the positive scalar γ_0 is defined in (10.27).

10.2.2 A Sliding Mode Control Law

Next, a sliding mode controller will be designed based on the system in (10.30) with respect to the virtual control $\hat{\nu}$. The proposed control law is given by $\hat{\nu}(t) = \hat{\nu}_l(t) + \hat{\nu}_n(t)$ where

$$\hat{\nu}_l(t) := -\tilde{A}_{21}\hat{e}_1(t) - \tilde{A}_{22}s(t) + \nu_m(t) \tag{10.40}$$

and $\nu_m(t)$ is defined in (10.17). The nonlinear component is defined to be

$$\hat{\nu}_n(t) := -\big(\rho(t) + \eta\big)\frac{s(t)}{\|s(t)\|} \quad \text{for } s(t) \ne 0 \tag{10.41}$$

where $s(t) = \hat{S}\hat{x}(t)$ and η is a positive scalar. The modulation function $\rho(t)$ represents an adaptive term which will be defined explicitly later in the section.

It follows that the actual control signal which is sent to the actuators is resolved from the 'virtual control law' $\nu(t)$ (from (10.40)–(10.41)), using (10.4), (10.5), (10.11) and (10.24). Therefore $u(t)$ is defined as

$$u(t) = W B_2^{\mathrm{T}} \big(B_2 W^2 B_2^{\mathrm{T}}\big)^{-1} \hat{\nu}(t) \tag{10.42}$$

In this chapter, provided the stability condition (10.39) is satisfied, the sliding mode controller for the 'virtual' system proposed above, can handle total actuator failures in the original system provided that $\det(B_2 W B_2^{\mathrm{T}}) \neq 0$.

In a fault-free situation it is not necessary and indeed is not advisable to have a large gain on the switched term—therefore, ideally the term $\rho(t)$ should only adapt to the onset of a fault and react accordingly. It is easy to see from (10.40) that, if $y_d(t)$ is bounded, $\hat{\nu}_l(t)$ is bounded by

$$\left\|\hat{\nu}_l(t)\right\| < l_1 \left\|e(t)\right\| + l_2 \tag{10.43}$$

where l_1 and l_2 are known positive constants. The adaptive gain from (10.41) is defined to be

$$\rho(t) = r(t)\big(l_1 \left\|e(t)\right\| + l_2\big) \tag{10.44}$$

The scalar variable $r(t)$ varies according to

$$\dot{r}(t) = a\big(l_1 \left\|e(t)\right\| + l_2\big) D_\epsilon\big(\left\|s(t)\right\|\big) - br(t) \tag{10.45}$$

where $r(0) = 0$ and the a and b are positive design constants. The function $D_\epsilon : \mathbb{R} \mapsto \mathbb{R}$ is the nonlinear function

$$D_\epsilon\big(\|s\|\big) = \begin{cases} 0 & \text{if } \|s\| < \epsilon \\ \|s\| & \text{otherwise} \end{cases} \tag{10.46}$$

where ϵ is a positive scalar. Here, ϵ is set to be small and helps define a boundary layer about the surface $\mathcal{S}$, inside which an acceptably close approximation to ideal sliding takes place. Provided the states evolve with time inside the boundary layer, no adaptation of the switching gains takes place. If a fault occurs, which starts to make the sliding motion degrade so that the states evolve outside the boundary layer i.e., $\|s(t)\| > \epsilon$, then the dynamic coefficients $r(t)$ increase in magnitude, (according to (10.45)), to force the states back into the boundary layer around the sliding surface.

As in Sect. 8.2.1, the choice of the design parameters η, a, b and ϵ depends on the closed-loop performance specifications and requires some design iteration. The choice of these design parameters will be discussed further in Sect. 10.5. Analogously to Proposition 8.1 in Chap. 8, the following proposition will show that $r(t)$ is bounded and motion inside a boundary layer around $\mathcal{S}$ is obtained.

Proposition 10.1 *Consider the potentially faulty error system represented by* (10.15) *with the control law in* (10.40)–(10.41); *then the adaptive gain $r(t)$ remains bounded and the switching states $s(t)$ enter a boundary layer around $\mathcal{S}$ in finite time.*

Proof Define a scalar

$$\zeta := 1/(1 - \gamma_1\gamma_0) > 0 \tag{10.47}$$

This is guaranteed to exist, since in the requirements of (10.27), the inequality $\gamma_1\gamma_0 < 1$ must hold. Consider as a candidate Lyapunov function

$$V = \frac{1}{2}\left(\|s\|^2 + \frac{1}{a}(1 - \gamma_1\gamma_0)\big(r(t) - \zeta\big)^2\right) \tag{10.48}$$

where a is the positive scalar from (10.45). Clearly $V(\cdot)$ is positive definite with respect to $\|s\|$, the adaptive gain error $r(t)-\zeta$, and is radially unbounded. Taking derivatives along trajectories

$$\dot{V} = s^{\mathrm{T}}\dot{s} + \frac{1}{a}(1-\gamma_1\gamma_0)\big(r(t)-\zeta\big)\dot{r}(t) \tag{10.49}$$

From (10.30) and using (10.40)

$$\begin{aligned}\dot{s}(t) &= \tilde{A}_{21}\hat{e}_1(t) + \tilde{A}_{22}s(t) + \big(\hat{\nu}(t) - Fx_m(t) - Gr(t)\big) + \big(\mathcal{K}B_1B_2^N B_2^+\big)\hat{\nu}(t)\\ &= \big(I + \mathcal{K}B_1B_2^N B_2^+\big)\hat{\nu}_n(t) + \big(\mathcal{K}B_1B_2^N B_2^+\big)\hat{\nu}_l(t)\end{aligned} \tag{10.50}$$

and so

$$\begin{aligned}s^{\mathrm{T}}\dot{s} &= s^{\mathrm{T}}\big(\mathcal{K}B_1B_2^N B_2^+\big)\hat{\nu}_n(t) + s^{\mathrm{T}}\big(\mathcal{K}B_1B_2^N B_2^+\big)\hat{\nu}_l(t) - \big(\rho(t)+\eta\big)\|s\|\\ &\le \|s\|\big(\rho(t)+\eta\big)\big\|\mathcal{K}B_1B_2^N B_2^+\big\| + \|s\|\big\|\mathcal{K}B_1B_2^N B_2^+\big\|\big\|\hat{\nu}_l(t)\big\| - \|s\|\big(\rho(t)+\eta\big)\\ &\le \|s\|\Big(\big(\rho(t)+\eta\big)\gamma_1\gamma_0 + \gamma_1\gamma_0\big\|\hat{\nu}_l(t)\big\| - \big(\rho(t)+\eta\big)\Big)\end{aligned} \tag{10.51}$$

Using the fact that $\gamma_1\gamma_0 = 1-(1-\gamma_1\gamma_0)$, the inequality above can be written as follows:

$$s^{\mathrm{T}}\dot{s} \le -\|s\|(1-\gamma_1\gamma_0)\big(\rho(t)+\eta\big) - \|s\|(1-\gamma_1\gamma_0)\big\|\hat{\nu}_l(t)\big\| + \|s\|\big\|\hat{\nu}_l(t)\big\| \tag{10.52}$$

In turn, using (10.47), the inequality above can be written as

$$s^{\mathrm{T}}\dot{s} \le -\|s\|(1-\gamma_1\gamma_0)\big(\eta + \big\|\hat{\nu}_l(t)\big\|\big) - \|s\|(1-\gamma_1\gamma_0)\big(\rho - \big\|\hat{\nu}_l(t)\big\|\zeta\big) \tag{10.53}$$

since $(1-\gamma_1\gamma_0)\zeta = 1$. Therefore, using (10.43)

$$s^{\mathrm{T}}\dot{s} \le -\|s\|(1-\gamma_1\gamma_0)\big(\eta + \big\|\hat{\nu}_l(t)\big\|\big) - \|s\|(1-\gamma_1\gamma_0)\big(\rho - \big(l_1\|e(t)\| + l_2\big)\zeta\big) \tag{10.54}$$

Substituting from (10.44) into the above yields

$$s^{\mathrm{T}}\dot{s} \le -\|s\|(1-\gamma_1\gamma_0)\big(\eta + \big\|\hat{\nu}_l(t)\big\|\big) - \|s\|(1-\gamma_1\gamma_0)\big(l_1\big\|e(t)\big\| + l_2\big)\big(r(t)-\zeta\big) \tag{10.55}$$

Finally, substituting (10.45) and (10.55) into (10.49) yields

$$\begin{aligned}\dot{V} &\le -\|s\|(1-\gamma_1\gamma_0)\big(\eta + \big\|\nu_l(t)\big\|\big) - \|s\|(1-\gamma_1\gamma_0)\big(l_1\big\|e(t)\big\| + l_2\big)\big(r(t)-\zeta\big)\\ &\quad + \frac{1}{a}(1-\gamma_1\gamma_0)\big(r(t)-\zeta\big)a\big(l_1\big\|e(t)\big\| + l_2\big)D_\epsilon\big(\big\|s(t)\big\|\big)\\ &\quad - \frac{1}{a}(1-\gamma_1\gamma_0)\big(r(t)-\zeta\big)br(t)\end{aligned} \tag{10.56}$$

If $\|s\| > \epsilon$ then $D_\epsilon(\|s\|) = \|s\|$ and so substituting in (10.56) and simplifying terms yields

$$\dot{V} \le -\|s\|(1-\gamma_1\gamma_0)\big(\eta + \big\|\nu_l(t)\big\|\big) - \frac{b}{a}(1-\gamma_1\gamma_0)\big(r(t)-\zeta\big)r(t) \tag{10.57}$$

Notice by construction $0 \le \gamma_1\gamma_0 < 1$ and $r(t) \ge 0$. Further manipulation of (10.57) and using (10.47) yields

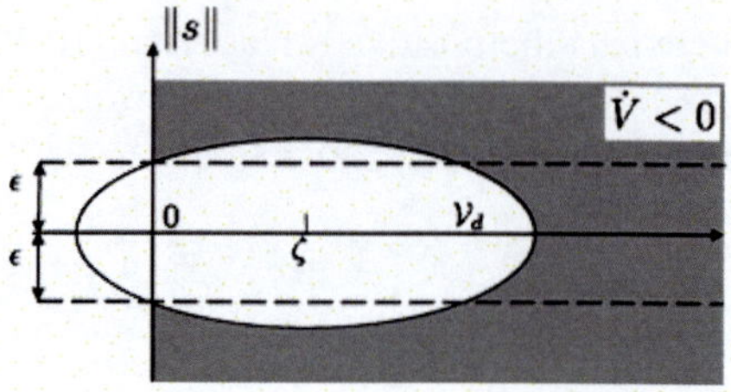

Fig. 10.2 Level set of the Lyapunov functions V

$$\dot{V} \le -\|s\|(1-\gamma_1\gamma_0)\big(\eta + \|\nu_l(t)\|\big) - \frac{b}{a}(1-\gamma_1\gamma_0)\left(\frac{1}{2}\zeta - r(t)\right)^2 + \frac{b}{4a(1-\gamma_1\gamma_0)} \tag{10.58}$$

since expanding the quadratic term on the right-hand side of (10.58) gives the right-hand side of (10.57). If $\|s\| > \epsilon$, then $\|s\|(1-\gamma_1\gamma_0)\eta \ge (1-\gamma_1\gamma_0)\epsilon\eta$. The quantities ϵ, η, a and b are design parameters and so if they are chosen to satisfy

$$\epsilon\eta \ge \frac{b}{4a(1-\gamma_1\gamma_0)^2} \tag{10.59}$$

then

$$\dot{V} \le -\|s\|(1-\gamma_1\gamma_0)\|\nu_l(t)\| - \frac{b}{a}(1-\gamma_1\gamma_0)\left(\frac{1}{2}\zeta - r(t)\right)^2 \le 0$$

If $\|s\| < \epsilon$ then $D_\epsilon(\|s\|) = 0$ and so substituting in (10.56) and simplifying terms yields

$$\dot{V} \le -\|s\|(1-\gamma_1\gamma_0)\big(\eta + \|\nu_l(t)\|\big) - \|s\|(1-\gamma_1\gamma_0)\big(l_1\|e(t)\| + l_2\big)\big(r(t) - \zeta\big) - \frac{b}{a}(1-\gamma_1\gamma_0)\big(r(t) - \zeta\big)r(t) \tag{10.60}$$

Notice again by construction $\gamma_1\gamma_0 < 1$ and $r(t) \ge 0$ and therefore for $\|s\| < \epsilon$ and $r(t) > \zeta$, it follows $\dot{V} < 0$. Define a rectangle in $\mathbb{R}^2$ as

$$\mathcal{R} = \big\{\big(\|s\|, r\big) \mid \|s\| \le \epsilon,\ 0 \le r \le \zeta\big\} \tag{10.61}$$

Also define $\mathcal{R}_+ \in \mathbb{R}^2$ as $\mathcal{R}_+ = \{(\|s\|, r) \mid r \ge 0\}$. By construction of the adaptive gains, $r(t) \ge 0$ for all time and so the trajectory of $(\|s(t)\|, r(t)) \in \mathcal{R}_+$ for all time, and so outside the set $\mathcal{R} \cap \mathcal{R}_+ = \mathcal{R}$, from (10.58) and (10.60), the derivative of the Lyapunov function $\dot{V} < 0$. Let $\mathcal{V}_d$ denote the truncated ellipsoid

$$\mathcal{V}_d = \big\{\big(\|s\|, r\big) \mid V\big(\|s\|, r\big) \le d\big\} \cap \mathcal{R}_+$$

where $V(\cdot)$ is defined in (10.48). Because $\mathcal{R}$ in (10.61) is a compact set, there exists a unique $d_0 > 0$ such that $d_0 = \min\{d \in \mathbb{R}_+ \mid \mathcal{R} \subset \mathcal{V}_d\}$ and in fact $d_0 = \frac{1}{2}(\epsilon^2 + \frac{\zeta}{a})$. As shown in Fig. 10.2, since $\mathcal{R} \subset \mathcal{V}_{d_0}$, it follows outside $\mathcal{V}_{d_0}$ the derivative of the Lyapunov function $\dot{V} < 0$ and so $\mathcal{V}_{d_0}$ is an invariant set which is entered in finite time t_0. Since $\mathcal{V}_{d_0}$ is entered in finite time, $V(\|s\|, r) \le d_0$ for all $t > t_0$ which implies $\|s\| \le \sqrt{2d_0}$ for all time $t > t_0$, and hence s enters and remains in a boundary layer of size $\sqrt{2d_0}$ around the ideal sliding surface $\mathcal{S}$. □

From the arguments above, for an appropriate choice of a, b and ϵ, close approximation to ideal sliding can be maintained in the presence of faults.

Remark 10.1 As discussed in Sect. 8.2.1, if $\epsilon = 0$ and $b = 0$, then ideal sliding can be guaranteed since it follows from (10.57) that

$$\dot{V}(s) \leq -\|s\|(1-\gamma_1\gamma_0)\big(\eta + \|\nu_l(t)\|\big)$$

and therefore ideal sliding can be attained and maintained in finite time. However this scheme is not practical since $r(t)$ may become unbounded in the presence of noise.

10.3 Fixed Control Allocation

The analysis in this section is similar to that in Sect. 10.2 above. The difference here is the assumption that there is no FDI or actuator effectiveness estimation available. Here the CA will be fixed, and $\Omega = I$ will be used in (10.5) instead of $\Omega = W = I - K$ as proposed in Sect. 10.2.

The effect of choosing the weighting matrix to be $\Omega(t) = W(t)$ as in Sect. 10.2 is that $u(t)$ in (10.4) depends explicitly on $K(t)$. Here instead, and perhaps more conventionally,

$$\Omega := I \tag{10.62}$$

will be considered. With this choice of weighting matrix, (10.4) becomes

$$u(t) = B_2^{\dagger}\nu(t) = B_2^{\mathrm{T}}\underbrace{\big(B_2B_2^{\mathrm{T}}\big)^{-1}}_{I}\nu(t) = B_2^{\mathrm{T}}\nu(t) \tag{10.63}$$

then it can be shown that (10.1) can be written as

$$\dot{x}(t) = Ax(t) + \underbrace{\begin{bmatrix} B_1B_2^{\mathrm{T}} \\ I \end{bmatrix}}_{B_\nu}\nu(t) - \underbrace{\begin{bmatrix} B_1KB_2^{\mathrm{T}} \\ B_2KB_2^{\mathrm{T}} \end{bmatrix}}_{\bar{B}_k}\nu(t) \tag{10.64}$$

Note that the last term in (10.64) is different from the last term in (10.12). Consider a reference model as defined in (10.13), then using the definition of the error signal in (10.14), from (10.64) and (10.13) the error system

$$\dot{e}(t) = Ae(t) + (A - A_m)x_m(t) + B_\nu\nu(t) - \bar{B}_k\nu(t) - B_m y_d(t) \tag{10.65}$$

where $e(t)$ is defined in (10.14), B_ν and $\bar{B}_k$ are defined in (10.64), and the reference model matrices A_m and B_m are given in (10.16). Define

$$\nu(t) = \nu_l(t) + \nu_n(t) \tag{10.66}$$

then using (10.16) and (10.17), (10.65) can be rewritten as

$$\dot{e}(t) = Ae(t) - \bar{B}_K \nu(t) + B_\nu \big(\underbrace{\nu(t) - F x_m(t) - G y_d(t)}_{-\nu_m(t)}\big) \tag{10.67}$$

where ν_m is defined in (10.17). As in Sect. 10.2, a coordinate transformation $e \mapsto T_r e(t) = \hat{e}(t)$ is introduced to obtain 'regular form', where T_r is defined in (10.20). By construction the matrix $B_2 B_2^{\mathrm{T}} = I_l$, and it is easy to check that (10.67) becomes

$$\dot{\hat{e}}(t) = \hat{A}\hat{e}(t) + \underbrace{\begin{bmatrix} 0 \\ I \end{bmatrix}}_{\hat{B}_\nu} \big(\nu(t) - \nu_m(t)\big) - \begin{bmatrix} -B_1 B_2^N (I-K) B_2^{\mathrm{T}} \\ I - B_2(I-K) B_2^{\mathrm{T}} \end{bmatrix} \nu(t) \tag{10.68}$$

where $\hat{A} := T_r A T_r^{-1}$ and B_2^N is defined in (10.22). The fact that $B_2^N B_2^{\mathrm{T}} = (I - B_2^{\mathrm{T}} B_2) B_2^{\mathrm{T}} = 0$ has also been used to obtain the top partition of the last term in (10.68). The last term in (10.68) is zero in the fault-free case ($K = 0$), but is treated as (unmatched) uncertainty when $K \neq 0$. Define

$$B_2^\ddagger := W B_2^{\mathrm{T}} \big(B_2 W B_2^{\mathrm{T}}\big)^{-1} \tag{10.69}$$

where W is defined in (10.7). It is shown in Sect. 10.2 there is an upper bound on the norm of the pseudo-inverse $B_2^\ddagger$ in (10.69) which is independent of W (as proven in Proposition 9.1 in Chap. 9); therefore

$$\|B_2^\ddagger\| = \big\|W B_2^{\mathrm{T}} \big(B_2 W B_2^{\mathrm{T}}\big)^{-1}\big\| = \big\|W^2 B_2^{\mathrm{T}} \big(B_2 W^2 B_2^{\mathrm{T}}\big)^{-1}\big\| < \gamma_0 \tag{10.70}$$

for all $W = \operatorname{diag}(w_1, \ldots, w_m)$ such that $0 < w_i \le 1$.

In the $\hat{e}(t)$ coordinates, a suitable choice for the sliding surface matrix is given by (10.28). Introduce another transformation $(\hat{e}_1, \hat{e}_2) \mapsto (\hat{e}_1, s)$, where T_s is defined in (10.29). Therefore (10.68) becomes

$$\begin{bmatrix} \dot{\hat{e}}_1(t) \\ \dot{s}(t) \end{bmatrix} = \begin{bmatrix} \tilde{A}_{11} & \tilde{A}_{12} \\ \tilde{A}_{21} & \tilde{A}_{22} \end{bmatrix} \begin{bmatrix} \hat{e}_1(t) \\ s(t) \end{bmatrix} + \begin{bmatrix} 0 \\ I \end{bmatrix} \big(\nu(t) - \nu_m(t)\big) - \begin{bmatrix} -B_1 B_2^N W B_2^{\mathrm{T}} \\ I - \mathcal{K} B_1 B_2^N W B_2^{\mathrm{T}} - B_2 W B_2^{\mathrm{T}} \end{bmatrix} \nu(t) \tag{10.71}$$

where $\tilde{A}_{11}$ and $\tilde{A}_{21}$ are defined in (10.31) and (10.32), respectively. Note that (10.71) has a similar structure to the one in (10.30) which uses the online CA in Sect. 10.2. One clear difference is the last term of both equations. As in Sect. 10.2, if (A, B_ν) is controllable, from (10.28), $\mathcal{K}$ can always be chosen to make $\tilde{A}_{11}$ stable. If a control law can be designed to induce a sliding motion, then during sliding $\dot{s}(t) = s(t) = 0$ and the equivalent control necessary to maintain sliding is obtained from solving for $\nu_{\mathrm{eq}}(t)$ from the lower equations of (10.71) to give

$$\nu_{\mathrm{eq}}(t) = \big(B_2 W B_2^{\mathrm{T}}\big)^{-1} \big(I + \mathcal{K} B_1 B_2^N B_2^\ddagger\big)^{-1} \big(-\tilde{A}_{21} \hat{e}_1(t) + \nu_m(t)\big) \tag{10.72}$$

where B_2^N is defined in (10.22) and $B_2^\ddagger$ is defined in (10.69). Substituting (10.72) into the top partition of (10.71), yields the reduced order system (which governs the sliding motion) given as

$$\dot{\hat{e}}_1(t) = \big(\tilde{A}_{11} - B_1 B_2^N B_2^{\ddagger}\big(I + \mathcal{K} B_1 B_2^N B_2^{\ddagger}\big)^{-1} \tilde{A}_{21}\big)\hat{e}_1(t) + B_1 B_2^N B_2^{\ddagger}\big(I + \mathcal{K} B_1 B_2^N B_2^{\ddagger}\big)^{-1} \nu_m(t) \tag{10.73}$$

Note that (10.73) has the same structure as the one in (10.34). The difference is the term $B_2^{\ddagger}$ is replaced by B_2^{+} in (10.34). Since γ_1 from (10.38) and γ_2 from (10.37) correspond to the same elements in the reduced order systems (10.34) and (10.73), and the fact that $\|B_2^{\ddagger}\| = \|B_2^{+}\| < \gamma_0$, the closed-loop system described in (10.73) will be stable if (10.39) is satisfied. Although the CA strategy proposed in this section is different from the one in Sect. 10.2, the stability analysis of the reduced order sliding motion is the same.

10.3.1 A Sliding Mode Control Law

In this section, a sliding mode controller will be designed based on the nominal no fault 'virtual' system in (10.71) with respect to ν. Here, the proposed control law is given by (10.66), where

$$\nu_l(t) := -\tilde{A}_{21}\hat{e}_1(t) - \tilde{A}_{22}s(t) + \nu_m(t) \tag{10.74}$$

and $\nu_m(t)$ is defined in (10.17). The nonlinear component is defined to be

$$\nu_n(t) := -\big(\rho(t) + \eta\big)\frac{s(t)}{\|s(t)\|} \quad \text{for } s(t) \neq 0 \tag{10.75}$$

where $s(t) = \hat{S}\hat{e}(t)$ and η is a positive scalar. The choice of the varying gain $\rho(t)$ will be discussed next.

It follows that the actual control which is sent to the actuators is resolved from the 'virtual control law' $\nu(t)$ defined in (10.3). Therefore $u(t)$, as defined in (10.63), is

$$u(t) = B_2^{\mathrm{T}}\nu(t)$$

and the control signal distribution is independent of the effectiveness gain K.

As argued in Sect. 10.2.2, $\nu_l(t)$ in (10.74) is bounded by

$$\big\|\nu_l(t)\big\| < l_1\big\|e(t)\big\| + l_2 \tag{10.76}$$

where l_1 and l_2 are known positive constants as in (10.43). The gain $\rho(t)$ from (10.75) is defined as in (10.44), where the scalar variable $r(t)$ varies according to (10.45) and (10.46).

Let $\mathcal{W}$ be the set of faults such that

$$\mathcal{W} = \big\{(w_1, \ldots, w_m) \in \underbrace{[0,1] \times [0,1] \times \cdots \times [0,1]}_{m \text{ times}} \mid \underline{\lambda}\big(B_2 W B_2^{\mathrm{T}}\big) := w > 0\big\} \tag{10.77}$$

where w is a strictly positive scalar and $\underline{\lambda}(B_2 W B_2^{\mathrm{T}})$ is the smallest eigenvalue of $(B_2 W B_2^{\mathrm{T}})$. Notice that $(w_1, \ldots, w_m) \in \mathcal{W} \Rightarrow \det(B_2 W B_2^{\mathrm{T}}) \neq 0$.

Using a similar analysis to that in Proposition 8.1, and in Sect. 10.2.2, the following lemma will show that $r(t)$ is bounded.

Proposition 10.2 *Consider the faulty system represented by* (10.1) *with the control law in* (10.74)–(10.75); *then the adaptive gain* $r(t)$ *from* (10.44)–(10.46) *remains bounded, and the switching states* $s(t)$ *enter a boundary layer around* $\mathcal{S}$ *in finite time for any fault condition* $(w_1, \dots, w_m) \in \mathcal{W}$.

Proof Define a scalar

$$\varsigma := \frac{(2+\gamma_1)}{w^2(1-\gamma_1\gamma_0)} \tag{10.78}$$

The expression for ς in (10.78) is guaranteed to be positive, since in the requirements of (10.39), the inequality $\gamma_1\gamma_0 < 1$ must hold. Assuming that $\dot{K}(t) = 0$ almost always, this implies $\dot{W}(t) = 0$ almost always and so only isolated abrupt step changes in the effectiveness are considered here. Using the fact that $(B_2 W B_2^{\mathrm{T}}) > 0$, the candidate Lyapunov function

$$V = \frac{1}{2}\left(s^{\mathrm{T}}(B_2 W B_2^{\mathrm{T}})s + \frac{1}{a}\underline{\lambda}(B_2 W B_2^{\mathrm{T}})^2(1-\gamma_1\gamma_0)\big(r(t)-\varsigma\big)^2\right) \tag{10.79}$$

where a is the positive scalar from (10.45), is positive definite with respect to s, the adaptive gain error $r(t) - \varsigma$, and is radially unbounded. Taking derivatives along trajectories

$$\dot{V} = s^{\mathrm{T}}(B_2 W B_2^{\mathrm{T}})\dot{s} + \frac{1}{a}\underline{\lambda}(B_2 W B_2^{\mathrm{T}})^2(1-\gamma_1\gamma_0)\big(r(t)-\varsigma\big)\dot{r}(t) \tag{10.80}$$

where from (10.71) (and using (10.66) and (10.74)),

$$\begin{aligned}\dot{s}(t) &= \tilde{A}_{21}\hat{x}_1(t) + \tilde{A}_{22}s(t) + \big(\nu(t) - \nu_m(t)\big) - \big(I - \mathcal{K}B_1B_2^N W B_2^{\mathrm{T}} - B_2 W B_2^{\mathrm{T}}\big)\nu(t)\\ &= \big(I + \mathcal{K}B_1B_2^N B_2^{\ddagger}\big)\big(B_2 W B_2^{\mathrm{T}}\big)\nu_n(t) - \big(I - \mathcal{K}B_1B_2^N W B_2^{\mathrm{T}} - B_2 W B_2^{\mathrm{T}}\big)\nu_l(t)\end{aligned}$$

Using the fact that

$$s(t)^{\mathrm{T}}\big(B_2 W B_2^{\mathrm{T}}\big)\big(B_2 W B_2^{\mathrm{T}}\big)s(t) = \big\|B_2 W B_2^{\mathrm{T}} s\big\|^2$$

where $\|(B_2 W B_2^{\mathrm{T}})\| \le \|B_2 B_2^{\mathrm{T}}\| = 1$, and $\|W B_2^{\mathrm{T}}\| \le \|W\|\|B_2^{\mathrm{T}}\| \le 1$ for all $(w_1, \dots, w_m) \in \mathcal{W}$, it follows that when $s \neq 0$

$$\begin{aligned}s^{\mathrm{T}}\big(B_2 W B_2^{\mathrm{T}}\big)\dot{s} &= -\frac{(\rho+\eta)}{\|s\|}\big\|B_2 W B_2^{\mathrm{T}} s\big\|^2\\ &\quad - (\rho+\eta)s^{\mathrm{T}}\big(B_2 W B_2^{\mathrm{T}}\big)\big(\mathcal{K}B_1B_2^N B_2^{\ddagger}\big)\big(B_2 W B_2^{\mathrm{T}}\big)\frac{s}{\|s\|}\\ &\quad - s^{\mathrm{T}}\big(B_2 W B_2^{\mathrm{T}}\big)\big(I - \mathcal{K}B_1B_2^N W B_2^{\mathrm{T}} - B_2 W B_2^{\mathrm{T}}\big)\nu_l(t)\\ &\le -\frac{(\rho+\eta)}{\|s\|}\big\|B_2 W B_2^{\mathrm{T}} s\big\|^2 + \frac{(\rho+\eta)}{\|s\|}\big\|B_2 W B_2^{\mathrm{T}} s\big\|^2\big\|\big(\mathcal{K}B_1B_2^N B_2^{\ddagger}\big)\big\|\\ &\quad + \big\|B_2 W B_2^{\mathrm{T}} s\big\|\big\|\big(I - \mathcal{K}B_1B_2^N W B_2^{\mathrm{T}} - B_2 W B_2^{\mathrm{T}}\big)\big\|\big\|\nu_l(t)\big\|\\ &\le \big\|B_2 W B_2^{\mathrm{T}} s\big\|\bigg(-\frac{(\rho+\eta)}{\|s\|}\big\|B_2 W B_2^{\mathrm{T}} s\big\|(1-\gamma_1\gamma_0)\\ &\quad + (2+\gamma_1)\big\|\nu_l(t)\big\|\bigg)\end{aligned} \tag{10.81}$$

since $\|\mathcal{K}B_1B_2^N B_2^\ddagger\| \leq \|\mathcal{K}B_1B_2^N\|\|B_2^\ddagger\| \leq \gamma_0\gamma_1$, and

$$\|I - \mathcal{K}B_1B_2^N W B_2^{\mathrm{T}} - B_2 W B_2^{\mathrm{T}}\| \leq 1 + \|\mathcal{K}B_1B_2^N W B_2^{\mathrm{T}}\| + \|B_2 W B_2^{\mathrm{T}}\| \leq 2 + \gamma_1$$

From the Rayleigh principle, $-\|B_2 W B_2^{\mathrm{T}} s\|^2 \leq -\underline{\lambda}(B_2 W B_2^{\mathrm{T}})^2\|s\|^2 = -w^2\|s\|^2$, and using the fact that $\bar{\lambda}(B_2 W B_2^{\mathrm{T}}) = 1$, inequality (10.81) implies

$$\begin{aligned} s^{\mathrm{T}}(B_2 W B_2^{\mathrm{T}})\dot{s} &\leq -w^2\|s\|(\rho+\eta)(1-\gamma_1\gamma_0) + \|s\|(2+\gamma_1)\|\nu_l(t)\| \\ &= w^2\|s\|(1-\gamma_1\gamma_0)\big(-(\rho+\eta) + \varsigma\|\nu_l(t)\|\big) \end{aligned} \tag{10.82}$$

where ς is defined in (10.78). Using (10.76) and (10.44), the inequality above can be written as

$$\begin{aligned} s^{\mathrm{T}}(B_2 W B_2^{\mathrm{T}})\dot{s} \leq{} & -w^2\|s\|(1-\gamma_1\gamma_0)\eta \\ & - w^2\|s\|(1-\gamma_1\gamma_0)\big(l_1\|x(t)\| + l_2\big)\big(r(t) - \varsigma\big) \end{aligned} \tag{10.83}$$

Finally, substituting (10.45) and (10.83) into (10.80) yields

$$\begin{aligned} \dot{V} \leq{} & -w^2\|s\|(1-\gamma_1\gamma_0)\eta - w^2\|s\|(1-\gamma_1\gamma_0)\big(l_1\|x(t)\| + l_2\big)\big(r(t) - \varsigma\big) \\ & + w^2(1-\gamma_1\gamma_0)\big(r(t) - \varsigma\big)\big(l_1\|x(t)\| + l_2\big)D_\epsilon\big(\|s(t)\|\big) \\ & - \frac{b}{a}w^2(1-\gamma_1\gamma_0)\big(r(t) - \varsigma\big)r(t) \end{aligned} \tag{10.84}$$

If $\|s\| > \epsilon$ then $D_\epsilon(\|s\|) = \|s\|$ and so substituting in (10.84) and simplifying terms yields

$$\dot{V} \leq -w^2\|s\|(1-\gamma_1\gamma_0)\eta - \frac{b}{a}w^2(1-\gamma_1\gamma_0)\big(r(t) - \varsigma\big)r(t) \tag{10.85}$$

By construction $0 \leq \gamma_1\gamma_0 < 1$ and $r(t) \geq 0$. Further manipulation of (10.85), and using (10.78) yields

$$\dot{V} \leq -w^2\|s\|(1-\gamma_1\gamma_0)\eta - \frac{b}{a}w^2(1-\gamma_1\gamma_0)\left(\frac{1}{2}\varsigma - r\right)^2 + \frac{b}{4a}\frac{(2+\gamma_1)^2}{w^2(1-\gamma_1\gamma_0)} \tag{10.86}$$

since expanding the quadratic term on the right-hand side of (10.86) gives the right-hand side of (10.85). If $\|s\| > \epsilon$, then $w^2\|s\|(1-\gamma_1\gamma_0)\eta \geq w^2(1-\gamma_1\gamma_0)\eta\epsilon$. The quantities ϵ, η, a and b are design parameters and so if they are chosen to satisfy

$$\epsilon\eta \geq \frac{b}{4a}\frac{(2+\gamma_1)^2}{w^4(1-\gamma_1\gamma_0)^2} = \frac{b}{4a}\varsigma^2 \tag{10.87}$$

then from (10.86)

$$\dot{V} \leq -\frac{b}{a}w^2(1-\gamma_1\gamma_0)\left(\frac{1}{2}\varsigma - r\right)^2 \leq 0$$

If $\|s\| < \epsilon$ then $D_\epsilon(\|s\|) = 0$ and so substituting in (10.84) and simplifying terms yields

$$\dot{V} \le -w^2\|s\|(1-\gamma_1\gamma_0)\eta - w^2\|s\|(1-\gamma_1\gamma_0)\big(l_1\|x(t)\| + l_2\big)\big(r(t)-\varsigma\big) \\ - \frac{b}{a}w^2(1-\gamma_1\gamma_0)\big(r(t)-\varsigma\big)r(t) \tag{10.88}$$

Notice by construction $\gamma_1\gamma_0 < 1$ and $r(t) \ge 0$ and therefore for $\|s\| < \epsilon$ and $r(t) > \varsigma$, it follows $\dot{V} < 0$. Define a rectangle in $\mathbb{R}^2$ as

$$\mathcal{R} = \big\{\big(\|s\|, r\big) \mid \|s\| \le \epsilon,\ 0 \le r \le \varsigma\big\} \tag{10.89}$$

Also define $\mathcal{R}_+ \in \mathbb{R}^2$ as $\mathcal{R}_+ = \{(\|s\|, r) \mid r \ge 0\}$. By construction of the adaptive gains, $r(t) \ge 0$ for all time and so the trajectory of $(\|s(t)\|, r(t)) \in \mathcal{R}_+$ for all time, and so outside the set $\mathcal{R} \cap \mathcal{R}_+ = \mathcal{R}$, from (10.86) and (10.88), the derivative of the Lyapunov function $\dot{V} < 0$. Let $\mathcal{V}_d$ denote the truncated ellipsoid

$$\mathcal{V}_d = \big\{\big(\|s\|, r\big) \mid V\big(\|s\|, r\big) \le d\big\} \cap \mathcal{R}_+$$

where $V(\cdot)$ is defined in (10.79). Because $\mathcal{R}$ in (10.89) is a compact set, for a given $w > 0$, there exists a unique $d_0 > 0$ such that $d_0 = \min\{d \in \mathbb{R}_+ \mid \mathcal{R} \subset \mathcal{V}_d\}$. As shown in Fig. 10.2, since $\mathcal{R} \subset \mathcal{V}_{d_0}$, it follows outside $\mathcal{V}_{d_0}$ the derivative of the Lyapunov function $\dot{V} < 0$ and so $\mathcal{V}_{d_0}$ is an invariant set which is entered in finite time t_0. Since $\mathcal{V}_{d_0}$ is entered in finite time, $V(\|s\|, r) \le d_0$ for all $t > t_0$ which implies $\|s\| \le \sqrt{2d_0/w}$ for $t > t_0$, and hence s enters and remains inside a boundary layer of size $\sqrt{2d_0/w}$ around the ideal sliding surface $\mathcal{S}$. □

Remark 10.2 Close approximation to ideal sliding can be maintained even in the presence of faults for an appropriate choice of a, b and ϵ. If $\epsilon = 0$ and $b = 0$, it follows from (10.85) that $\dot{V} \le -w^2\|s\|(1-\gamma_1\gamma_0)\eta$, which means that ideal sliding can be attained and maintained in finite time. However, in the presence of noise $r(t)$ may become unbounded.

10.4 Adaptive Reference Model

One benefit of using a model-reference framework is it allows the possibility of changing the reference model when a fault or failure occurs. This can be done using multiple pre-designed reference models or alternatively by adapting the reference model online. In some fault conditions, to reduce the demands on the damaged actuators, a 'slower' reference model is desirable. In this chapter, the reference model will be allowed to adapt to faults. Assume that two reference models have been designed, one will be the nominal reference model designed for a fault-free situation represented by (10.13), and the other will be the conservative reference model for the faulty situation. The idea is to blend the two models so that

$$A_m = \lambda(A + B_\nu F_1) + (1-\lambda)(A + B_\nu F_2) \tag{10.90}$$

$$B_m = \lambda B_\nu G_1 + (1-\lambda)B_\nu G_2 \tag{10.91}$$

where $\lambda \in [0\ 1]$ and is a function of the w_i from W in (10.7). The matrices F_1 and G_1 are associated with the nominal ideal model, while F_2 and G_2 are associated with the conservative one. In the fault-free case, the nominal reference model will be used; but when significant failures occur, the 'slow' model will be employed. When partial faults occur, a mix of both reference models will be taken.

10.5 ADMIRE Simulations: Online Control Allocation

10.5.1 Controller Design

The linear model from Sect. 6.4.2.1 has been used here and is associated with a low speed flight condition of Mach 0.22 at an altitude of 3000 m. The states are angle of attack (AoA) (rad), sideslip angle (rad), roll rate (rad/s), pitch rate (rad/s) and yaw rate (rad/s). The controlled outputs are AoA, sideslip and roll rate. The control surfaces are $\delta = [\delta_c\ \delta_{\mathrm{re}}\ \delta_{\mathrm{le}}\ \delta_r]^{\mathrm{T}}$, which represent the deflections (rad) of the canard, right elevon, left elevon and rudder, respectively.

The linearised (A, B) matrices are given in (6.135) and (6.136) where the partition of the B matrix shows the terms B_1 and B_2 (although a further change of coordinates is necessary to scale B_2 to ensure $B_2 B_2^{\mathrm{T}} = I$). The feed-back matrices for the ideal model from (10.16) have been designed using an LQR approach for the triple (A, B_ν, C_c) where

$$C_c = [\, I_3 \quad 0_{3\times 2}\,] \tag{10.92}$$

is the distribution matrix associated with the controlled outputs. The state weighting matrix has been chosen as $\mathbf{Q}_m = C_c \mathbf{Q}_c C_c$. For this particular problem

$$\mathbf{Q}_c = \mathrm{diag}(1, 5000, 0.1)$$

in the fault-free case, and the control penalty weight is $2I_3$. In the faulty case,

$$\mathbf{Q}_c = \mathrm{diag}(0.01, 50, 0.001)$$

and I_3 has been chosen as the control penalty. Alternatively, other methods such as eigenstructure assignment [90, 170] could also be used to design the ideal models. The feed-forward matrix G has been designed using the inverse steady-state gain for the virtual system (A, B_ν, C_c), specifically

$$G = -\big(C_c(A + B_\nu F)^{-1} B_\nu\big)^{-1}$$

Based on the above, the feed-back matrices from (10.90) and (10.91) are

$$F_1 = \begin{bmatrix} 0.0006 & -30.4857 & -1.1726 & 0.0003 & 4.8569 \\ -1.5365 & -0.0276 & -0.0006 & -0.8933 & 0.0022 \\ -0.0003 & 36.3096 & 1.0527 & -0.0002 & -4.9790 \end{bmatrix}$$

$$F_2 = \begin{bmatrix} -0.0003 & -3.8030 & -0.3749 & -0.0002 & 1.7039 \\ -1.3459 & -0.0072 & -0.0002 & -0.8304 & 0.0010 \\ 0.0006 & 3.4436 & 0.3142 & 0.0004 & -1.4496 \end{bmatrix}$$

and the feed-forward matrices are given by

$$G_1 = \begin{bmatrix} -0.0007 & 32.7235 & 0.2213 \\ 1.1753 & 0.0128 & 0.0001 \\ -0.0003 & -37.8584 & 0.1885 \end{bmatrix}$$

$$G_2 = \begin{bmatrix} 0.0005 & 5.7154 & 0.1294 \\ 0.9441 & -0.0069 & 0.0001 \\ -0.0015 & -4.6283 & 0.1371 \end{bmatrix}$$

A quadratic optimal design (similar to the one in Sect. 3.4.1) has been used to obtain the sliding surface matrix S. The symmetric positive definite weighting matrix has been chosen as $\mathbf{Q} = \text{diag}([30, 30, 1, 1, 1])$. This results in a hyperplane

$$\mathcal{K} = \begin{bmatrix} 0.0002 & 0.2906 \\ 1.9922 & -0.0019 \\ -0.0088 & -5.5060 \end{bmatrix}$$

In the simulations the discontinuity in the nonlinear control term in (10.41) has been smoothed by using a sigmoidal approximation from (3.46), where the scalar δ has been chosen as $\delta = 0.001$. This removes the discontinuity and introduces a further degree of tuning to accommodate the actuator limits.

During the design stage, and based on analysis from Sect. 10.2.2, it was found $\text{rank}(B_2 W B_2^{\mathrm{T}}) < 3$ when the rudder completely fails or any two surfaces from the set consisting of the canard and the left and right elevons completely fail. This is an expected result since there is no redundancy for the rudder to provide yaw; and when two actuators fail from either the canard or elevons, it means that there is no redundancy left in the system. From (10.27), it can be verified that $\gamma_0 = 2.0913$. Simple calculations show that $\gamma_1 = 0.1549$, therefore $\gamma_1\gamma_0 = 0.3239 < 1$ and one of the stability requirements in (10.39) is satisfied. Also for this particular choice of sliding surface $\|\tilde{G}(s)\|_\infty = \gamma_2 = 0.1277$. Therefore, from (10.39) the expression

$$\frac{\gamma_2\gamma_0}{1-\gamma_1\gamma_0} = 0.3951 < 1$$

which shows that the closed-loop system is stable for all $0 < w_i \le 1$. The variables related to the adaptive nonlinear gain (Sect. 10.2.2) have been chosen as follows: it was found that choosing $l_1 = 0$ and $l_2 = 1$ gave sufficiently good performance. This removes the dependance of $r(t)$ on $e(t)$ and simplifies the implementation. The parameter η from (10.41) was chosen as $\eta = 1$. The adaptation parameters from (10.45) have been chosen as $a = 1000$, $b = 0.1$ and $\epsilon = 2 \times 10^{-5}$. The parameter ϵ was chosen to be able to tolerate the variation in $\|s(t)\|$ associated with normal changes in flight condition but small enough to be sensitive to deviation from zero resulting from faults or failures. Here a has been chosen to be large to enable small changes in $\|s(t)\|$ to cause significant changes in the gain, so that the control system reacts quickly to a fault. The parameter b on the other hand dictates the rate at which $\rho(t)$ will decrease after $\|s(t)\|$ has returned below the threshold ϵ. For practical reasons, the adaptive gain is limited at a maximum of $\rho_{\max} = 5$. From (10.59), $(1 - \gamma_1\gamma_0)\epsilon\eta = 3 \times 10^{-5}$ and $b/4a(1-\gamma_1\gamma_0) = 2.5 \times 10^{-5}$ and therefore the condition in Proposition 10.1 is satisfied.

10.5.2 Actuator Fault Estimation Using Least Squares

In (10.7), the weighting matrix W which is used for the control signal redistribution, depends on information about the actuator fault. In this chapter, it will be assumed that a measurement of the actual actuator deflection is available. This is not an unrealistic assumption in aircraft systems [35]. Information provided by the actual actuator deflection can be compared with the signals from the controller to indicate the effectiveness of the actuator. The idea is to use a 'least squares' method to estimate the coefficients w_i and c_i in a relationship of the form

$$u_{(i,a)} = w_i u_i + c_i$$

where $u_{(i,a)}$ represents the actual deflection and u_i represents the demanded deflection i.e., the controller output. The scalars w_i and c_i are obtained from a least squares optimisation and $W = \mathrm{diag}(w_1, \dots, w_m)$. If the ith actuator is working perfectly, $w_i = 1$ and $c_i = 0$. If $w_i < 1$ then a fault is present. In the Admire simulation, 10 data samples from a 'moving window', collected at 100 Hz, are used to compute the w_i and c_i.

10.5.3 Simulation Results

In the following simulations, the aircraft model undertakes a 'α roll' manoeuvre. In this section a step demand of magnitude 10 deg is applied to AoA during 1–7 s and a step of 150 deg/s for roll rate is applied during 3–9 s. (There is no reference command for sideslip angle—see Fig. 10.3.) Figure 10.3 shows the response of the closed-loop system and compares the nominal no fault condition (dashed line) and a situation when the canard jams at 2 s (solid line). Note that some of the solid line overlaps the dashed line. It can be seen that the control signal is systematically rerouted to the right and left elevon to maintain the required performance. The tracking responses show no degradation in performance and the commanded state responses show that the controller tracks the reference model 'perfectly'. Figure 10.3 also shows accurate canard effectiveness estimation (w_c) from the least squares method.

Figure 10.4 shows a non-symmetric actuator failure where the left elevon locks with an offset at 1.6 s (i.e., a lock at a non-trim position). This scenario is much more difficult compared to the one considered in [122]. Initially in Fig. 10.4, a control signal is sent to the left elevon. After the failure has been detected, the weight w_i for the left elevon (w_{le}) in the control allocation is changed and the control signal sent to the left elevon is 'switched off' and redistributed to the canard and right elevon. Figure 10.4 shows much more significant nonlinear gain adaptation (both with and without reference model adaptation) when the fault occurs. Figure 10.4 also shows comparisons between the controller with (solid line) and without (dashed line) reference model adaptation. It can be seen that in the event of a severe failure, by 'slowing down' the reference model the control surface deflections become less aggressive compared with the non-adapted reference model. Therefore (as observed in the literature; see, e.g., [142]), a much slower response is desirable to provide a safe level of degraded performance and to prevent further damage to the aircraft.

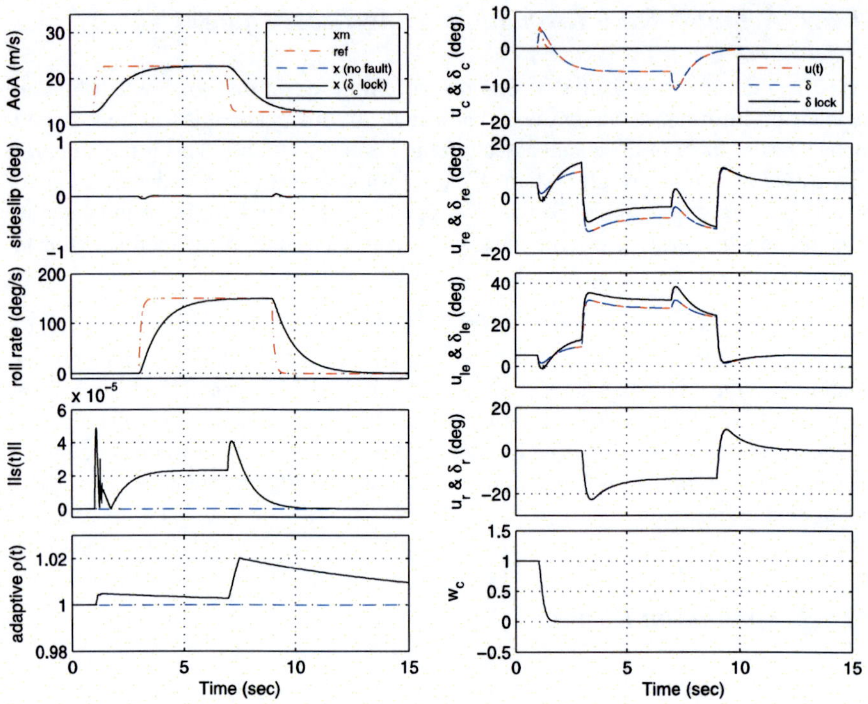

Fig. 10.3 Online CA: nominal no fault condition (*dashed line* --) vs. canard lock failure (*solid line* —): system responses

10.6 ADMIRE Simulations: Fixed Control Allocation

10.6.1 Controller Design

Note that in both the actuator effectiveness based CA scheme and the fixed and equally distributed CA scheme, the controller is designed based on the system in (10.21) and (10.68) when $K = 0$: i.e., based on the pair $(\hat{A}, \hat{B}_\nu)$. The same sliding surface design as in Sect. 10.5.1 is used here. The only difference is the actual control signal being sent to the actuators. In Sect. 10.5.1, the control signals sent to the actuators are based on (10.42) while in this section, the control signals sent to the actuators are based on (10.63). Note also, as mentioned in Sect. 10.3, although the CA strategy proposed in this section is different from the one in Sect. 10.2, the stability analysis of the reduced order sliding motion must satisfy the same condition i.e., (10.39). This is due to the fact that γ_1 from (10.38) and γ_2 from (10.37) correspond to the same elements in the reduced order systems (10.34) and (10.73), and the fact that $\|B_2^{\ddagger}\| = \|B_2^{+}\|$. Therefore, the analysis in Sect. 10.5.1 is also valid for the case of the fixed CA scheme.

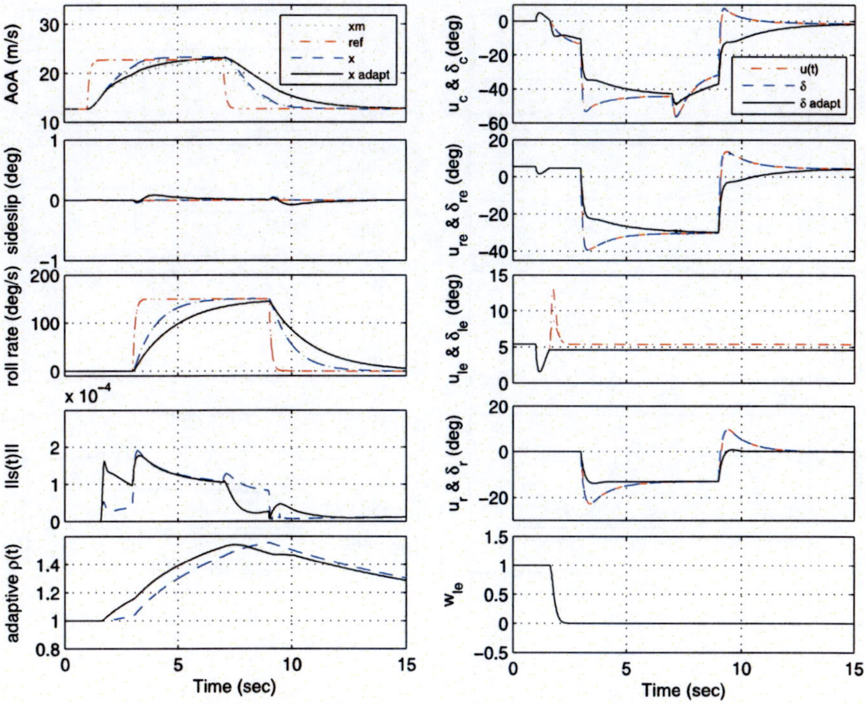

Fig. 10.4 Online CA: left elevon lock with offset: without (*dashed line* ––) and with adaptive reference model (*solid line* —)

The variables related to the adaptive nonlinear gain in Sect. 10.3.1 have been chosen as follows: again choosing $l_1 = 0$ and $l_2 = 1$ gave sufficiently good performance. The parameter η from (10.75) was chosen as $\eta = 1$. The adaptation parameters from Sect. 10.3.1 have been chosen as $a = 3200$, $b = 0.0001$ and $\epsilon = 3 \times 10^{-5}$. As in Sect. 10.5, the parameter ϵ was chosen to be able to tolerate the variation in $\|s(t)\|$ due to normal changes in flight condition but small enough to enable the adaptive gain to react to deviations from zero due to faults or failures. Here a has been chosen to be large to enable small changes in $\|s(t)\|$ to cause significant changes in the gain, so that the control system reacts quickly to a fault. The parameter b on the other hand dictates the rate at which $\rho(t)$ will decrease, after $\|s(t)\|$ has returned below the threshold ϵ. As in Sect. 10.5.1, $\text{rank}(B_2 W B_2^{\mathrm{T}}) < 3$ when the rudder completely fails or any two surfaces from the set consisting of the canard and the left and right elevons completely fail. Using the fact that, as in Sect. 10.5.3, only a single actuator failure is considered, w from (10.77) has been found to be $w = 0.2286$. Therefore from (10.78), $\varsigma = 60.9629$ (where γ_1 and γ_0 are defined in Sect. 10.5.1). From (10.87), $\epsilon\eta = 3 \times 10^{-5}$ and $b\varsigma^2/4a = 2.903 \times 10^{-5}$ and therefore the condition in Proposition 10.2 is satisfied.

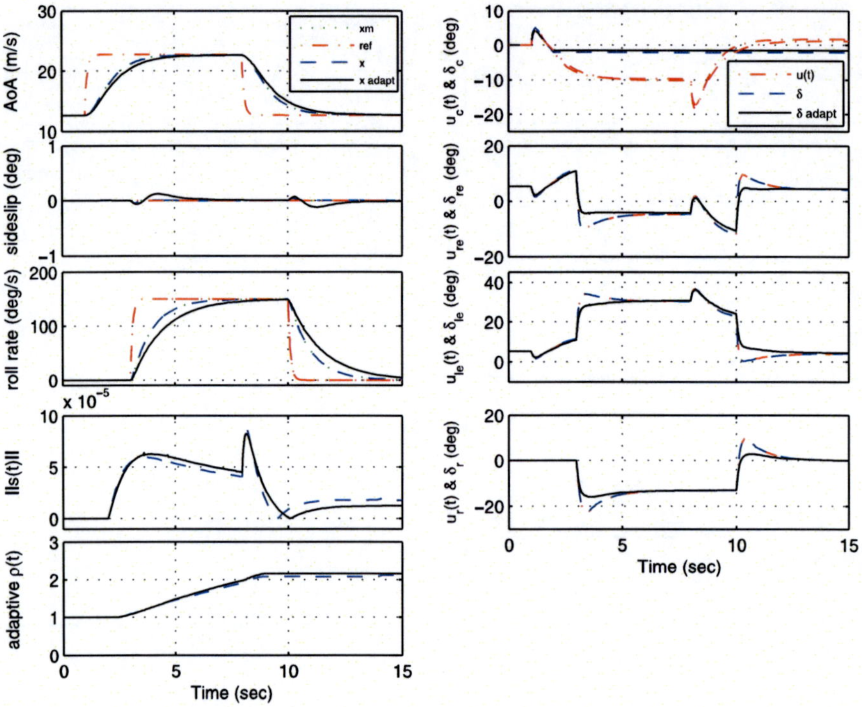

Fig. 10.5 Fixed CA: canard lock with offset: without (*dashed line* --) and with adaptive reference model (*solid line* —)

10.6.2 Simulation Results

The same manoeuvre and failure test conditions as in Sect. 10.5.3 is considered here. Figure 10.5 shows the aircraft responses when the canard fails at 2 s, while Fig. 10.6 shows when the left elevon fails at 1.6 s. Figure 10.5 shows the response of the closed-loop system and compares the controller with (solid line) and without (dashed line) reference model adaptation. It can be seen that the tracking responses show no degradation in performance and the commanded state responses show that the controller tracks the reference model 'perfectly'. Similar responses to those in Sect. 10.5.3 have been obtained. The controller with the adaptive reference model (solid line) shows much less aggressive control surface deflection compared to the one with a fixed reference model (thin line). There are a few differences in Fig. 10.5 compared to Fig. 10.3 associated with the signal $\|s(t)\|$, the adaptive gain $\rho(t)$, and the control signal to the canard $u_c(t)$. Figure 10.5 shows that in the absence of the actuator effectiveness estimation, the fixed CA scheme continues to send signals to the canard even after it has failed. The absence of the actuator effectiveness estimate gives much larger deviations from the sliding surface compared to the one in Fig. 10.3. As a consequence of the larger $\|s(t)\|$, the adaptive gain $\rho(t)$ is much higher than the one in Fig. 10.3. The higher gain $\rho(t)$ gives larger control signals

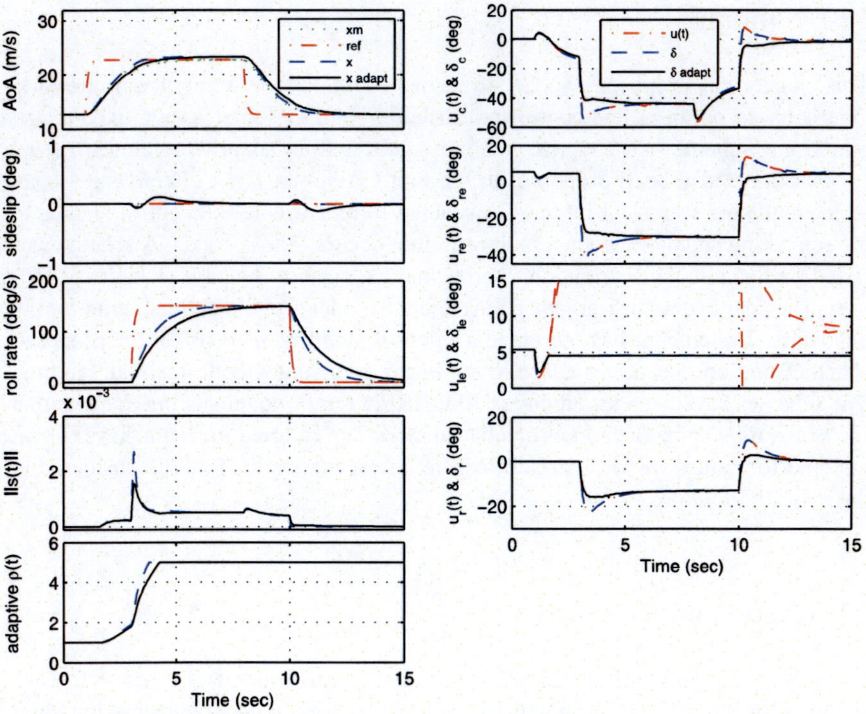

Fig. 10.6 Fixed CA: left elevon lock with offset: without (*dashed line* ---) and with adaptive reference model (*solid line* —)

than in the nominal condition, which are sent to all actuators to compensate for the jammed canard. This replicates the condition of passive FTC using sliding modes without any FDI as shown in Chap. 8. The difference in this section is that total actuator failures can be handled. The slow rate at which $\rho(t)$ is decreasing in Fig. 10.5 is due to the choice of b for the adaptive gain.

Figure 10.6 shows a non-symmetric type of actuator failure where the left elevon locks with an offset. This figure also shows a similar trend to Fig. 10.5. The control signal sent to the jammed left elevon is not shut off (because of the absence of an FDI scheme). The deviation from the sliding surface (the $\|s(t)\|$ signal) is much higher compared to the one in Fig. 10.4, and this causes the adaptive gain $\rho(t)$ to reach the maximum set gain. The large adaptive gain increases the magnitude of the control signals sent to all the actuators and thus enables the desired AoA, sideslip and roll rate performance to be maintained. Again, the slow rate at which $\rho(t)$ is decreasing in Fig. 10.6 is due to the choice of b for the adaptive gain. Figure 10.6 also shows comparisons between the controller with (solid line) and without (dashed line) reference model adaptation. By 'slowing down' the reference model, especially in the event of a severe failure, the control surface deflections become less aggressive compared to the non-adapted reference model.

10.7 Summary

This chapter has proposed two CA strategies using a model-reference framework. One is based on an online control redistribution and the other uses a fixed distribution CA scheme. An adaptive nonlinear gain and an adaptive reference model for FTC have also been proposed. In the first CA strategy, the effectiveness level of the actuators was used by the CA scheme to redistribute the control signals to the remaining actuators when a fault or failure occurs. The second CA scheme was based on fixed equal distribution to all actuators, even in the case of faults or failures. The adaptive nonlinear gain and reference model provide online tuning for the controller. The chapter has presented a rigorous stability analysis for the proposed schemes and proposed two adaptive nonlinear gain strategies to maintain sliding. The schemes have been tested on the ADMIRE aircraft model and have shown that faults and total actuator failures can be handled. The simulation results have shown good performance even in more difficult failure scenarios than those considered in the existing literature.

10.8 Notes and References

References [142] and [59] describe some recent research that uses model-reference schemes for active FTC. The popularity of the model-reference framework for adaptation and FTC is due to several advantageous features. Many performance specifications are given in the time domain e.g., rise time, damping ratio, decoupling effects etc. These can be represented in terms of an ideal transfer function response, which become the reference signals the closed-loop system must follow for tracking purposes. Another advantage of using a model-reference framework for FTC is that it allows the reference model to be changed online to cope with changes in the operational conditions especially during faults or failure.

In the SMC literature, model-reference schemes have been used for tracking (as seen in Sect. 3.5.2 and [17, 60, 70, 239, 301]). In terms of FTC, work such as [59] has investigated combining model-reference and SMC.

Chapter 11
SIMONA Implementation Results

The previous chapters proposed and analysed new fault tolerant control schemes using a combination of CA and SMC. This chapter builds on the previous ones and considers the practical issues arising from implementing such schemes for controlling both the longitudinal and lateral axes of a 6-DOF flight motion simulator configured to represent a B747 aircraft. Here, a 'proof of concept' controller will be presented to highlight the practicality of the sliding mode controllers for real-time application.

11.1 Introduction

The combination of sliding modes and control allocation developed in Chap. 9 provides a powerful tool for the development of simple, robust fault tolerant flight controllers that work for a wide range of faults and failures without requiring any reconfiguration (provided there is still enough redundancy in the system). The work in Shtessel et al., [226] and Wells and Hess [269] provides practical examples of the combination of SMC and CA for FTC. The work by Shin et al., [222] uses CA ideas, but formulates the problem from an adaptive controller point of view. However none of these papers provide a detailed stability analysis and discuss sliding mode controller design issues when using control allocation. The scheme in Chap. 9 uses a control law which depends on (an estimate of) the 'efficiency/effectiveness' of the actuators. In this chapter, these ideas are extended and the potential of SMC and CA is demonstrated through an implementation of these ideas on a six degree of freedom (6-DOF) research flight simulator called SIMONA (SImulation, MOtion and NAvigation) running a high-fidelity nonlinear aircraft model based on FTLAB747 V6.5/7.1/2006b [234].

H. Alwi et al., *Fault Detection and Fault-Tolerant Control Using Sliding Modes*, Advances in Industrial Control,
DOI 10.1007/978-0-85729-650-4_11,

11.2 Test Facilities (SIMONA)

Typically a newly developed controller is first designed and tested on a benchmark model, and then tested on a flight simulator before being certified by actual flight testing. Flight testing is very expensive. For the study of faults and failures, a high-fidelity nonlinear aircraft model can simulate closely real-life conditions and the performance of an aircraft with high accuracy, and apart from the cost saving, is much safer. Flight simulators are also used before an actual flight for training and to get useful feedback from pilots on the effectiveness of the controller systems. The work in this chapter is based on a high-fidelity nonlinear B747 model for design and desktop simulations, and has subsequently been implemented on a flight simulator. The details of the high-fidelity nonlinear B747 model and the flight simulator are described in the subsections below.

11.2.1 The SIMONA Research Simulator

The SIMONA Research Simulator (SRS) in Fig. 11.1 is a research project of Delft University of Technology. During its design and fabrication the university employed new techniques and insights from various fields to optimise performance and operational flexibility. The resulting flight simulator provides researchers with a powerful tool that can be adapted to various uses [243]. In the years since it has been operational, the SRS has been used for research into human (motion) perception [125, 258, 288], aircraft handling qualities [91, 113], fly-by-wire control algorithms and flight deck displays [163, 194], flight procedures [65, 66] and air traffic control [264]. The flexible software architecture and high-fidelity cueing environment allows the integration of the B747 model from Smaili et al., [234], complete with failures and the assessment of the controller in a realistic aircraft environment.

The flight deck of the SRS provides the two pilots with simulated instruments that match the aircraft under investigation (Fig. 11.1(b) and 11.1(c). The pilots can interface with the aircraft through a conventional control column or a sidestick controller, a centre pedestal with engine controls and a Mode Control Panel (MCP) for the autopilot. The windows give a wide view on a virtual environment and a motion system moves the entire cabin to simulate aircraft accelerations.

A modular network of personal computers (PCs) provides the processing power to run the simulator. Each PC has a specific task, e.g., driving the pilot controls, generating the instrument graphics, running the aircraft model or logging data. A high-speed fibre-optic network provides synchronisation and communication services for all the computers. The modular approach makes it easy to exchange for example the aircraft model for another, without affecting the rest of the simulation software. In particular, the software is able to interface with SIMULINK® models.

Fig. 11.1 SIMONA research simulator (picture courtesy of SIMONA, Delft University of Technology)

(a) Outside view

(b) Flight deck - centre view

(c) Flight deck - right view

11.2.2 Benchmark V2.2—FTLAB747 V6.5/7.1/2006b

Over the years the FTLAB747 software has been developed for the study of FTC and FDI schemes [183]. The amount of redundancy available on this aircraft model [183] makes it suitable to test new fault tolerant schemes. Recently this software has been upgraded to V6.5/7.1/2006b by Smaili et al., [234] to allow all the control surfaces to be controlled independently offering more degrees of control flexibility especially during faults or failures. This model is the basis for the results in this chapter. This 'modified' aircraft is essentially a fly-by-wire aircraft [35] where all the control surfaces are controlled electronically compared to the 'classical' B747 aircraft which uses mechanical linkages which therefore limit the usability of some of the control surfaces in fault or failure conditions. In this chapter, two sliding mode controllers for lateral and longitudinal control have been designed and tested under multiple fault and failure conditions before being implemented on the 6-DOF SIMONA flight simulator to show the capabilities of the proposed method.

To be able to fly with a pilot in the loop, the benchmark B747 model [234] (from FTLAB747) was slightly adapted from the offline model. The aircraft model was isolated from peripheral utility functions such as the autopilot, to follow the reference scenario and MATLAB® logging functions. Its inputs and outputs were standardised to fit in the SRS software environment and the SIMULINK® model was converted to C code using the REAL-TIME WORKSHOP®. Finally the model was integrated with the pilot controls, aircraft instruments and cueing devices of the SRS. Section 11.4 describes in more detail similar steps that were taken for the controller described in the next section, which was subsequently integrated in the SRS and coupled with the B747 model.

11.3 Controller Design

The 12 rigid body states of the B747 aircraft can be divided into six longitudinal axis states and six lateral and directional axis states which are all determined from the 6-degree of freedom equations of motion. The states are given by $x = [p\ q\ r\ V_{\text{tas}}\ \alpha\ \beta\ \phi\ \theta\ \psi\ h_e\ x_e\ y_e]^{\mathrm{T}}$. For the longitudinal axis, the states are pitch rate q, true airspeed V_{tas}, angle of attack α, pitch angle θ and altitude h_e. Meanwhile for the lateral and directional axes, the states are roll rate p, yaw rate r, sideslip angle β, roll angle ϕ and yaw angle ψ. The control surfaces comprise four ailerons (inboard and outboard on each wing), 12 spoilers (two inboard spoilers and four outboard spoilers on each wing), two rudders (upper and lower), four elevators (an inboard and outboard on each left and right elevator), a horizontal stabiliser and four engine thrusts (which are controlled through engine pressure ratios (EPR)).

In this chapter both lateral and longitudinal control is considered. One of the controller design objectives considered here is to bring a faulty aircraft to a near landing condition. This can be achieved by a change of direction through a 'banking turn' manoeuvre [36], followed by a decrease in altitude and speed. This can be achieved by tracking appropriate roll angle (ϕ) and sideslip angle (β) commands using the lateral controller, and tracking flight path angle (FPA) and airspeed (V_{tas}) commands using the longitudinal controller. For lateral control, the settling time when there is no fault/failure should be approximately 20 s for ϕ and 20 s for β. If a fault/failure occurs, the tracking requirement is 25 s for ϕ and β. These specifications are chosen to ensure that there is almost zero side force and therefore passenger comfort is maintained (page 233 of Bryson [36]). For longitudinal control, the settling time when there is no failure should be 20 s for FPA and 45 s for V_{tas}. If a failure occurs, the tracking requirement is 30 s for FPA with no difference in the V_{tas} tracking. These specifications are taken from Ganguli et al., [105].

A linearisation has been obtained around an operating condition of 263,000 kg, 92.6 m/s true airspeed, and an altitude of 600 m at 25.6% of maximum thrust and at a 20 deg flap position. The result is a 12th order linear model (separated into two sixth order models) associated with the lateral and longitudinal states. For design purposes, only the first four longitudinal ($x_{\text{long}} = [q\ V_{\text{tas}}\ \alpha\ \theta]^{\text{T}}$) and lateral states ($x_{\text{lat}} = [p\ r\ \beta\ \phi]^{\text{T}}$) have been retained. For lateral control, the four individual engine pressure ratios (EPR) and the four individual ailerons have been used. The 10 spoilers[1] have been aggregated to produce two control inputs on each wing (spoilers 1–4, 5, 8 and 9–12 have been grouped, respectively). The other input represents rudder deflection (the upper and lower rudder has been aggregated to produce a single control signal). For longitudinal control, the four elevators have been aggregated to produce one control input while the four longitudinal EPRs can be controlled independently. The other input represents horizontal stabiliser deflection. The following state-space system pairs represent the lateral and longitudinal systems about the trim condition:

$$A_{\text{lat}} = \begin{bmatrix} -1.0579 & 0.1718 & -1.6478 & 0.0004 \\ -0.1186 & -0.2066 & 0.2767 & -0.0019 \\ 0.1014 & -0.9887 & -0.0999 & 0.1055 \\ 1.0000 & 0.0893 & 0 & 0 \end{bmatrix} \tag{11.1}$$

$$B_{\text{lat}} = \left[\begin{array}{ccccccc} -0.0832 & 0.0832 & -0.2285 & 0.2285 & -0.2625 & -0.0678 & 0.0678 \\ -0.0154 & 0.0154 & -0.0123 & 0.0123 & -0.0180 & -0.0052 & 0.0052 \\ \hline 0 & 0 & 0 & 0 & 0.0017 & 0.0006 & -0.0006 \\ 0 & 0 & 0 & 0 & 0 & 0 & 0 \end{array}\right.$$

$$\left.\begin{array}{cccccc} 0.2625 & 0.1187 & 0.0246 & 0.0140 & -0.0140 & -0.0246 \\ 0.0180 & -0.2478 & 0.1269 & 0.0724 & -0.0724 & -0.1269 \\ \hline -0.0017 & 0.0174 & 0.0005 & 0.0005 & -0.0005 & -0.0005 \\ 0 & 0 & 0 & 0 & 0 & 0 \end{array}\right] \begin{array}{l} \left.\vphantom{\begin{array}{c}0\\0\end{array}}\right\} B_{\text{lat},2} \\ \left.\vphantom{\begin{array}{c}0\\0\end{array}}\right\} B_{\text{lat},1} \end{array} \tag{11.2}$$

[1]Spoilers 6 and 7 are ground spoilers and are not used during flight [119].

and

$$A_{\text{long}} = \begin{bmatrix} -0.5137 & 0.0004 & -0.5831 & 0 \\ 0 & -0.0166 & 1.7171 & -9.8046 \\ 1.0064 & -0.0021 & -0.6284 & 0 \\ 1.0000 & 0 & 0 & 0 \end{bmatrix} \tag{11.3}$$

$$B_{\text{long}} = \left[\begin{array}{cccccc} -0.6228 & -1.3578 & 0.0082 & 0.0218 & 0.0218 & 0.0082 \\ 0 & -0.1756 & 1.4268 & 1.4268 & 1.4268 & 1.4268 \\ \hline -0.0352 & -0.0819 & -0.0021 & -0.0021 & -0.0021 & -0.0021 \\ 0 & 0 & 0 & 0 & 0 & 0 \end{array}\right] \begin{array}{l} \big\} B_{\text{long},2} \\ \\ \big\} B_{\text{long},1} \end{array} \tag{11.4}$$

where the states represent $x_{\text{lat}} = [p\ r\ \beta\ \phi]^{\text{T}}$ and $x_{\text{long}} = [q\ V_{\text{tas}}\ \alpha\ \theta]^{\text{T}}$. The lateral control surfaces are

$$\delta_{\text{lat}} = [\delta_{\text{air}}\ \delta_{\text{ail}}\ \delta_{\text{aor}}\ \delta_{\text{aol}}\ \delta_{sp1-4}\ \delta_{sp5}\ \delta_{sp8}\ \delta_{sp9-12}\ \delta_r\ e1_{\text{lat}}\ e2_{\text{lat}}\ e3_{\text{lat}}\ e4_{\text{lat}}]^{\text{T}}$$

which represent aileron deflection (right and left: inboard and outboard) (rad), spoiler deflections (left: 1–4 and 5, right: 8 and 9–12) (rad), rudder deflection (rad) and lateral engine pressure ratios (EPR). The longitudinal control surfaces are $\delta_{\text{long}} = [\delta_e\ \delta_s\ e1_{\text{long}}\ e2_{\text{long}}\ e3_{\text{long}}\ e4_{\text{long}}]^{\text{T}}$, which represent elevator deflection (rad), horizontal stabiliser deflection (rad), and longitudinal EPR. The partition of B in (11.2) and (11.4) shows the terms B_1 and B_2 (although a further change of coordinates is necessary to obtain the form in (9.9) to scale B_2 to ensure $B_2 B_2^{\text{T}} = I$). The controlled output distribution matrices are

$$C_{c_{\text{lat}}} = \begin{bmatrix} 0 & 0 & 1 & 0 \\ 0 & 0 & 0 & 1 \end{bmatrix}, \qquad C_{c_{\text{long}}} = \begin{bmatrix} 0 & 0 & -1 & 1 \\ 0 & 1 & 0 & 0 \end{bmatrix}$$

which represent the states ϕ and β for lateral control and flight path angle (FPA) and V_{tas} for longitudinal control. These linear models will be used to design the control schemes which will be described in the next sections.

To include a tracking facility, integral action (as discussed in Sect. 3.5.1) has been included for both longitudinal and lateral control. For the generic system in (9.7), let $x_r(t)$ represent integral action states:

$$\dot{x}_r(t) = r(t) - C_c x(t) \tag{11.5}$$

where $C_c \in \mathbb{R}^{l \times n}$ is the distribution matrix associated with the controlled outputs, and the differentiable (filtered reference) signal $r(t)$ satisfies

$$\dot{r}(t) = \Gamma\big(r(t) - r_c\big) \tag{11.6}$$

In the above, $\Gamma \in \mathbb{R}^{l \times l}$ is a stable design matrix and r_c is a piecewise constant demand vector. Augmenting the states from (11.1)–(11.4) with the integral action states and defining $x_a(t) = \text{col}(x_r(t), x(t))$ it follows that

$$\dot{x}_a(t) = A_a x_a(t) + B_a u(t) + B_r r(t) \tag{11.7}$$

where

$$A_a = \begin{bmatrix} 0 & -C_c \\ 0 & A \end{bmatrix}, \qquad B_a = \begin{bmatrix} 0 \\ B \end{bmatrix}, \qquad B_r = \begin{bmatrix} I_l \\ 0 \end{bmatrix} \tag{11.8}$$

If (A, B) is controllable and (A, B, C_c) does not have any zeros at the origin then (A_a, B_a) is controllable. Define a switching function $s_a(t) : \mathbb{R}^{(n+l)} \to \mathbb{R}^l$ to be

$$s_a(t) = S_a x_a(t) \tag{11.9}$$

where $S_a \in \mathbb{R}^{l\times(n+l)}$ and $S_a B_a = I_l$. As in Equation (9.42)–(9.43), the proposed 'virtual control law' comprises two components $\hat{\nu}(t) = \hat{\nu}_l(t) + \hat{\nu}_n(t)$. Now because of the reference signal $r(t)$, the linear component has a feed-forward reference term and so $\hat{\nu}_l(t) = L_{x_a} x_a(t) + L_r r(t)$ where the gain $L_{x_a} = -\hat{S}_a \hat{A}_a$ and $L_r = -\hat{S}_a \hat{B}_r$. Here $\hat{A}_a$, $\hat{B}_r$ and $\hat{S}_a$ are the matrices from (11.8) and (11.9) after a transformation to achieve the regular form in (9.19) has been performed. The nonlinear component is defined as

$$\hat{\nu}_n(t) = -\rho(t, x_a)\frac{s_a(t)}{\|s_a(t)\|} \quad \text{for } s_a(t) \neq 0 \tag{11.10}$$

and from (9.47)

$$u(t) = W B_2^{\mathrm{T}} \left(B_2 W^2 B_2^{\mathrm{T}}\right)^{-1} \hat{\nu}(t) \tag{11.11}$$

Notice that the control sent to the actuators is dependent on the effectiveness gains k_i (through the diagonal weighting matrix W).

11.3.1 Lateral Controller Design

In normal operation, the ailerons will be the primary control surface for ϕ tracking, whilst the spoilers introduce redundancy. Meanwhile for β tracking, the rudder will be the primary control surface and differential engine thrust is the associated redundancy. It will be assumed that at least one of the control surfaces for both ϕ and β tracking will be available when a fault or failure occurs (i.e., one of either the four ailerons or the four spoilers will be available and one of either the rudder or the four engine thrusts are available). Based on these assumptions, it can be verified from a numerical search that $\gamma_{0_{\text{lat}}}$ from (9.27) is $\gamma_{0_{\text{lat}}} = 8.1314$.

The matrix which defines the hyperplane must now be synthesised so that the conditions of (9.36) are satisfied. A quadratic optimal design (as in Sect. 3.4.1) has been used to obtain the sliding surface $S_{a_{\text{lat}}}$ which depends on the matrix $\mathcal{K}_{\text{lat}}$ in (9.28). Here the symmetric positive definite state weighting matrix has been chosen as $\mathbf{Q}_{\text{lat}} = \text{diag}(0.005, 0.1, 6, 6, 1, 1)$. The first two terms of $\mathbf{Q}_{\text{lat}}$ are associated with the integral action states and are less heavily weighted. The third and fourth term of $\mathbf{Q}_{\text{lat}}$ are associated with the equations of the angular acceleration in roll and yaw (i.e., $B_{\text{lat},2}$ term partition in (9.9)) and thus weight the virtual control term. Thus by analogy to a more typical LQR framework, they affect the speed of response of the closed-loop system. Here, the third and fourth terms of $\mathbf{Q}_{\text{lat}}$ have been heavily weighted compared to the last two terms to reflect a reasonably fast closed-loop system response. The poles associated with the reduced order sliding motion are $\{-0.0707, -0.3867, -0.3405 \pm 0.1484i\}$. Based on this value of $\mathcal{K}_{\text{lat}}$, simple calculations from (9.32) show that $\gamma_{1_{\text{lat}}} = 0.0145$, therefore $\gamma_{0_{\text{lat}}} \gamma_{1_{\text{lat}}} = 0.1180 < 1$ and so the requirements of (9.36) are satisfied. Also for this particular choice of sliding surface, $\|\tilde{G}_{\text{lat}}(s)\|_{\infty} = \gamma_{2_{\text{lat}}} = 0.0764$ from (9.35). Therefore from (9.36),

$$\frac{\gamma_{2_{\text{lat}}} \gamma_{0_{\text{lat}}}}{1 - \gamma_{1_{\text{lat}}} \gamma_{0_{\text{lat}}}} = 0.7043 < 1$$

which shows that the system is stable for $0 < w_i \leq 1$. The pre-filter matrix from (11.6) has been designed to be $\Gamma_{\text{lat}} = \text{diag}(-0.5, -0.5)$. This may be viewed as representing the ideal response in the ϕ and the β channels. For implementation, the discontinuity in the nonlinear control term in (11.10) has been smoothed by using a sigmoidal approximation described in Sect. 3.2.2 where the scalar $\delta_{\text{lat}} = 0.05$. This removes the discontinuity and introduces a further degree of tuning to accommodate the actuator rate limits—especially during actuator fault or failure conditions.

To emulate a real aircraft flight control capability, an outer loop heading control was designed based on a proportional plus derivative controller, to provide a roll command to the inner loop sliding mode controller. In the implementation, this outer loop heading control can be activated by a switch in the cockpit. The proportional gain was set as $K_{p_{\text{lat}}} = 0.5$ and the derivative gain was set as $K_{d_{\text{lat}}} = 0.1$

11.3.2 Longitudinal Controller Design

In normal operation, the elevators will be the primary control surface for FPA tracking, whilst the horizontal stabiliser introduces redundancy. For V_{tas} tracking, the collective thrust (from the four engines) will be the actuator. It will be assumed that at least one of the control surfaces for FPA tracking will still be available when a fault or failure occurs. It is also assumed that at least one of the four engines is available for V_{tas} tracking. Based on these assumptions, it can be verified from a numerical search that $\gamma_{0_{\text{long}}} = 8.2913$ from (9.27).

As in the lateral controller, a quadratic optimal design has been used to obtain the sliding surface matrix (and therefore the matrix $\mathcal{K}_{\text{long}}$). The weighting matrix has been chosen as $\mathbf{Q}_{\text{long}} = \text{diag}(0.1, 0.1, 10, 50, 1, 1)$. Again, similar to the lateral controller design, the first two terms of $\mathbf{Q}_{\text{long}}$ are associated with the integral action states and are less heavily weighted. The third and fourth terms of $\mathbf{Q}_{\text{long}}$ are associated with the $B_{\text{long},2}$ term partition in (9.9) (i.e., states q and V_{tas}) which weight the virtual control term, and have been heavily weighted compared to the last two terms. The poles associated with the reduced order sliding motion are $\{-0.7066, -0.2393 \pm 0.1706i, -0.0447\}$. Based on this value of $\mathcal{K}_{\text{long}}$, simple calculations from (9.32) show that $\gamma_{1_{\text{long}}} = 1.9513 \times 10^{-4}$: therefore

Table 11.1 SIMONA hardware interconnections

Input	Source	Output	Destination
Ref. signals e.g., ψ_{cmd}	MCP	Actuator commands	Aircraft model
Aircraft states e.g., ψ	Aircraft model	Controller data	Data logger
Config. switches e.g., ψ or ϕ	Pilot switches	Configuration choice	Controller

$\gamma_{0_{long}}\gamma_{1_{long}} = 0.0016 < 1$ and so the requirements of (9.36) are satisfied. For this choice of sliding surface, $\|\tilde{G}_{long}(s)\|_\infty = \gamma_{2_{long}} = 0.0112$ from (9.35). Therefore from (9.36),

$$\frac{\gamma_{2_{long}}\gamma_{0_{long}}}{1-\gamma_{1_{long}}\gamma_{0_{long}}} = 0.0931 < 1$$

which shows that the system is stable for $0 < w_i \leq 1$. The pre-filter matrix from (11.6) has been designed to be $\Gamma_{long} = \text{diag}(-0.5, -0.125)$. The discontinuity in the nonlinear control term in (11.10) has been smoothed by using a sigmoidal approximation where the scalar $\delta_{long} = 0.05$.

An outer loop altitude control scheme was designed based on a proportional plus derivative controller to provide a FPA command to the inner loop sliding mode controller. In the SIMONA implementation, this outer loop altitude control can be activated by a switch in the cockpit. The proportional and the derivative gains were set as $K_{p_{long}} = 0.001$ and $K_{d_{long}} = 0.05$, respectively.

Note that both the lateral and longitudinal controller manipulate EPR. For lateral control, differential EPR is required as a secondary 'actuator' for β tracking; whilst for longitudinal control, collective EPR is used for V_{tas} tracking. In the final implementations, 'control mixing' was employed, whereby the signals from both the lateral controller ($e_{1_{lat}}$, $e_{2_{lat}}$, $e_{3_{lat}}$ and $e_{4_{lat}}$) and longitudinal controller ($e_{1_{long}}$, $e_{2_{long}}$, $e_{3_{long}}$ and $e_{4_{long}}$) were added together before being applied into each of the engines. This is similar to the control strategy used for the NASA propulsion control aircraft described in Burcham et al., [43].

11.4 SIMONA Implementations

The controller was implemented as a SIMULINK® (version 2006b) model with appropriate inputs and outputs to connect it with the aircraft model and the SIMONA hardware, as described in Table 11.1 and Fig. 11.2.

The controller/aircraft model combination contains an algebraic loop, with the controller requiring input from the model whilst producing the required input for the aircraft model. In the SRS this was solved by the controller module using the aircraft state data from the previous time step. All data are time stamped, ensuring consistency across different modules within the simulation, even when they are on physically different processors.

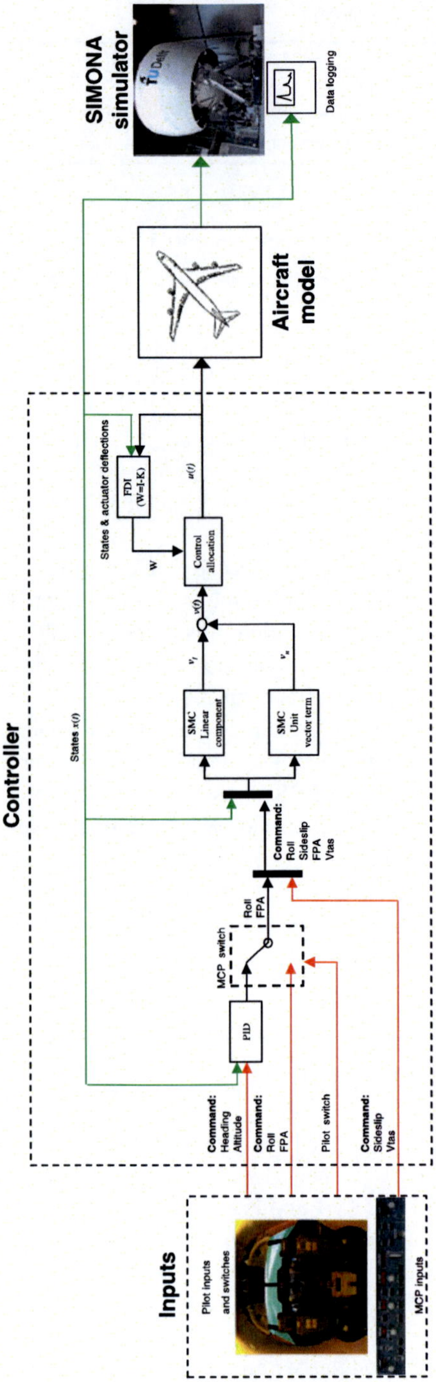

Fig. 11.2 SIMONA interconnections

Fig. 11.3 Mode control panel (MCP)

The controller was set up to work with an Ode4 solver with a fixed time step of 0.01 s. Using the REAL-TIME WORKSHOP®, the controller block diagram from the SIMULINK® environment was converted to C-code and integrated into the SRS, where it runs on a dual Intel® Pentium® III 1 GHz processor, together with the aircraft model and motion control software. The available processing power is sufficient to run the controller in real-time, i.e., within 10 ms per time step.

A connection with the Mode Control Panel (MCP) on the flight deck (Fig. 11.3) enables the selection of 'control modes' e.g., altitude hold, heading select and reference values. The simulator trials were performed with the speed, altitude and heading select modes active. The pilot commands new headings, speeds or altitudes by adjusting the controls on the MCP.

11.5 Results

The results presented in this chapter are all from the 6-DOF SIMONA research simulator based on the full 77 state nonlinear B747 model. For passenger comfort during turning manoeuvres, the reference command for ϕ was limited to 25 deg and a 0 deg reference applied to β, to force the nose of the aircraft to point towards the heading angle. It was assumed that the aircraft has recently taken off and reached an altitude of 600 m. After a few seconds of straight and level flight, failures occur on the actuators. The immediate action requested by the pilot is to change the heading to 180 deg and to head back to the runway. The altitude is then changed from 600 m (1967.2 ft) to 30.5 m (100 ft) before the V_{tas} is reduced from 92.8 m/s (180 kts) to 82.3 m/s (160 kts), to approximate a landing manoeuvre.

Five different control surface failures have been tested on the simulator: all elevators jam with a 3 deg offset, all ailerons jam with a 3 deg offset, a stabiliser runaway, all rudders runaway, and finally both rudders detach from the vertical fin [234]. All the trials have been done with and without wind and turbulence. However, only the most significant results are shown in this chapter.

11.5.1 No Fault

Figures 11.4(a)–11.4(d) show the fault-free responses of the controller. Figure 11.4(a) shows that there is a small amount of coupling between roll and sideslip

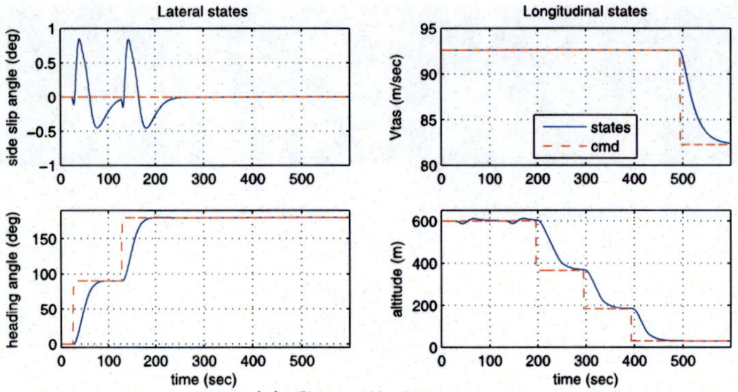

(a) Controlled states

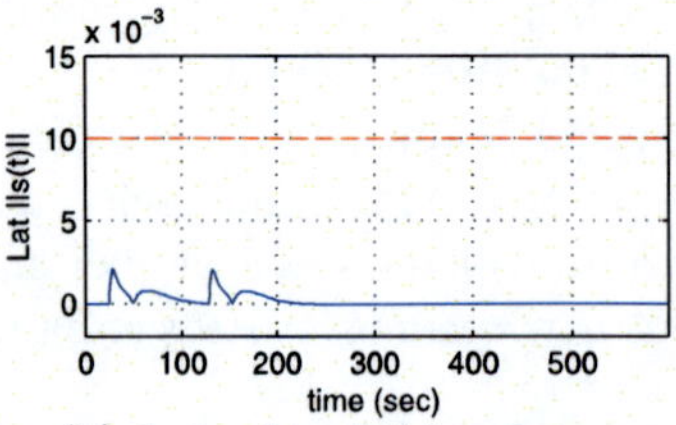

(b) Lateral switching func

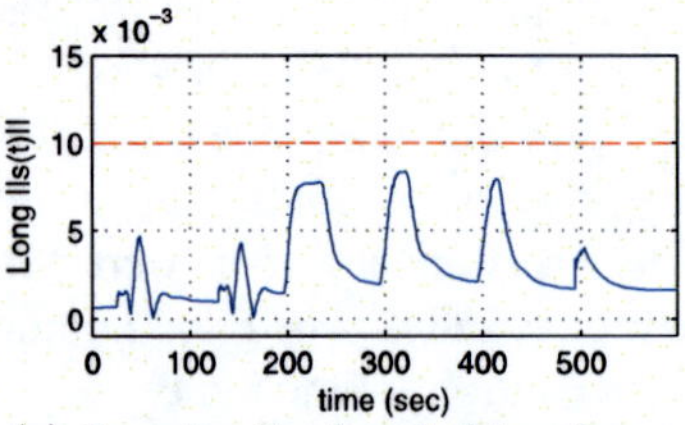

(c) Longitudinal switching func

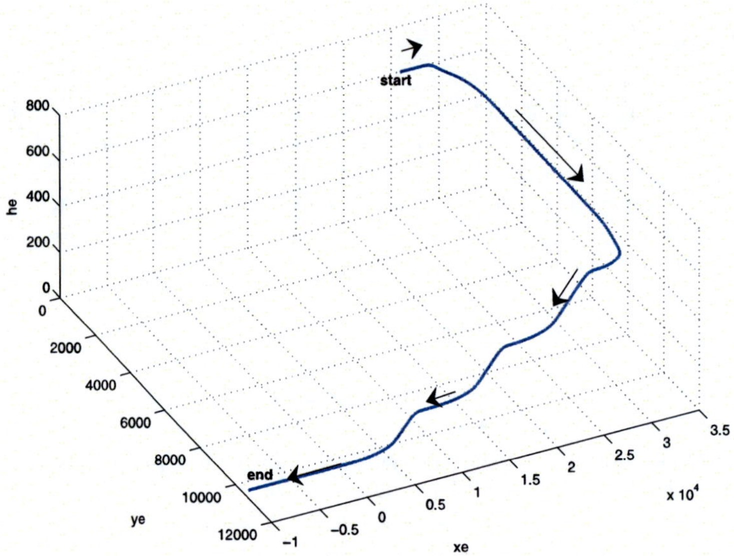

(d) Flight trajectory

Fig. 11.4 No fault conditions

during a heading change. There is also a small change in altitude during heading change. The heading is changed by means of two 90 deg step inputs followed by a change in altitude from 600 m to 30 m in three steps: 600 m to 366 m to 183 m and finally to 30 m above the runway. Figure 11.4(a) shows good tracking by the states of the command signals. Figures 11.4(b) and 11.4(c) show the nominal variations in the norm of the switching function signals. Finally, Fig. 11.4(d) shows the overall trajectory of the aircraft in 3D. Here, the change in heading and altitude can be seen more clearly.

11.5.2 Actuator Effectiveness

In (11.11), the weighting matrix W depends on information about the actuator fault. As in Sect. 10.5.2, it will be assumed in this chapter that a measurement of the actual control surface deflection is available. The idea is to use a 'least squares' method to estimate the actuator effectiveness W by comparing the actual control surface deflection with the demanded deflection i.e., the controller output. In the SIMONA implementation, 10 data samples from a 'moving window', collected at 100 Hz are used to compute the W. In the SIMONA implementation, both the lateral and longitudinal controller have their own fault estimation block.

11.5.3 Stabiliser Runaway

Figures 11.5(a)–11.5(d), showing a stabiliser runaway failure, are comparable to those in Figs. 11.4(a)–11.4(d). As before, Fig. 11.5(a) shows no visible degradation in performance. The only difference that can be seen is in the switching function shown in Fig. 11.5(d). Here the switching function exceeds the nominal condition briefly after the failure occurs but immediately returns and remains close to zero. Figure 11.5(b) shows the stabiliser runaway to a maximum positive deflection of 3 deg at its maximum deflection rate of 0.5 deg/s [119]. During the stabiliser runaway, Fig. 11.5(b) shows the elevator moves to the negative deflection to counteract the effect of stabiliser runaway. Figure 11.5(b) also shows that to counteract the stabiliser deflection change of about 6 deg (from a trim deflection of -3 deg to maximum deflection of 3 deg), the elevator deflection has offset approximately 12 deg from its trim deflection of approximately 2 deg. Finally, Fig. 11.5(c) shows that the effectiveness of the stabiliser has been successfully estimated.

11.5.4 Elevator Jam with Offset

Figures 11.6(a)–11.6(d) show the system responses when the elevator jams with an offset in the presence of wind and gusts. Figure 11.6(a) shows that the states maintain the required performance (as in the no fault condition in Fig. 11.4(a)) throughout the manoeuvre. A small change in altitude during the elevator jam is

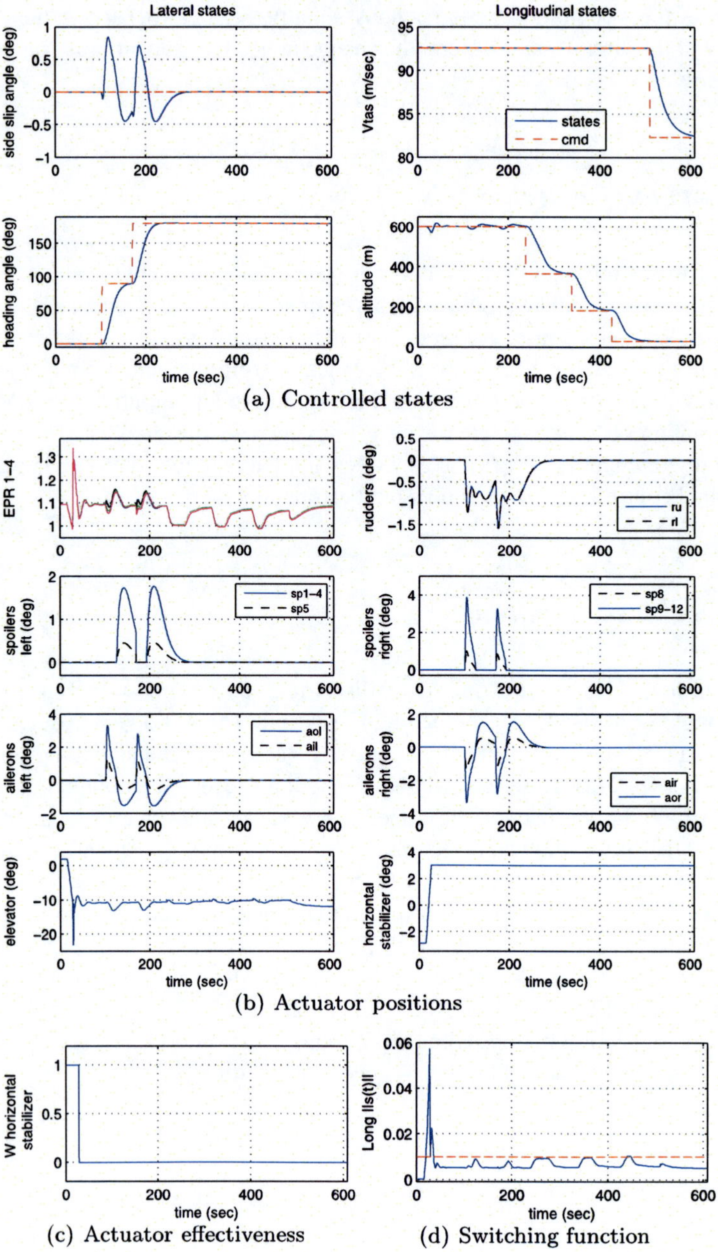

(a) Controlled states

(b) Actuator positions

(c) Actuator effectiveness

(d) Switching function

Fig. 11.5 Stabiliser runaway

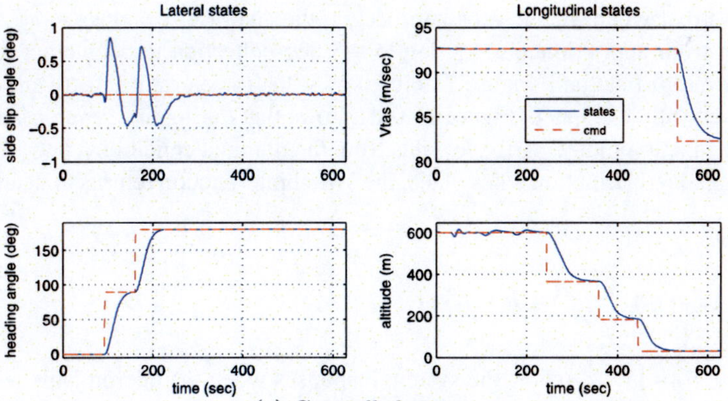

(a) Controlled states

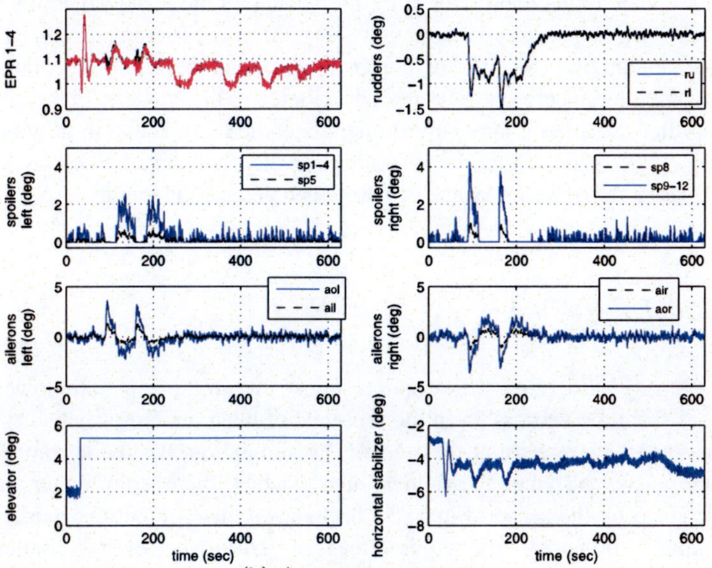

(b) Actuator positions

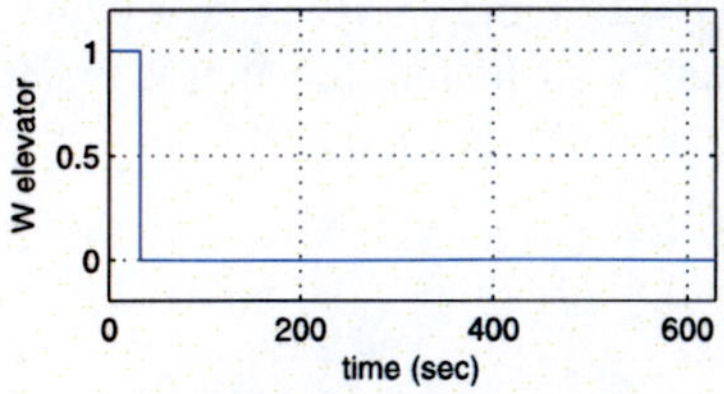

(c) Actuator effectiveness

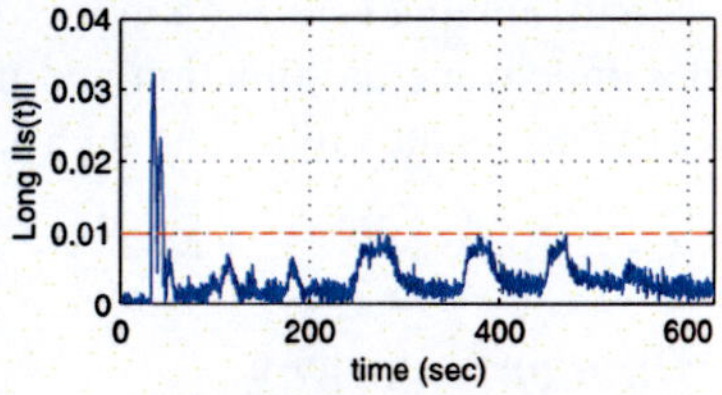

(d) Switching function

Fig. 11.6 Elevator jam with offset (wind and gust)

visible at approximately 50 s, since the effect of the elevator offset to 5 deg from the trim deflection (see Fig. 11.6(b)) creates an unwanted pitch moment. Figure 11.6(b) also shows that when the elevator jam occurs, the stabiliser becomes more active and offset from approximately -3 deg to -5 deg deflection to counteract the effect of elevator offset jam. Figure 11.6(c) shows that the elevator effectiveness has been successfully estimated. Figure 11.6(d) shows that during the elevator jam, the switching function momentarily deviates from the nominal conditions; but once the control signal redistribution takes place, the switching function returns to near zero.

11.5.5 Aileron Jam with Offset

Figures 11.7(a)–11.7(d) show the system responses when the aileron jams with an offset. Figure 11.7(a) shows no visible difference in terms of tracking performance compared to the no fault condition. Figure 11.7(b) shows the offset deflection of the aileron from 0 deg to an offset deflection of 3 deg. After the aileron jams, the aileron effectiveness estimation (Fig. 11.7(c)) drops to zero, and therefore the control signals sent to the ailerons are shut off and redistributed to the spoilers. Figure 11.7(c) shows that after the aileron jams, the spoilers become more active to provide roll. Finally, Fig. 11.7(d) shows that the switching function is maintained close to zero, indicating that nominal performance is maintained despite the failure.

11.5.6 Rudder Missing

Figures 11.8(a)–11.8(d) show the system responses when the upper and lower rudder detaches from the vertical fin in the presence of wind and gusts. This is shown clearly in Fig. 11.8(b) where at the start of the simulation, the rudder moves due to wind and gusts, and when the rudders are detached, there is no longer any deflection detected by the sensor. Figure 11.8(a) shows that without the rudders, the aircraft manages to maintain the required level of performance even in challenging wind and gust conditions. There is visually no difference in the sideslip performance compared to the nominal situation in Fig. 11.4(a). Finally, Fig. 11.8(c) shows accurate rudder effectiveness estimation and that the switching function is maintained close to zero despite the failure and the presence of wind and gusts.

11.5.7 Rudder Runaway

Figure 11.9(b) shows that the upper and lower rudders runaway to the 5 deg position. Not only does the rudder runaway cause a tendency to turn to one side (and therefore affecting the lateral performance), it also creates difficulties in the longitudinal

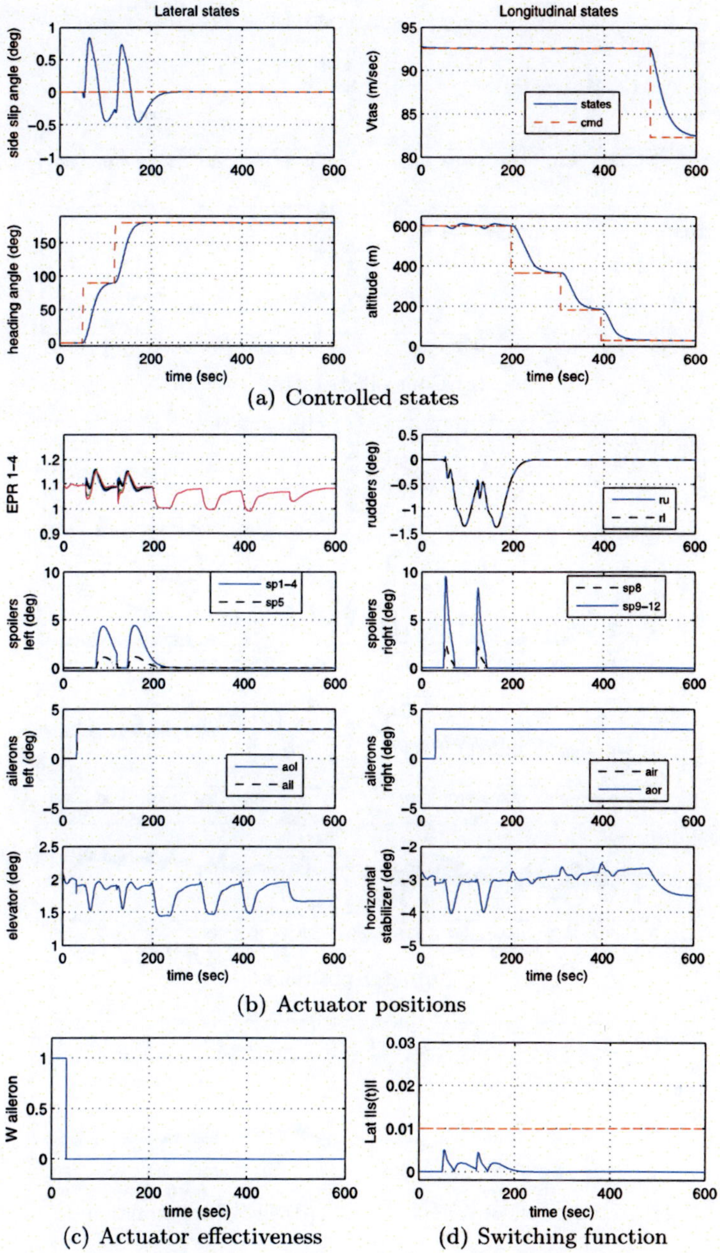

(a) Controlled states

(b) Actuator positions

(c) Actuator effectiveness

(d) Switching function

Fig. 11.7 Aileron jam with offset

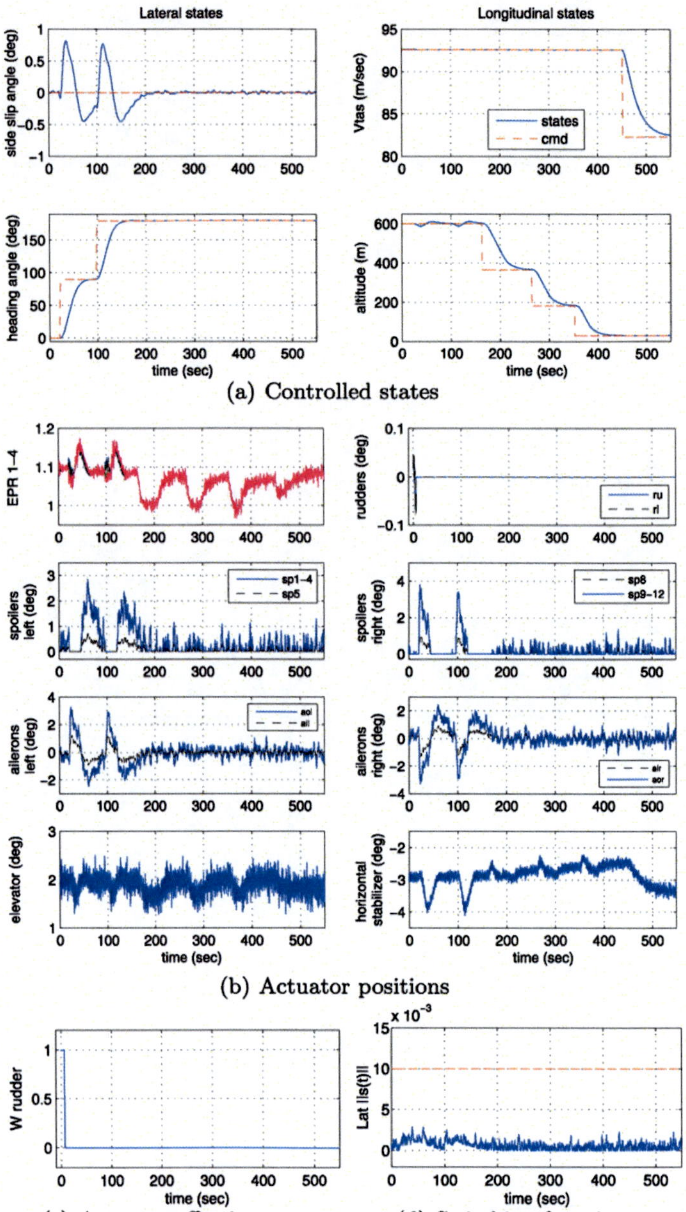

(a) Controlled states

(b) Actuator positions

(c) Actuator effectiveness

(d) Switching function

Fig. 11.8 Rudder missing with wind and gust

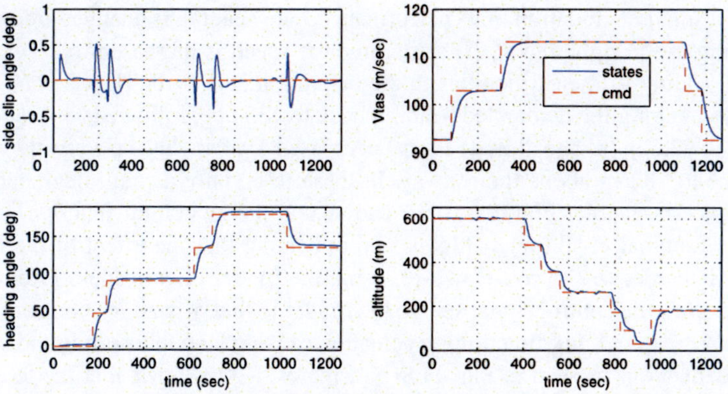

(a) Controlled states

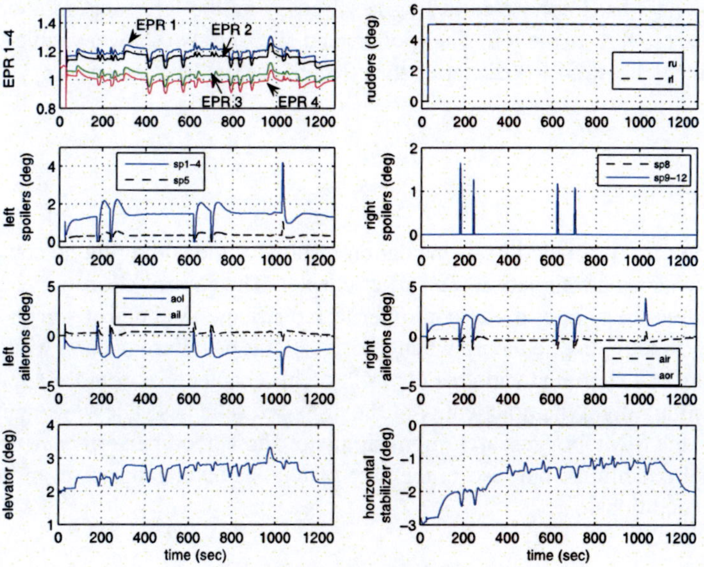

(b) Actuator positions

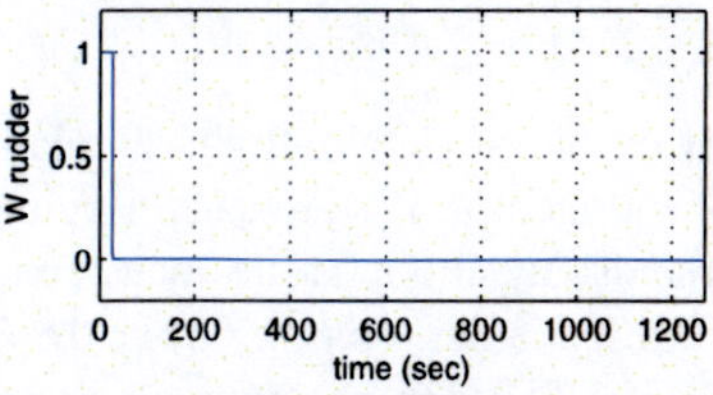

(c) Actuator effectiveness

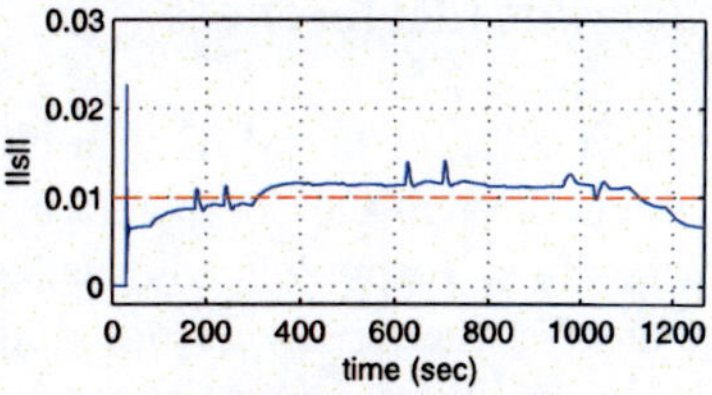

(d) Switching function

Fig. 11.9 Rudder runaway

axis and results in a tendency to pitch up. Figure 11.9(a) shows that the controller is tested on a slightly different manoeuvre. The sideslip command is kept at 0 deg and has only a small degradation in its performance. The heading is changed by 180 deg by banking to the right and at the same time the speed is increased to 113.18 m/s (220 kts) adding further difficulties to the banking manoeuvre. Then a bank left is tested by changing the demanded heading back to 135 deg, followed by a reduction in speed to 92.6 m/s. The altitude is also decreased to 30 m, before a small increase in altitude to 182 m above the runway. In these tests, only a small degradation in performance is visible. Figure 11.9(d) shows that the switching function just exceeds the nominal condition at high speed indicating that the effect of the rudder runaway is harder to control. However, using the rudder effectiveness information in Fig. 11.9(c), the control signal sent to the rudder is shut off and the control signals are sent to the remaining functioning actuators causing a visible split in the control surface deflections as seen in Fig. 11.9(b). Figure 11.9(b) shows that the four EPR have split to counteract the effect of the banking turn. Engines 3 and 4 on the right wing show less EPR compared to Engines 1 and 2 on the left wing, to counteract the tendency to turn to the left. The spoilers and ailerons also show a visible split in terms of the deflections to counteract the effect of the rudder runaway.

11.6 Summary

This chapter has presented implementation results for the sliding mode control allocation schemes developed in the earlier chapter. The control allocation aspect is used to allow the sliding mode controller to redistribute the control signals to the remaining functioning actuators when a serious fault or failure occurs, without reconfiguring or switching to another controller. This chapter has provided a 'proof of concept' to highlight the practicality of the proposed schemes. The implementation results on the SIMONA research flight simulator show good performance not only in nominal conditions, but also in the case of total actuator failures, even in wind and gust conditions.

11.7 Notes and References

The implementation of the SMC controller on the SIMONA flight simulator was made possible through the co-operation of the faculty of Aerospace Engineering, Delft University of Technology, the Netherlands, under the Group for Aeronautical Research and Technology in Europe, Flight Mechanics, Action Group 16 (GARTEUR AG16) program. As part of the GARTEUR AG16 program, a collection of other FTC and FDI schemes were evaluated. The results for various types of actuator failures can be found in [80, 117, 144, 172, 173, 242, 261].

Chapter 12
Case Study: ELAL Bijlmermeer Incident

The flight simulator results shown in Chap. 11 concentrated on individual control surface failures (jams with offsets or runaways). In this chapter, the model-reference controller from Chap. 10 will be applied to the more challenging task of 'landing' the B747 aircraft based on the ELAL flight 1862 Bijlmermeer scenario, with the assumption that no actuator effectiveness estimate is available.

12.1 Introduction

An independent investigation by Delft University of Technology [235] into the ELAL flight 1862, which crashed into an apartment building in Bijlmermeer, Amsterdam, suggested that there was still some control and flying capability associated with the crippled aircraft. This is backed up by an early publication of FTC on the ELAL 1862 scenario in [177], which showed that it was possible to control the crippled aircraft (although in [177], an exact damage model is assumed to be available).

This chapter presents the flight simulator results based on the FTC scheme from Sect. 10.3 obtained by experienced pilots based on the ELAL flight 1862 (Bijlmermeer incident) scenario. The results in this chapter are the outcome of the controller evaluation 'flight testing' campaign and the GARTEUR AG16 final workshop at Delft University of Technology, The Netherlands, in November 2007. The results represent the successful real-time implementation of the SMC controller proposed in Sect. 10.3 on the SIMONA 6-DOF flight simulator with professional pilots flying and evaluating the controller. The simulator tests by the pilots were done with the ELAL flight 1862 scenario which was previously used in the independent investigation of the Bijlmermeer incident by Delft University of Technology [235].

The ELAL flight 1862 incident represents a challenging scenario for any FTC strategy. In this chapter, it will be assumed that the controller has no knowledge of the failure and damage to the airframe, and that there is no FDI or fault estimation available.

The controller that has been used is the model-reference sliding mode controller proposed in Chap. 10. Here, since there is no FDI and no actuator effectiveness

H. Alwi et al., *Fault Detection and Fault-Tolerant Control Using Sliding Modes*,
Advances in Industrial Control,
DOI 10.1007/978-0-85729-650-4_12,

Fig. 12.1 ELAL flight 1862: the aircraft (picture: Werner Fischdick (c))

estimation available, fixed control allocation will be used. In this situation, there is no control signal redistribution to the healthy control surfaces. Instead, the fixed and equally distributed control allocation scheme is sufficient to access the remaining available control surfaces and 'passively' control the aircraft while ensuring stability and some nominal performance.

An outer loop inertial landing system (ILS) based on a PID is used in order to provide an outer loop command (roll and flight path demand) to guide the aircraft to capture the localiser (LOC) and glide slope (GS), and therefore landing the aircraft using a typical landing procedure. This is a typical controller configuration used by the pilot to reduce the workload, although this setup can be changed from the Mode Control Panel (MCP—see Fig. 11.3) to allow the pilot to manually land the aircraft.

12.2 ELAL Flight 1862: The Incident

A brief discussion of the actual incident will be presented in this section. This summary is based on the actual incident report by The Netherlands Aviation Safety Board [8].

On the 4th October 1992, the ELAL flight 1862 freighter aircraft—a Boeing 747-200 (Fig. 12.1)—departed from Schiphol Airport, Amsterdam after refuelling. This was a scheduled flight from New York JFK airport to Tel Aviv Israel, stopping in

Amsterdam for refuelling and a crew change [8]. Two runways were in use that day, 01L for takeoff and 06 for landing.

The aircraft takeoff started from runway 01L at 17:21 with a gross weight of 338.3 metric tons. The centre of gravity (CG) for takeoff was 23.1 percent of the mean aerodynamic chord (MAC). The aircraft initial climb was normal. At about 17:27.30, as the aircraft reached an altitude of about 6500 ft, the pilots transmit an emergency call as the aircraft was turning to the right. The pilots found themselves operating under extreme workload conditions trying to control the aircraft. Straight and level flight required full positive rudder pedal deflection and 60% to 70% maximum lateral control (the wheel almost full to the left) [8]. At around 17:28.06, the air traffic controller (ATC) instructed the pilot to turn to a heading of 260 deg. At 17:28.17 the crew reported a fire in engine No. 3 and reported the loss of thrust in engines Nos. 3 and 4. At 17:28.57, the crew was initially informed that runway 06 was in use for landing but the crew requested runway 27 which was much closer to their current location in order to land quickly. However, the aircraft was only 7 miles away from runway 27 at an altitude of 5000 ft i.e., too high for a landing approach. The crew was instructed by the ATC to turn right to a heading of 360 deg and descend to 2000 ft. At 17:31.17 'flap 1' setting could be heard as part of the conversation in the cockpit. The ATC then instructed a heading change to 100 deg and asked for the current status of the aircraft. The crew reported that engines Nos. 3 and 4 were inoperative and reported a problem on the wing flaps. The aircraft passed through the required heading of 100 deg and maintained a heading of 120 deg with airspeed 260 knots in a gradual descent to 2000 ft. As the aircraft was still heading to the required localiser (associated with runway 27), no correction in heading was instructed by the ATC. The ATC then instructed another right turn to a heading of 270 deg to intercept the localiser and align for the final approach course. At this point the aircraft was at 4000 ft, 260 knots ground speed and at a heading of 120 deg, 3 nautical miles north of runway 27 centre line and 11 miles away from the runway. The heading change took about 30 s and it was realised that the aircraft was going to overshoot the localiser. The ATC instructed another right turn and a change of heading to 290 deg in an attempt to capture the localiser from the south. After 20 s the ATC instructed a heading and altitude change to 310 deg and 1500 ft, respectively. At 17:35.03 the flight crew acknowledged the instruction and immediately reported control difficulties. At approximately 17:35.28, the co-pilot reported that the aircraft was 'going down' while the pilot was raising all flaps and lowering the landing gear. At 17:35.42 the aircraft crashed 13 km east of Schiphol airport into an eleven floor apartment building in Bijlmermeer, a suburb of Amsterdam (Fig. 12.2). The aircraft was destroyed during impact and in the subsequent fire. Investigation of the crash site indicated that the impact was at a very steep flight path angle with bank angle slightly over 90 deg to the right and with the nose down approximately 70 deg [8]. All the 4 flight crew and approximately 43 other people on the ground were fatally injured. The trajectory of the ELAL flight 1862 can be seen in Fig. 12.3. Further details on the incident can be found in [8].

(a)

(b) (picture: Jos Wiersema via Wikipedia)

(c) (picture: Jos Wiersema via Wikipedia)

(d) (picture: bbc.co.uk)

Fig. 12.2 ELAL flight 1862: the impact area at the Bijlmermeer apartment building

12.3 ELAL Flight 1862: Aircraft Damage Analysis

This section provides a summarised version of the incident report [8] describing the actual damage on the aircraft:

Unknown to the flight crew, the inboard fuse pin[1] that held engine No. 3 to the pylon[2] broke due to fatigue. This caused No. 3 engine and its pylon to also separate from the right wing shortly after takeoff causing damage to the leading edge of the right wing. The shedding of engine No. 3 from the right wing in an outboard and rearward direction resulted in a collision with No. 4 engine (see Fig. 12.4), causing it and its pylon to separate from the wing. The damage was extensive to the right wing structure (see Fig. 12.4). Several parts of the leading edge flaps and leading edge structures such as the leading edge flap No. 18 drive unit, the top skin panel above pylon No. 3, the adjacent inboard top skin panel (located above the most outboard

[1]The role of the fuse pin is to allow the engine to separate from the wing under a strong impact load that occurs in the event of a crash or hard landing in order to protect the fuselage from engine fire.

[2]The pylon is the structural component connecting the jet engine to the main wing.

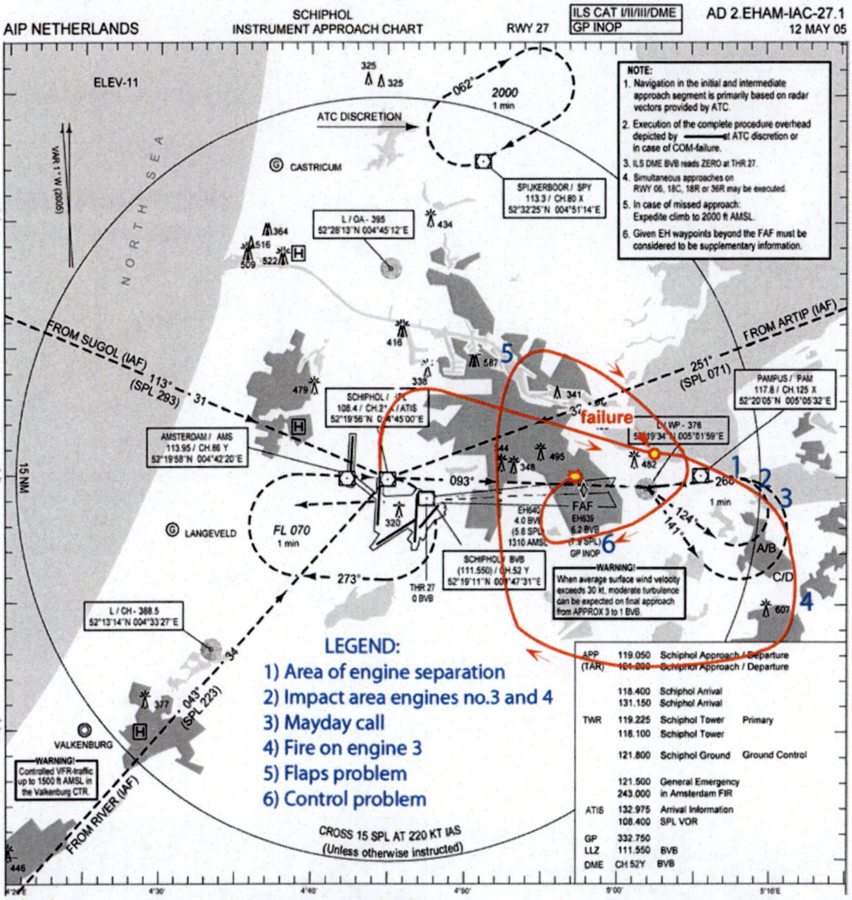

Fig. 12.3 ELAL flight 1862: flight trajectory

Krügerflap[3]) and the pneumatic duct of the bleed air system (which is normally located in the wing leading edge, between engines Nos. 3 and 4) detached from the wing when engines Nos. 3 and 4 were lost. It is assumed that due to the speed of the aircraft, the aerodynamic distortion, and turbulence, some other parts were also blown off the leading edge of the right hand wing up to the front spar. Right wing leading edge damage is assumed to have occurred up to the front spar of the right hand wing over an area approximately 1 metre left of pylon No. 3, to approximately 1 metre to the right of pylon No. 4. Figure 12.4 illustrates the estimated damage to the right wing. (The amount of damage on the wing leading edge after the separation of pylon No. 2 from a B747 accident at Anchorage on March 31, 1993 (Fig. 12.5), is indicative of the amount of damage probably inflicted on the El Al 1862 right wing leading edge [8].)

[3]The Krügerflap is the hinged flap on the leading edge.

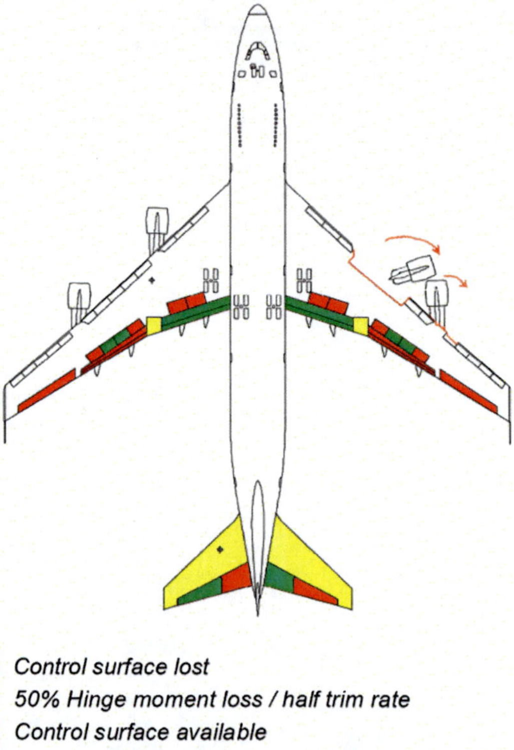

Fig. 12.4 ELAL flight 1862: actuator fault/failure and structural damage (figure adapted from [231])

Fig. 12.5 Wing damage due to separation of engine No. 2, Anchorage, 1993 (figure from [7])

In summary, the damage and the effect to the ELAL flight 1862 aircraft after the engines Nos. 3 and 4 separation are:

1. loss of thrust from engines Nos. 3 and 4;
2. right wing leading edge damage causing changes to the wing aerodynamic (higher drag and less lift due to the disrupted airflow over the damaged wing);

3. right inboard aileron and remaining spoilers 10 and 11 are less effective due to the airflow disruption resulting from the damage to the leading edge of the right wing;
4. limited roll control due to the loss of the outboard aileron (which is required during slow speed) and only partially available spoilers due to hydraulic Nos. 3 and 4 system loss;
5. weight loss of about 10 tonnes due to the separation of two engines;
6. hydraulic and pneumatic system Nos. 3 and 4 pressure loss;
7. loss and partial loss (half trim rate) of several control surfaces due to the loss of hydraulic systems Nos. 3 and 4 (see Fig. 12.4);
8. lateral CG displacement due to the loss of the engines;
9. degraded lateral control due to lower rudder lag as a result of hydraulic pressure loss;
10. positive yawing moment to the right due to asymmetric thrust from engines Nos. 1 and 2;
11. partial loss of the right wing leading edge flaps and the loss of outboard trailing edge flaps means that a high speed landing is necessary.

12.3.1 ELAL Flight 1862: Controllability and Performance

When the flight crew transmitted the emergency call at 17:27.30, the aircraft was turning to the right. In order to stabilise the aircraft at 260 knots and maintain straight and level flight, an almost full positive rudder deflection and almost maximum wheel (60%–70%) deflection was applied by the pilot [8]. The amount of corrections in order to obtain straight and level flight was something unexpected for the pilot, indicating something unusual had occurred. A few paragraphs from [234] best describe the expected effect when engine failure occurs (but with the engine still intact) compared to the ELAL 1862 scenario;

> The aircraft design and certification requirements state that there should be enough controllability to handle a multiple engine failure on one side in order to continue flight. The air minimum control speed (V_{mca}) is defined as the minimum speed during a failure of the most critical engine, at which aircraft control and a fixed heading can be maintained with full rudder and with sufficient lateral control authority to bank 5 degrees into the operating engine. The first sign of an engine failure will be a sudden roll of the aircraft. If directional control is not applied, or with a fixed rudder deflection, thrust asymmetry will cause the aircraft to yaw. Assuming a right multiple engine failure for a nominal case with no structural wing damage, the resulting yaw will create a negative side slip angle that creates a positive roll moment to the right. Instant control compensation in an engine failure flight condition consists of applying a rudder pedal input to counteract the yawing moment, a control wheel deflection to counteract the rolling moment or applying a thrust reduction on the remaining engines to decrease the yawing moment.
>
> For the case of ElAl Flight 1862, the wing damage caused an additional lift loss and drag increase on the right wing. Because these effects are a function of angle of attack, an increase of angle of attack will create an additional rolling moment and yawing moment into the direction of the dead engines. This in turn will require more opposite control wheel deflection, especially to counteract bank steepening during manoeuvring. Banking into the

> dead engines will increase the minimum control speed and therefore reduce the available controllability.
> The Flight 1862 accident aircraft was designed to have enough rudder authority to keep the control wheel almost neutral with two engines inoperative on one side. However, in the case of Flight 1862, the DFDR (digital flight data recorder) indicates that control wheel deflections between 20 to 60 degrees to the left were needed for lateral control and straight flight. The aerodynamic effects due to the wing damage and degraded effectiveness of the right wing inboard aileron, required larger left wing down control wheel deflections than in the nominal case.

The extensive damage to the right wing caused severe disruption of the airflow at the leading edge of the right wing, causing major aerodynamic changes thus reducing its lift generating capability and increasing drag [8].

> At small angles of attack the lift on both wings was essentially equal, at higher angles of attack the increase in lift on the damaged wing was less than the increase of lift on the undamaged wing. An increase in angle of attack therefore generates a roll moment. In the case of El Al 1862 this increase caused bank steepening during the right turns in the direction of the damaged wing. This effect was confirmed by the DFDR data.

Later studies [235] managed to estimate the total drag using the reconstruction of the ELAL 1862 aircraft in simulation. When compared with the DFDR a reasonable match was obtained. It is further stated in [235] that a 10% drag increase at low angle of attack was estimated compared to the undamaged wing. At higher angles of attack, a 20% to 30% increase in drag was estimated, due to flow separation behind the damaged leading edge. The extensive damage to the leading edge of the right wing (which caused flow separation and turbulence) meant that the right inboard elevator and the remaining spoilers 10 and 11 became less effective. This required more aileron deflection on the left inboard wing just to maintain straight and level flight. This is reflected in the almost maximum wheel deflection to the left applied by the pilot.

It is further reported in [8] that

> An energy analysis was performed based upon altitude and airspeed data from the DFDR. It should be realised that this method does not allow extrapolation of performance capabilities in other conditions than those encountered during this flight. Based on this analysis the following conclusions were made:
>
> - Marginal level flight capability was available at 270 knots and go-around power with a limited manoeuvring capability
> - Performance degraded below about 260 knots at increased angles of attack. Deceleration to 256 knots resulted in a considerable sink rate.
>
> ...Until the last phase of the flight, aircraft control was possible but extremely difficult. The aircraft was in a right turn to intercept the localiser and the crew was preparing for the final approach and may have selected the leading edge flaps electrically. During the last minute, the following occurred, as derived from DFDR data: the aircraft decelerated when the pitch attitude was increased probably to reduce the rate of descent. The associated increase in angle of attack caused increased drag. The additional drag from a side slip and possibly extended leading edge flaps resulted in a further speed decay. This speed decay was probably the reason for increasing thrust on the two remaining engines. All this generated an increased roll moment to the right by
>
> 1. asymmetric lift generation at an increased angle of attack;

2. high thrust asymmetry;
3. loss of aerodynamic efficiency of the right hand inboard aileron at increased angle of attack;
4. possible asymmetric lift due to leading edge flaps operation.

The resulting roll moment exceeded the available roll control.

This contributed to the excessive roll, loss of altitude and loss of control during the last phase of the flight. A similar effect was also seen during the GARTEUR AG16 flight evaluation test by three experienced professional pilots in the SIMONA motion flight simulator when using the classical/existing B747 controller (see the discussions in Sect. 12.6).

In one of the conclusions of the incident report [8], it was mentioned that in the case of ELAL 1862

> Performance and controllability were so severely limited that the airplane was marginally flyable . . . Because of the marginal controllability a safe landing became highly improbable, if not virtually impossible.

The independent investigation undertaken later in Delft University of Technology [232, 235], which also looked into the performance capabilities using simulations, suggests that there was still some control and flying capability and that the aircraft was still recoverable. It is further elaborated in [234] that

> from a technical point of view, the accident aircraft was recoverable if unconventional control strategies were used.

Further studies using [177] showed that it is possible to control the crippled aircraft (although in this FTC paper, an exact model of the damaged aircraft was assumed to be available).

One of the main findings in [232, 235] is that

> a significant improvement in available performance and controllability was available at lower weight if more fuel had been jettisoned

It is also interesting to highlight that, for the duration of the incident, the flight crew was unaware engines Nos. 3 and 4 had separated from the wing despite reporting loss of thrust from both engines. In [8]

> Information regarding the separation of both engines did not reach the ATC controllers concerned with the emergency, and was therefore not relayed to the crew. Although it remains questionable if, when relayed, this knowledge would have changed the course of events, it could have given the crew at least a better understanding of the unusual situation.

This is the motivation for the tests in this chapter to be carried out under the assumption that the type of failure is unknown and in the absence of any FDI or fault reconstruction strategy.

12.4 Controller Design

In this chapter both lateral and longitudinal control is considered. The main objective of the controller design is to bring the damaged ELAL 1862 aircraft to a near landing condition on Runway 27 at Schiphol airport (through a proper landing approach using localiser (LOC) and glide slope (GS) capture procedures if possible). It is assumed that no FDI or fault reconstruction is available to replicate the actual ELAL 1862 scenario—indeed the flight crew were even unaware that engine Nos. 3 and 4 had detached from the right wing.

As in Chap. 11, a linearisation of the nominal aircraft has been obtained around an operating condition of 263,000 kg, 92.6 m/s true airspeed, and an altitude of 600 m at 25.6% of maximum thrust and at a 20 deg flap position. The result is a 12th order linear model (separated into two sixth order models) associated with the lateral and longitudinal states. For design purposes, only the first four longitudinal and lateral states have been retained. The state-space system pairs, representing the lateral and longitudinal systems about the trim condition, can be found in (11.1), (11.2), (11.3) and (11.4) in Chap. 11. The states represent $x_{\text{lat}} = [p\ r\ \beta\ \phi]^{\text{T}}$ and $x_{\text{long}} = [q\ V_{\text{tas}}\ \alpha\ \theta]^{\text{T}}$. The lateral control surfaces are $\delta_{\text{lat}} = [\delta_{\text{air}}\ \delta_{\text{ail}}\ \delta_{\text{aor}}\ \delta_{\text{aol}}\ \delta_{sp1-4}\ \delta_{sp5}\ \delta_{sp8}\ \delta_{sp9-12}\ \delta_r\ e1_{\text{lat}}\ e2_{\text{lat}}\ e3_{\text{lat}}\ e4_{\text{lat}}]^{\text{T}}$ which represent aileron deflection (right and left: inboard and outboard) (rad), spoiler deflections (left: 1–4 and 5, right: 8 and 9–12) (rad), rudder deflection (rad) and lateral engine pressure ratios (EPR). The longitudinal control surfaces are $\delta_{\text{long}} = [\delta_e\ \delta_s\ e1_{\text{long}}\ e2_{\text{long}}\ e3_{\text{long}}\ e4_{\text{long}}]^{\text{T}}$, which represent elevator deflection (rad), horizontal stabiliser deflection (rad), and longitudinal EPR.

As in Chap. 11, the controlled output distribution matrices are

$$C_{c\text{lat}} = \begin{bmatrix} 0 & 0 & 1 & 0 \\ 0 & 0 & 0 & 1 \end{bmatrix}, \qquad C_{c\text{long}} = \begin{bmatrix} 0 & 0 & -1 & 1 \\ 0 & 1 & 0 & 0 \end{bmatrix}$$

which represent the states ϕ and β for lateral control, and flight path angle (FPA) and V_{tas} for longitudinal control. These linear models of the *nominal damage free aircraft* will be used to design the control schemes which will be described in the next sections. This is a major difference compared to [177] where the MPC controller is designed based on the exact knowledge of the damaged aircraft.

In the original coordinates, the control law can be summarised as

$$\nu_l(t) = L_e e(t) + F x_m(t) + G y_d(t)$$

where $L_e = -SA$ since SB_ν has been scaled so that $SB_\nu = I$. The nonlinear term $\nu_n(t)$ is given in (10.75), where the nonlinear gain $\rho(t)$ is based on the adaptive law (10.44)–(10.46). The final control law is given in (10.63) i.e., a fixed CA scheme which does not depend on estimates of $K(t)$.

12.4.1 Lateral Controller Design

The feedback matrices for the ideal lateral model from (10.17) have been designed using eigenstructure assignment [170]. The eigenvalues were chosen as $\{-0.3500 \pm 0.1500i, -0.5000, -0.4000\}$ and the desired and obtained eigenstructure are, respectively:

$$\underbrace{\begin{bmatrix} \star + *i & \star - \star i & \star & 0 \\ 0 & 0 & 0 & 0 \\ \star + \star i & \star - \star i & 0 & 0 \\ 1 + \star i & 1 - \star i & 1 & 1 \end{bmatrix}}_{\text{desired}}$$

$$\Longrightarrow \underbrace{\begin{bmatrix} 0.3195 - 0.1369i & 0.3195 + 0.1369i & 0.4498 & 0.3748 \\ -0.0000 - 0.0000i & -0.0000 + 0.0000i & -0.0430 & -0.0526 \\ 0.1619 + 0.1412i & 0.1619 - 0.1412i & 0.0182 & 0.0275 \\ -0.9127 & -0.9127 & -0.8919 & -0.9252 \end{bmatrix}}_{\text{obtained}}$$

which yields

$$F_{\text{lat}} = \begin{bmatrix} 0.5592 & -0.8808 & -0.6384 & 0.1010 \\ 0.0823 & 1.3729 & 2.5265 & -0.5851 \end{bmatrix}$$

As in Sect. 10.5.1, the feed-forward matrix G_{lat} has been designed using the inverse steady-state gain for the virtual triple $(A_{\text{lat}}, B_{\nu_{\text{lat}}}, C_{c_{\text{lat}}})$: specifically

$$G_{\text{lat}} = -\left(C_{c_{\text{lat}}}(A_{\text{lat}} + B_{\nu_{\text{lat}}} F_{\text{lat}})^{-1} B_{\nu_{\text{lat}}}\right)^{-1}$$

which yields

$$G_{\text{lat}} = \begin{bmatrix} -0.3078 & 0.0651 \\ 0.7310 & 0.3891 \end{bmatrix}$$

As in Chap. 11, it will be assumed that at least one of the control surfaces for both ϕ and β tracking will be available when a fault or failure occurs (i.e., one of either the four ailerons or the four spoilers will be available, and one of either the rudder or the four engine thrusts are available). Based on these assumptions, it can be verified from a numerical search that $\gamma_{0_{\text{lat}}}$ from (10.70) is $\gamma_{0_{\text{lat}}} = 8.1314$.

The matrix which defines the hyperplane must now be synthesised so that the conditions of (10.39) are satisfied. A quadratic optimal design (as discussed in Sect. 3.4.1) has been used to obtain the sliding surface S_{lat} which depends on the matrix $\mathcal{K}_{\text{lat}}$ in (10.28). The symmetric positive definite state weighting matrix has been chosen as $\mathbf{Q}_{\text{lat}} = \text{diag}(2, 2, 1, 1)$. The first and second terms of $\mathbf{Q}_{\text{lat}}$ are associated with the equations of angular acceleration in roll and yaw (i.e., $B_{\text{lat},2}$ term partition in (11.2)) and thus weight the virtual control term. By analogy to typical LQR frameworks, they affect the speed of response of the closed-loop system. Here, the first and second terms of $\mathbf{Q}_{\text{lat}}$ have been more heavily weighted compared to the last two terms, to give a reasonably fast closed-loop system response. The poles associated with the reduced order sliding motion are $\{-0.7136 \pm 0.0522i\}$, where

$$\mathcal{K}_{\text{lat}} = \begin{bmatrix} 0.0813 & -1.9138 \\ 1.3455 & 0.1854 \end{bmatrix}$$

Based on this value of $\mathcal{K}_{lat}$, calculations using (10.38) show that $\gamma_{1_{lat}} = 0.0230$, therefore $\gamma_{0_{lat}}\gamma_{1_{lat}} = 0.1870 < 1$ and so the requirements of (10.39) are satisfied. Also for this particular choice of sliding surface, $\|\tilde{G}_{lat}(s)\|_\infty = \gamma_{2_{lat}} = 0.0563$ from (10.37). Therefore, from (10.39),

$$\frac{\gamma_{2_{lat}}\gamma_{0_{lat}}}{1 - \gamma_{1_{lat}}\gamma_{0_{lat}}} = 0.5627 < 1$$

which shows that the system is stable for all $0 < w_i \le 1$. For implementation, the discontinuity in the nonlinear control term in (10.41) has been smoothed by using a sigmoidal approximation as in Sect. 3.2.2 where the scalar $\delta_{lat} = 0.05$. This introduces a further degree of tuning to accommodate the actuator rate limits—especially during actuator fault or failure conditions.

For simplicity, the variables related to the adaptive nonlinear gain have been chosen as $l_{1_{lat}} = 0$ and $l_{2_{lat}} = 1$, as in the last chapter. The parameter η_{lat} from (10.41) was chosen as $\eta_{lat} = 1$. In practice, a maximum limit ρ_{max} for the adaptive nonlinear gain in (10.44) has been imposed to avoid the actuators becoming too aggressive. Here, the maximum gain was set at $\rho_{max_{lat}} = 5$. The adaptation parameters from (10.45) have been chosen as $a_{lat} = 100$, $b_{lat} = 0.01$ and $\epsilon_{lat} = 5 \times 10^{-2}$. As in the last chapter, the parameter ϵ_{lat} was chosen to be able to tolerate variations in $\|s_{lat}(t)\|$ due to normal changes in flight condition, but small enough to enable the adaptive gain to be sensitive enough to deviation from zero due to faults or failures. Here a_{lat} has been chosen to be large to enable small changes in $\|s_{lat}(t)\|$ to cause significant changes in the gain, so that the control system reacts quickly to a fault. The parameter b_{lat} on the other hand dictates the rate at which $\rho_{lat}(t)$ will decrease after $\|s_{lat}(t)\|$ has returned below the threshold defined by ϵ_{lat}.

In order to emulate real aircraft flight control capability, an outer loop heading control law was designed based on a PID, to provide a roll command to the inner loop sliding mode controller. In the SIMONA implementation, this outer loop heading control can be activated by a switch in the cockpit. The proportional gain was chosen as $K_{p_{lat}} = 3$, the integrator gain was set as $K_{i_{lat}} = 0.1$ and the derivative gain was set as $K_{d_{lat}} = 3$. Note that the integrator component is only activated when the heading angle error is less than 5 deg to remove unwanted oscillation during manoeuvres but to still eliminate steady-state error.

12.4.2 Longitudinal Controller Design

As in the lateral controller, the feedback matrices for the ideal longitudinal model from (10.17) have been designed using eigenstructure assignment [170]. The eigenvalues were chosen as $\{-0.2400 \pm 0.1700 \; -0.7000 \; -0.1250\}$ and the desired and obtained eigenstructures are

$$\underbrace{\begin{bmatrix} 0.5+\star i & 0.5-\star i & 0 & 0 \\ 0 & 0 & 0 & 1 \\ 0.5+\star i & 0.5-\star i & 0 & 0 \\ 0 & 0 & 1 & 0 \end{bmatrix}}_{\text{desired}}$$

$$\Longrightarrow \underbrace{\begin{bmatrix} 0.1812-0.1283i & 0.1812+0.1283i & -0.1057 & 0.0001 \\ -0.0020+0.0015i & -0.0020-0.0015i & -0.0060 & 1.0000 \\ 0.3220-0.5264i & 0.3220+0.5264i & 0.9829 & -0.0037 \\ -0.7549 & -0.7549 & 0.1510 & -0.0012 \end{bmatrix}}_{\text{obtained}}$$

respectively, which yields

$$F_{\text{long}} = \begin{bmatrix} -0.0012 & -0.0380 & -0.6113 & 3.4367 \\ -0.0523 & 0.0017 & 0.4395 & -0.2396 \end{bmatrix}$$

As in the lateral control design, the feed-forward matrix G_{long} has been designed using the inverse steady-state gain for the virtual triple so that

$$G_{\text{long}} = -\left(C_{c_{\text{long}}}(A_{\text{long}} + B_{\nu_{\text{long}}} F_{\text{long}})^{-1} B_{\nu_{\text{long}}}\right)^{-1}$$

Here, the lateral feed-forward matrix G_{lat} is given by

$$G_{\text{long}} = \begin{bmatrix} -0.0015 & 0.0438 \\ 0.0665 & -0.0024 \end{bmatrix}$$

As in Chap. 11, it will be assumed that at least one of the control surfaces for FPA tracking will still be available when a fault or failure occurs. It is also assumed that at least one of the four engines is available for V_{tas} tracking. Based on these assumptions, it can be verified from a numerical search that $\gamma_{0_{\text{long}}} = 8.2913$ from (10.70).

As in the lateral controller, a quadratic optimal design has been used to obtain the sliding surface matrix. The weighting matrix has been chosen as $\mathbf{Q}_{\text{long}} = \text{diag}(2, 2, 1, 1)$. Again, the first two terms of $\mathbf{Q}_{\text{long}}$ are associated with the $B_{long,2}$ term partition in (10.2) (i.e., the q and V_{tas} states) which weight the virtual control term. These have been more heavily weighted compared to the last two terms. The poles associated with the reduced order sliding motion are $\{-1.1157, -0.3737\}$ where

$$\mathcal{K}_{\text{long}} = \begin{bmatrix} -0.0124 & -0.0037 \\ 0.4786 & 0.1247 \end{bmatrix}$$

Based on this value of $\mathcal{K}_{\text{long}}$, simple calculations from (10.38) show that $\gamma_{1_{\text{long}}} = 3.0160 \times 10^{-4}$. Therefore $\gamma_{0_{\text{long}}}\gamma_{1_{\text{long}}} = 0.0025 < 1$ and so the requirements of (10.39) are satisfied. For this choice of sliding surface, $\|\tilde{G}_{\text{long}}(s)\|_\infty = \gamma_{2_{\text{long}}} = 0.0066$ from (10.37). Therefore from (10.39),

$$\frac{\gamma_{2_{\text{long}}}\gamma_{0_{\text{long}}}}{1-\gamma_{1_{\text{long}}}\gamma_{0_{\text{long}}}} = 0.0551 < 1$$

which shows that the system is stable for all $0 < w_i \leq 1$. The discontinuity in the nonlinear control term in (10.41) has been smoothed by using a sigmoidal approximation where the scalar $\delta_{\text{long}} = 0.05$.

As in the lateral design, the variables related to the adaptive nonlinear gain have been chosen as $l_{1_{\text{long}}} = 0$ and $l_{2_{\text{long}}} = 1$. This was also found to give sufficiently good performance and removes the dependence of $r(t)$ on $x(t)$ and simplifies the implementation. The parameter η_{long} from (10.41) was chosen as $\eta_{\text{long}} = 1$. In practice, a maximum limit $\rho_{\max}$ for the adaptive nonlinear gain in (10.44) is imposed to avoid the actuators becoming too aggressive. Here, the maximum gain was set as $\rho_{\max_{\text{long}}} = 2$. The adaptation parameters from (10.45) have been chosen similar to those in the lateral design: $a_{\text{long}} = 100$, $b_{\text{long}} = 0.01$ and $\epsilon_{\text{long}} = 5 \times 10^{-2}$.

Again, to emulate real aircraft flight control capability, an outer loop altitude control law was designed based on a PID, to provide a FPA command to the inner loop sliding mode controller. In the SIMONA implementation, this outer loop altitude control can be activated by a switch in the cockpit. The proportional gain was set as $K_{p_{\text{long}}} = 0.001$, the integrator gain was set as $K_{i_{\text{long}}} = 0.00004$ and the derivative gain was set as $K_{d_{\text{long}}} = 0.02$. Note that the integrator component is only activated when the altitude error is less than 15 m to remove unwanted oscillation during manoeuvres but to eliminate steady-state error.

Note that both the lateral and longitudinal controllers manipulate the engine EPRs. For lateral control, differential engine EPR is required as a secondary 'actuator' for β tracking; whilst for longitudinal control, collective EPR is used for V_{tas} tracking. In the trials, 'control mixing' was employed, where the signals from both the lateral controller ($e_{1_{\text{lat}}}$, $e_{2_{\text{lat}}}$, $e_{3_{\text{lat}}}$ and $e_{4_{\text{lat}}}$) and longitudinal controller ($e_{1_{\text{long}}}$, $e_{2_{\text{long}}}$, $e_{3_{\text{long}}}$ and $e_{4_{\text{long}}}$) were added together before being applied to each of the engines.

12.5 SIMONA Implementation

The designed controller was implemented on the SIMONA flight simulator. The command input from the pilot is through the MCP. The new additional element in this control scheme is the implementation of the APP (approach) button which is engaged in order to intercept the LOC[4] (localiser) and GS[5] (glide slope) for the desired runway.

The controller was implemented as a SIMULINK® (version 2006b) model with appropriate inputs and outputs to connect it with the SIMONA hardware, as described in Fig. 12.6. In Fig. 12.6, one major difference compared to the implemen-

[4]A localiser is one component of an Instrument Landing System (ILS). The localiser provides runway centre-line guidance to an aircraft. The localiser is placed at the far end of the approached runway. It covers a distance of up to 46.3 km in a cone of up to 10 deg either side of the course.

[5]The glide slope provides vertical guidance to the aircraft during descent. The standard glide slope path is 3 deg. The glide slope signal is emitted by an antenna, located near the end of the runway. The glide slope provides the precise required altitude leading to the touchdown zone of the runway.

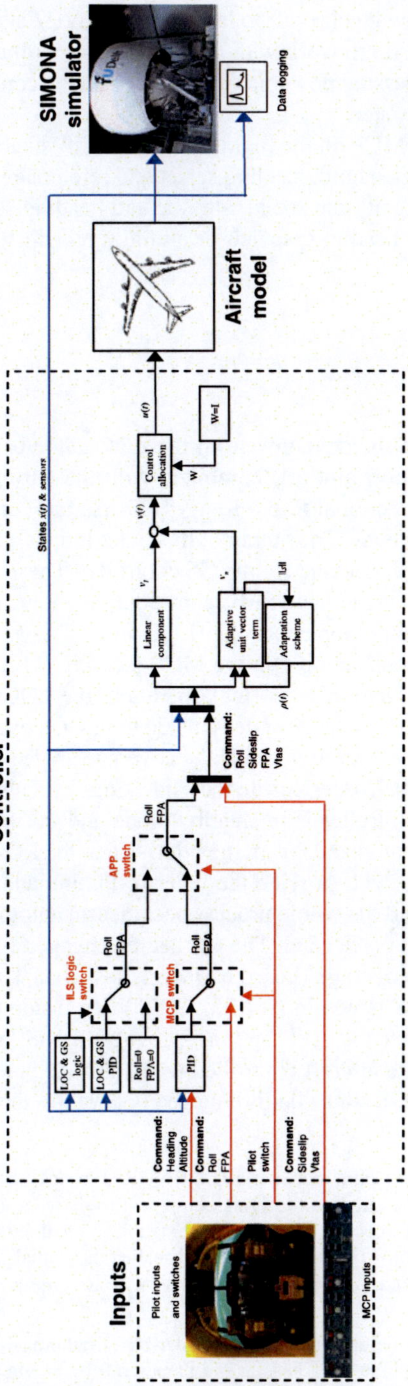

Fig. 12.6 SIMONA interconnections

tation in Chap. 11 is the addition of the ILS landing capability which has been added to the control loop to allow the aircraft to land on Runway 27 at Schiphol.

The controller was set up to work with an Ode4 solver with a fixed time step of 0.01 s. The available processing power is sufficient to run the controller in real-time, i.e., within 10 ms per time step.

A connection with the MCP on the flight deck (Fig. 11.3) enables the selection of 'control modes' e.g., altitude hold, heading select and reference values. The simulator trials were performed with the speed, altitude and heading select modes active. The pilot commands new headings, speeds or altitudes by adjusting the controls on the MCP.

12.5.1 ILS Landing

A sensor which measures the deviation from the LOC angle/beam error (which is available in typical transport aircraft) combined with the current aircraft heading and VOR (VHF Omni-directional Radio Range[6]) course radial is used for aligning the aircraft towards the runway. The output of this outer loop is a roll demand for the LOC controller and an FPA demand for the GS controller. These demand signals replace the pilot commands to the main SMC controller, to allow an almost automatic landing procedure. The outer loop controller (LOC and GS) is armed by the pilot by engaging the APP (approach) button on the MCP (see Fig. 11.3) when the aircraft is near the LOC signal coverage. In normal operation, the LOC will be the first to be engaged (LOC valid) when the aircraft is inside the LOC coverage (the DME[7] is less than 46.3 km when the aircraft is inside the coverage angle of ± 10 deg from the LOC beacon and $(-7, -0.75)$ deg inside the glide slope (GS) beacon). During the armed phase, the LOC controller is in standby mode and the aircraft is controlled either by heading or roll commands from the pilot. When the LOC is engaged (LOC valid), the LOC controller will provide the inner roll command to the core lateral sliding mode controller and the whole process becomes an automatic landing mode: i.e., no input from the pilot is needed. The GS is then engaged (GS valid) when the aircraft is inside the GS coverage (i.e., the DME is less than 18.5 km, the LOC is within ± 8 deg and the GS is within $(-1.35, -5.25)$ deg inside coverage). The GS is in an armed phase (after the APP button is engaged), and the GS controller is in a standby mode with the aircraft controlled using altitude or via FPA commands from the pilot. When the GS controller is engaged (GS valid), the GS controller will

[6]VOR is a type of radio navigation system for aircraft. VORs broadcast a VHF composite radio signal and data that allows the airborne receiving equipment to derive a magnetic bearing from the station to the aircraft (the direction from the VOR station in relation to the earth's magnetic North at the time of installation). This line of position is called the 'radial' in VOR parlance. The intersection of two radials from different VOR stations on a chart allow a 'fix' of the specific position of the aircraft.

[7]DME (Distance Measuring Equipment) is a transponder-based radio navigation technology that measures the distance between the LOC beacon and the aircraft by timing the propagation delay of VHF or UHF radio signals.

provide the FPA command to the core longitudinal SMC controller: again no input from the pilot is needed. If for some reason, during the LOC and GS manoeuvre to the runway, the LOC or GS becomes invalid (i.e., if the aircraft goes outside the LOC and GS coverage 'cones'), then the LOC and GS controller provide zero roll and FPA commands, respectively. Then, the pilot can disengage the APP button to retake full control of the aircraft.

12.6 Pilotted SIMONA Flight Simulator Results

The controller has been flown by three different professional pilots with experience on B747, B767, A330 and Citation II aircraft. An experienced B767 and Citation II pilot, rigorously tested the controller during the flight evaluation campaign before the GARTEUR AG16 final workshop in November 2007. During the AG16 final workshop, an experienced B747 pilot, flew the damaged 'aircraft' on the SIMONA simulator, during a presentation to the general public, including the local Dutch press (TV news, radio and newspapers). The results presented here are from tests flown by an experienced A330 KLM (Royal Dutch Airlines) pilot and a test pilot for NLR (National Aerospace Laboratory, The Netherlands) during the pilot evaluation campaign in November 2007.

Note that even though the controller has been designed based on the linearisation using a weight of approximately 263,000 kg, the controller was tested with a heavy trim weight of 317,000 kg as per the actual ELAL 1862 aircraft. This removes the advantage of low weight and low speed manoeuvrability and higher performance and controllability compared to the heavy trim weight as discussed in Sect. 12.3, which was one of the main findings in [232]. The heavy trim weight for the flight test also replicates the actual ELAL 1862 scenario and fits with the assumption that the exact damage and condition of the aircraft, post faults, is unknown, thus making the challenge even harder.

The flight test was made as realistic as possible. As in the actual ELAL 1862 scenario, the aircraft flew in a northerly direction from runway 01L before starting to make a right turn. Immediately after the right turn, the ELAL failure scenario occurred (see Fig. 12.7) whereby engines Nos. 3 and 4 detached from the right wing and caused significant damage to the right wing. The chosen runway, Runway 27, faced west at an angle of approximately 269 deg from the north. Therefore in order for the aircraft to land, two 90 deg turns must be performed before aligning the aircraft on Runway 27. During the third right turn, the aircraft was required to capture a localiser signal which guides the heading of the aircraft to line it up with the runway. During this normal procedure for landing, the aircraft will also be required to intercept a glide slope signal to enable the aircraft to descend at about a 3 deg flight path angle, which will bring the aircraft to the landing target zone. The flare[8] and the actual landing of the aircraft were not carried out and the simulation was stopped at a point 50 ft above the ground level.

[8]Flare is the nose up manoeuvre of the aircraft, used at the final part of landing i.e., just before touch down to arrest the descent rate.

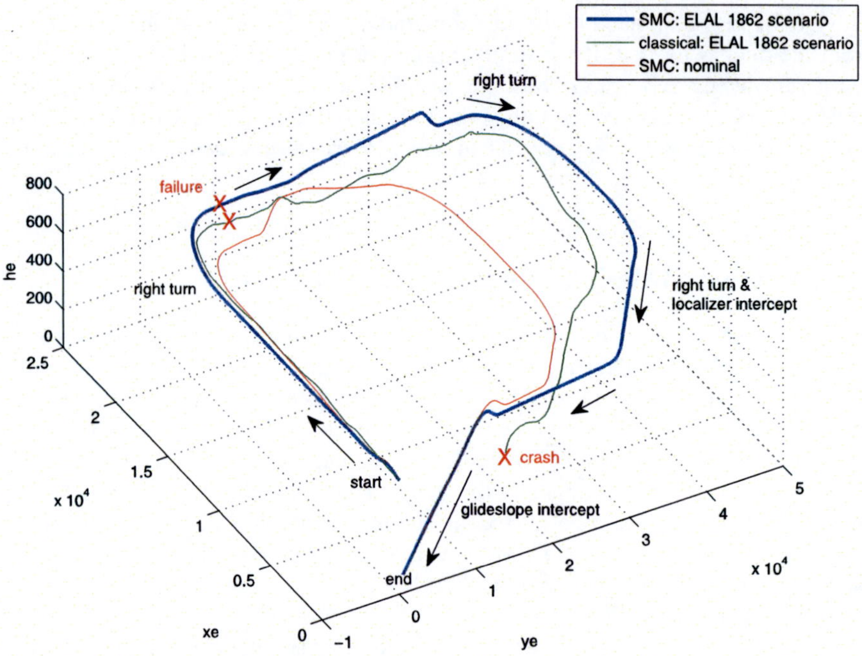

Fig. 12.7 Classical and SMC controller: 3D flight trajectory

12.6.1 Classical Controller

Figures 12.8–12.10 show the results of the piloted evaluation using the classical controller tested under the ELAL flight 1862 scenario conditions. The classical controller was tested by the pilots to give them some 'feel' and an idea of the severity of the actual ELAL scenario. The classical controller makes the pilot appreciate the controllability challenges and difficulties experienced during the failure, especially when compared to the FTC schemes.

Figure 12.7 shows an example of the flight trajectory of the piloted classical controller in the ELAL 1862 scenario. After the failure, the aircraft is still able to do right turns. Only during the final stage of the test flight does the aircraft lose control and crash before being able to line up with the runway.

The results shown in Figs. 12.8, 12.9 and 12.10 have been carried out with the heavy trim weight. Figure 12.8 shows the pilot control deflections. As described in Sect. 12.3, and in the incident studies in [8, 232], similar patterns appear. Immediately after the failure, the deflection of wheel, column and pedal increase in magnitude. As in the ELAL 1862 flight, almost maximum wheel deflection to the left to counter the right turn is visible. Also visible is the pedal deflection to counteract the yawing moment of the asymmetric thrust. Figure 12.8 also shows that close to the final stages of the test, a flap setting of 1 deg is selected to prepare for landing. At about 600 s, the power lever angle (throttle) is also reduced for landing. However,

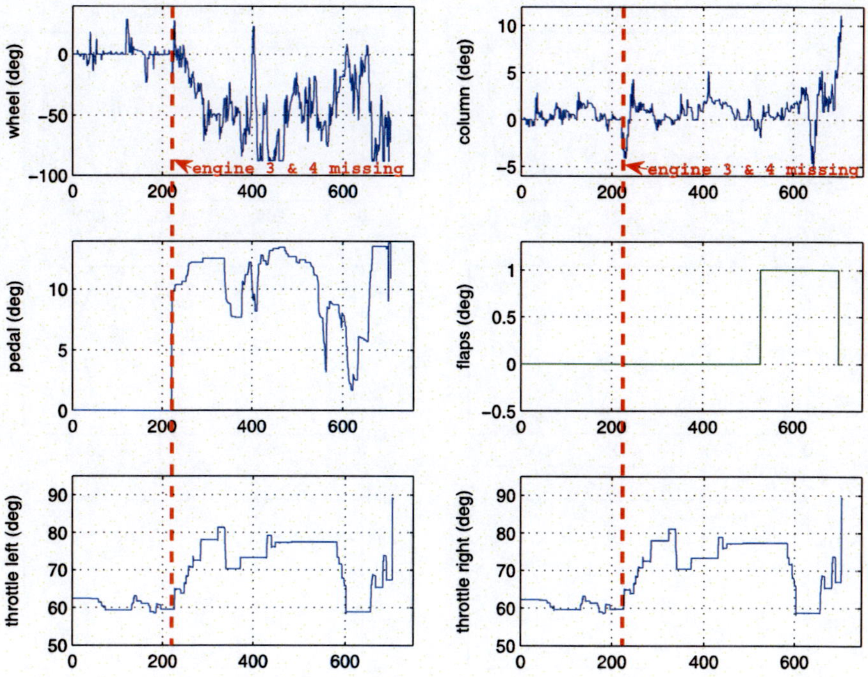

Fig. 12.8 ELAL 1862 scenario: classical controller: pilot deflection

when the speed reaches 110 m/s (approx 220 kts) near 700 s (Fig. 12.9), the aircraft becomes hard to control and banks to the right. Figure 12.8 shows that maximum left pilot wheel deflection is applied. Still unable to recover from the right bank, the flap is returned to a zero degree setting, and the throttle input is increased in order to regain control. However, the aircraft still rolls to the right and loses altitude and speed. The loss of altitude and FPA is tried to be compensated for by the high positive (pulled towards the pilot) column deflection. At this stage, all control is lost and the aircraft rolls at almost 80 deg right with the FPA nearing -40 deg and the pitch angle passing -20 deg. This is similar to what is described in the incident report in [8] when the ELAL 1862 aircraft hit the apartment building in Bijlmermeer, Amsterdam.

Analyzing the plots further, it can be seen that when the throttle is reduced in preparation for landing, the speed becomes low, and during descent, the angle of attack becomes high. As discussed in Sect. 12.3.1 and in [8, 232, 234, 235], the increase in the angle of attack causes high flow separation and turbulence behind the damaged right wing leading edge, resulting in the loss of lift and high drag (compared to the left wing). This increases the rolling and yawing moment to the right and results in a further drop in altitude and speed.

Figure 12.10 shows the control surface deflections of the classical controller. One major feature of the classical controller is that most of the control surfaces are me-

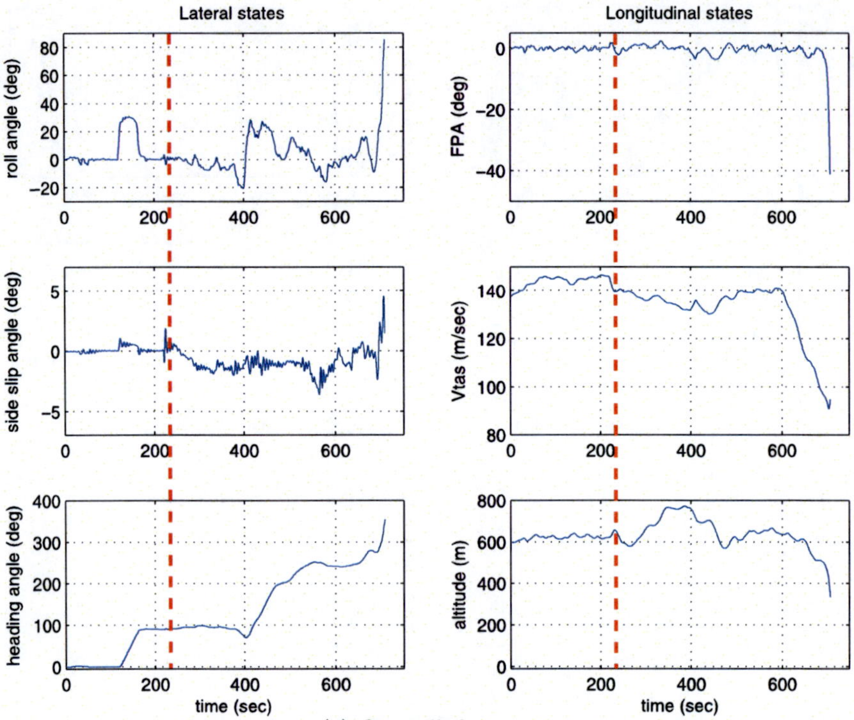

(a) Controlled states

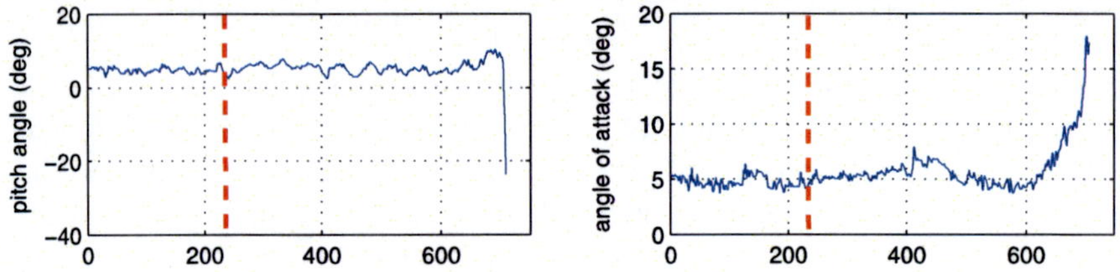

(b) Pitch angle and angle of attack

Fig. 12.9 ELAL 1862 scenario: classical controller: states

chanically linked. For example, the outer ailerons on the left and right wing are only fully active when a flap setting of more than 5 deg is used [118, 119]. This can be seen in Fig. 12.10, where the outboard aileron is inactive throughout the flight test. The high deflection of the left aileron and spoilers up to the saturation limits (−20 deg for the aileron and 40 deg for the spoiler) after the engine failure, shows that there is limited control even at a speed of 130–140 kts. Note that a positive deflection for the ailerons is a deflection down, and for the spoilers, positive is deflected up [118, 119]. The high deflections of the control surfaces on the left wing compensate for the loss of efficiency of the right inboard aileron and the remaining spoilers (10 and 11), and for the higher drag and loss of lift due to flow separation/turbulence caused by the damage to the leading edge of the right wing.

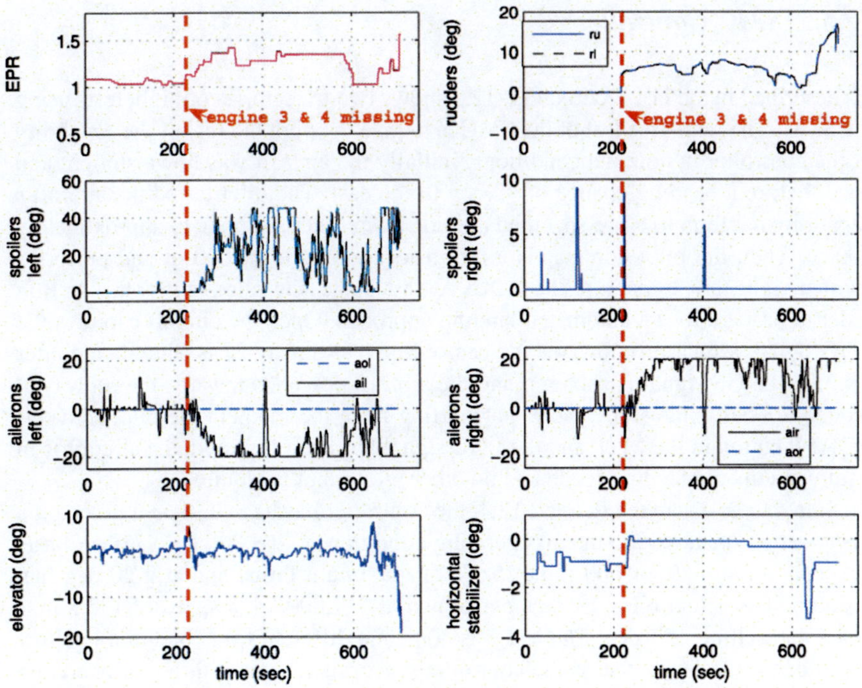

Fig. 12.10 ELAL 1862 scenario: classical controller: control surfaces deflection

This is also due to the fact that there is no direct authority to the outer ailerons due to mechanical linkages preventing the outboard aileron being functional in the classical controller setup [118, 119]. (This is one of the motivations for fly-by-wire (FBW) aircraft control.) Fig. 12.10 shows that the aileron deflections are most of the time at their saturation limits after the loss of engines 3 and 4 in order to obtain straight and level flight, and therefore most of the roll manoeuvre capability is assisted by the spoiler deflections. Shortly after the reduction in speed i.e., after approximately 600 s, the left aileron and spoilers saturate again, but due to the lower speed and higher angle of attack, the control surface deflections are insufficient to regain control as the aircraft has gone beyond the capability of the control surfaces to provide enough performance. Note that the general control surface deflections and behaviour in Figs. 12.8–12.10 closely follow the findings of the actual ELAL 1862 incident reported in [8].

Figure 12.7 shows the flight trajectory of the test. Three different trajectories are shown; the ELAL 1862 scenario with classical and SMC controllers and one with the SMC without any failure. With the classical controller, the pilot manages to maintain some performance and managed two banking turn manoeuvres. During the preparation for landing and capture of the localiser, the aircraft loses control and the simulation was stopped. The other two trajectories, associated with safe landings by the SMC controller, will be discussed in the next section.

12.6.2 SMC Controller

As described in the last section, Fig. 12.7 shows two trajectories of flight tests using the SMC. The fault-free test of the SMC is to give the pilot the feel of the capability of the controller in nominal conditions. Initially the aircraft was flown straight and level, before a heading change of 90 deg to the east. The pilot tested the aircraft's capability to climb to a pre-specified altitude from 600 m altitude to approximately 800 m. Then the pilot commands a return to an altitude of 600 m and performs another right turn to capture the LOC. At this stage, the pilot 'arms' the APP in order to prepare for an automated landing approach. Once the aircraft captures the LOC signal, a final turn towards the centre line of Runway 27 is started, and after a while, the GS signal is captured and the aircraft descends towards the runway at around a 3 deg glide slope. Note that starting from the moment the pilot activates the APP button in the MCP and the LOC signal has been captured, the aircraft is in a fully automated landing mode and no other pilot input is required.

Note that as discussed in Sect. 12.4, the controller has been designed based on a linearisation obtained around an operating condition of 263,000 kg, 92.6 m/s true airspeed, an altitude of 600 m at 25.6% of maximum thrust and at a 20 deg flap position. The actual pilot test was performed at 317,000 kg, a speed of 133.8 m/s and a flap setting of 1 deg. The test was done at a different trim condition to allow the pilot to rigorously access the controller performance under different operating conditions. Also note that the B747 aircraft setup in the GARTEUR AG16 program using the FTLAB747 software, has been modified to include a state of the art fly-by-wire capability, 'removing' mechanical links and locks from the classical B747 configuration. This allows more flexibility in the control strategy exploiting independent control of all available surfaces, thus increasing the ways redundant control surfaces can be used to achieve fault tolerant control.

Figure 12.7 also shows the trajectory of the SMC controller tested with the ELAL 1862 failure scenario. The same controller as that used in the nominal fault-free case is applied. In general, the controller performs the same right turn manoeuvres, LOC and GS intercept and lands on Runway 27. The SMC with the ELAL 1862 failure manages to bring the aircraft close to landing on the desired runway. Figure 12.11 shows the controlled states of the damaged aircraft with the SMC controller. Note that, at the beginning of the simulation, before the failure occurs at around 200 s, the FPA, V_{tas} and altitude show small steady-state errors due to the mismatch between the designed trim conditions and the test conditions as described earlier. This is due to the absence of integrators in the main SMC controller. The mismatch between the designed and test trim conditions demonstrate the controller coping with uncertainty and allows the pilot to rigorously test the controller outside its 'comfort zone'.

Figure 12.11 shows that after the failure occurs, at approximately 200 s, the climb capability of the aircraft is slightly degraded when the pilot requests an increase in altitude to 800 m (from 600 m). On the other hand, the more important descent capability of the SMC controller is not degraded as it is able to follow the glide slope of 3 deg towards the runway. This is shown in Fig. 12.12. The glide slope error is maintained below 0.5 deg. Figure 12.11 also shows that the side slip angle of the

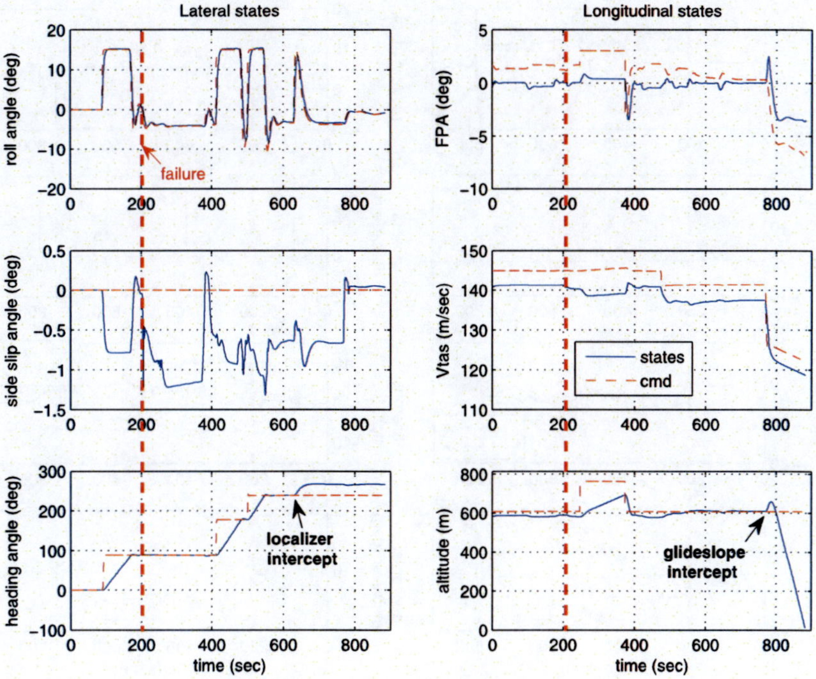

Fig. 12.11 ELAL 1862 scenario: SMC controller: controlled states

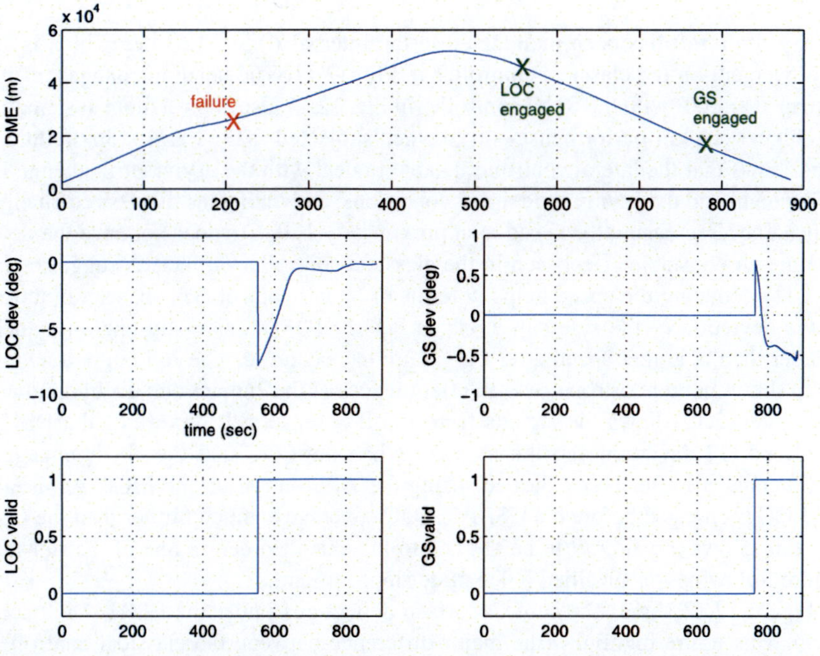

Fig. 12.12 ELAL 1862 scenario: SMC controller: LOC and GS deviation angle

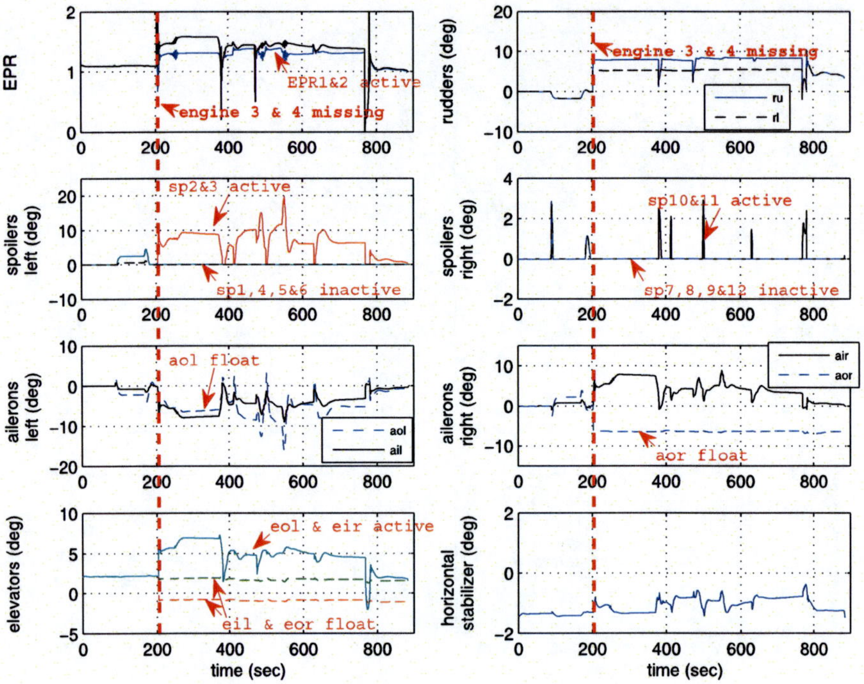

Fig. 12.13 ELAL 1862 scenario: SMC controller: control surfaces deflection

damaged aircraft has been maintained in the interval (0.5, −1.5) deg which is an improvement on the classical controller in Fig. 12.11. The heading changes of the damaged aircraft with the SMC controller in Fig. 12.11 also show a more systematic and higher level of performance—even when subjected to the ELAL 1862 failures. This shows that the lateral controller is able to deal with the asymmetric change in CG, weight and the asymmetric thrust conditions, and maintains the desired change in heading. Decreasing the speed to approximately 120 m/s does not have the devastating and unstable effect seen in the classical controller. In fact, as suggested in [8, 232], reducing the speed helps in terms of lateral control. This is seen in terms of the deviation of the side slip angle in Fig. 12.11. The side slip angle is much smaller than at higher speed after the failure has occurred. The roll angle tracking again shows good performance even after the loss of the engines and the hydraulics.

Figure 12.12 shows the signals from the ILS sensors. It represents the DME, LOC and GS deviation and the moment when the LOC and the GS is engaged (valid/engaged) after being 'armed' using the APP button on the MCP. As usual, the LOC is engaged before the GS. The LOC coverage is much further than the GS, and this allows the aircraft to be aligned to the extended centre line of the runway before following the specified 3 deg glide slope descent.

Figure 12.13 shows the control surface deflections under the ELAL 1862 scenario. This figure highlights the major difference between the classical controller (which is mechanically linked) and the FBW aircraft that has been provided by the

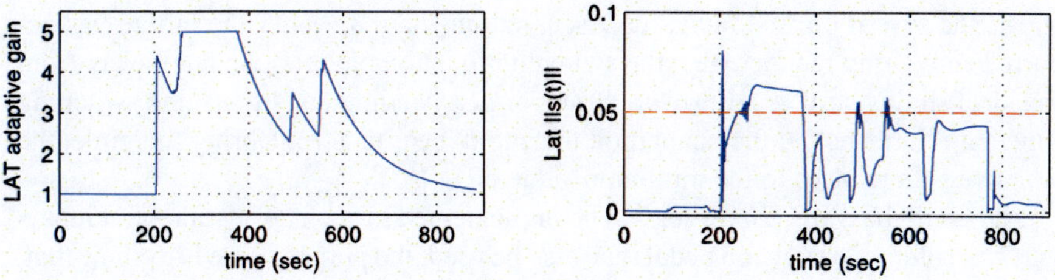

Fig. 12.14 ELAL 1862 scenario: SMC controller: lateral adaptive gain

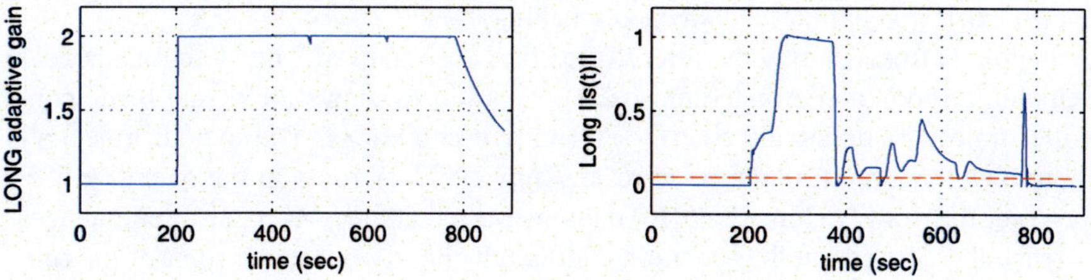

Fig. 12.15 ELAL 1862 scenario: SMC controller: longitudinal adaptive gain

GARTEUR AG16 modification [234]. In this figure, the outer ailerons can be seen to be independently mobile before the occurrence of the failure. After the failure, the outer ailerons (left and right) 'float' due to the loss of hydraulic systems 3 and 4. Independent control can also be seen in the spoilers, elevators, rudders and EPR. The effect of losing the hydraulic system can also be seen in the 'floating' of the inner left and outer right elevators (see Fig. 12.13) where a clear distinction between the control surface deflection can also be seen. The spoilers also show similar patterns. Before the loss of engines 3 and 4, all the spoilers seem to be moving independently; and when the failure occurs, only spoilers 2, 3, 10 and 11 are active, the rest remain at zero deflection. In general, the control surface deflections of the elevators, ailerons and spoilers are almost half of the ones from the classical controller (see Fig. 12.13). The control surface deflections from the SMC controller do not reach the saturation limits of the surfaces and the spoilers and the ailerons, are generally less aggressive. Engine EPR shows that differential thrust has been used to achieve the desired performance, obtaining a small side slip angle and roll angle. Note that all the surfaces are controlled independently by the CA SMC scheme. The pilot input only comes from supplying the higher level commands such as heading and altitude change (or roll or FPA commands through the MCP panel). This reduces the pilot's workload compared to the classical controller where the demand is high.

Figures 12.14 and 12.15 show the adaptive gain and the associated $\|s(t)\|$ signals that initiate the gain adaptation. Before the occurrences of the failure, the sliding signal $s(t)$ is below the selected threshold. Once the threshold is exceeded, the gain is adapted from a minimum of 1 up to the maximum of 5, and 2 for the lateral and longitudinal axes, respectively. High deviation from the sliding surface $s(t) = 0$

shows the severity of the faults (as discussed in Chap. 8). After the failure has occurred and during manoeuvres, the switching function plot $s(t)$ deviates away from the ideal sliding surface. However, during or near landing conditions, the switching function returns below the adaptation threshold, near to zero. During this time, the adaptive gain reduces to the minimum value of 1.

Although the SMC controller can be implemented in such a way that pilot inputs (such as column, wheel and pedal) can also be used, the purpose here is to show that, as a proof of concept, the SMC controller is more than able to handle all the rigorous tests and failures it is subjected to using the minimal amount of input from the pilot, thus lowering the workload during an emergency condition. This will allow pilots to concentrate on higher level decision making.

Figure 12.16(a) shows the aircraft position on a map of The Netherlands near Schiphol airport (and Amsterdam itself). This figure shows the actual SMC controller trajectory under the ELAL 1862 failure conditions. The overall trajectory shows that the aircraft manages to reach Runway 27. A zoom of the overall trajectory near the runway (Fig. 12.16(b)) demonstrates that the SMC controller manages to reach the desired landing position—although slightly out to the right of the runway.

12.7 Summary

This chapter has presented piloted flight simulator results associated with the ELAL flight 1862 (Bijlmermeer incident) scenario, which is one of the case studies of the GARTEUR action group AG16. The results represent the successful implementation of a FTC SMC controller on the SIMONA 6-DOF flight simulator configured to represent a B747, with pilots flying and testing the controller. The results from the proposed SMC controller show that the control surface deflections are much lower than that of the classical controller. A significant reduction in the pilot's workload, especially with the implementation of the ILS landing approach, has been shown from the SMC controller tests. The comments and feedback from the professional pilots after the simulator flight tests show that the proposed SMC scheme has the ability to help pilots 'land' the aircraft safely on the designated runway.

12.8 Notes and References

Other piloted evaluation results on the SIMONA flight simulator for the ELAL flight 1862 scenario can be found in [80]. Other studies which consider the ELAL flight 1862 scenario appear in [126, 177].

The incident in Sioux city in 1999 and more recently in Baghdad in 2003 [41] showed remarkable recoveries using only engine thrust after a major loss of hydraulics to all control surfaces. Another incident with a similar failure was flight 123 in Japan [109]. Learning from these past incidents, it is clearly possible that,

(a) Overall trajectory

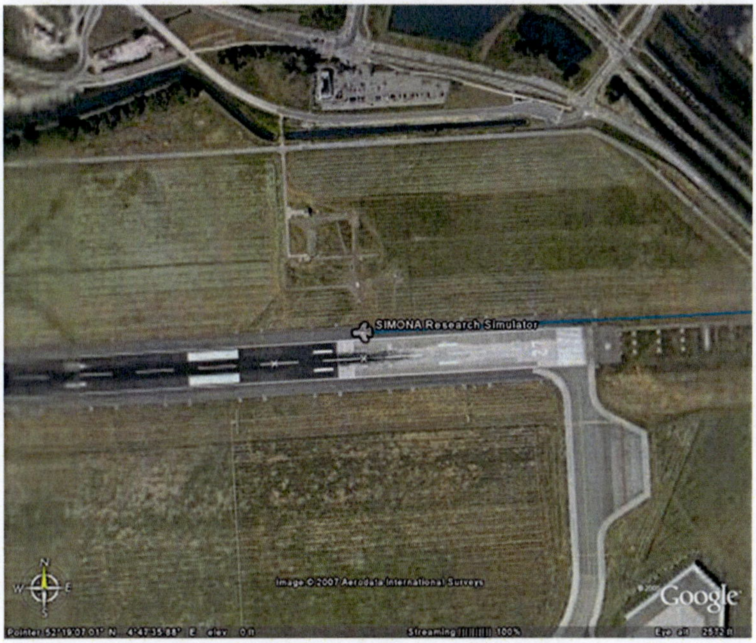

(b) Zoomed trajectory near the runway

Fig. 12.16 Piloted SIMONA flight trajectory of ELAL 1862 scenario with model-reference SMC controller with control allocation (picture: Google Earth (c)) using only the engines, a safe approach and landing can be achieved. After the incident in Sioux city, the National Transportation Safety Board (NTSB) recommended and encouraged the development of backup flight controllers to be used in such emergency situations [253]. NASA Dryden led by F.W. Burcham and C.G. Fullerton took up the challenge and conceived, developed and tested the first propulsion controlled aircraft system [253]. Impressive simulator and actual flight test results were obtained during the project. Tests were done on many types of aircraft ranging from fighter jets (F15 [253]) to large transport aircraft (DC11 and the B747 [40, 42–44, 253]). Based on one particular type of failure, the PCA scheme is considered as a MMST by [143]. Tests using the PCA scheme system and its variant by NASA Dryden have been undertaken on a real aircraft. The PCA scheme has shown its airworthiness during these flight tests. Even though the PCA considers one of the most catastrophic types of failure, it is not sufficient to solve general FTC problems [143]. In the case of airframe damage, loss of engines (as occurred in the ELAL flight 1862 Bijlmermeer incident [233]) and indeed the partial loss of the hydraulic systems, PCA might not be suitable. NASA reports on PCA can be found in [40–44, 253]. Other papers such as [120, 133] also consider propulsion control strategies. The two papers use a different type of controller design from the MMST, and in [133] an adaptive neural network with control allocation is used instead.

Chapter 13
Concluding Remarks

This book has described recent research in the area of sliding mode theory applied to fault detection and fault tolerant control problems, conducted over the previous decade. There is significant interest in the area of fault detection and fault tolerant control, as automation and autonomy are pursued by engineers working in a wide variety of application areas. Consequently there is an ever expanding literature describing different paradigms to address these issues. This book has focussed on sliding mode methods applied to these problems, and consequently describes a narrow focus of work tackling these issues. The results which have been described are all based on rigorous underpinning theory, but with the underlying requirement that they be practical, and hence implementable in real-time. The methods described in the preceding chapters represent what might be viewed as a 'classical' sliding mode approach, since the discontinuities associated with the switching term appear when examining the first derivative of the switching function. In recent years, so-called higher order sliding mode approaches have been extensively researched whereby the switching function and a finite number of its higher derivatives are forced to remain at zero. The application of these ideas to fault detection is only now beginning to emerge, and it will doubtless represent an area of future development. One example of a higher order sliding mode approach to FDI is given in [218].

One of the purposes of this book is to try to demonstrate the practicality of sliding mode methods for fault detection and fault tolerant control. One of the—at least in our opinion—popular misconceptions associated with sliding modes is that there is always 'chattering', which renders the method inapplicable for real engineering systems. We hope in the preceding chapters—and particularly the ones describing the implementation of the controllers and observers on the SIMONA research simulator (at Delft University of Technology)—that we have demonstrated that these methods can be developed further and could form part of health monitoring and control strategies in industrial engineering systems.

Currently, despite all the various sophisticated schemes for fault detection proposed by academia, these are not being taken up by industry. Many of the schemes used in, for example, the aerospace sector would be considered as 'basic' by academics and often rely on very simple condition monitoring checks (e.g., signal deviations from their nominal behaviour) or trend analysis. On-going work at the time

H. Alwi et al., *Fault Detection and Fault-Tolerant Control Using Sliding Modes*, Advances in Industrial Control,
DOI 10.1007/978-0-85729-650-4_13,

of writing is investigating the use of sliding mode schemes for industrial fault detection problems associated with modern civil aircraft through the European FP7 project ADDSAFE.[1] If successful, the project offers the possibility of testing such schemes on AIRBUS's 'Iron-Bird' facility at Toulouse. This would represent a significant step in terms of taking these ideas from an academic environment towards an industrial one. Watch this space ...

[1] For details see http://addsafe.deimos-space.com/

References

1. Civil aviation safety data. Technical report, Civil Aviation Authority of the Netherlands (CAA-NL) (2003)
2. In-flight upset; 240 km NW Perth, WA; Boeing Co 777-200, 9M-MRG. Aviation safety investigation report—Final 200503722, Australian Transport Safety Bureau (2005)
3. The transportation safety board of Canada issues safety recommendations to improve rudder inspections on Airbus aircraft. Communiqués TSB # A04/2006 A05F0047, Transportation Safety Board of Canada (2006)
4. An, Y.H.: A design of fault tolerant flight control systems for sensor and actuator failures using online learning neural networks. Ph.D. thesis, West Virginia University (1998)
5. Anand, M.D., Selvaraj, T., Kumanan, S., Janarthanan, J.: A hybrid fuzzy logic artificial neural network algorithm-based fault detection and isolation for industrial robot manipulators. Int. J. Manuf. Res. **2**(3), 279–302 (2007)
6. Anderson, B.D.O., Moore, J.B.: Optimal Control: Linear Quadratic Methods. Prentice Hall, New York (1989)
7. Anon: In-flight engine separation, Japan Airlines, Inc., Flight 46E, Boeing 747-121, N473EV, Anchorage, Alaska, March 31, 1993. Aircraft accident report NTSB/AAR-93/06, National Transportation Safety Board (1993)
8. Anon: EL AL flight 1862, aircraft accident report 92-11. Technical report, Netherlands Aviation Safety Board, Hoofddorp (1994)
9. Anon: Uncontrolled descent and collision with terrain, USAIR flight 427, Boeing 737-300, N513AU, near Aliquippa, Pennsylvania September 8, 1994. Aircraft accident report NTSB/AAR-99-01, National Transportation Safety Board (1994)
10. Aravena, J., Zhou, K., Li, X.R., Chowdhury, F.: Fault tolerant safe flight controller bank. In: Proceedings of the IFAC Symposium SAFEPROCESS '06, Beijing, China, pp. 807–812 (2006)
11. Armeni, S., Casavola, A., Mosca, E.: Robust fault detection and isolation for LPV systems under a sensitivity constraint. Int. J. Adapt. Control Signal Process. **23**, 55–72 (2009)
12. Åström, K.J., Hagglund, T.: Pid Controllers: Theory, Design and Tuning. Instrument Society of America, Research Triangle Park (1995)
13. Åström, K.J., Wittenmark, B.: Computer Controlled Systems: Theory and Design. Prentice Hall, Englewood Cliffs (1984)
14. Åström, K.J., Wittenmark, B.: Adaptive Control. Addison-Wesley, Reading (1989)
15. Baev, S., Shtessel, Y., Edwards, C.: HOSM observer for a class of non-minimum phase causal nonlinear MIMO systems. In: Proceedings of the IFAC World Congress, Seoul, pp. 4797–4802 (2008)
16. Balas, G.J.: Linear, parameter-varying control and its application to a turbofan engine. Int. J. Robust Nonlinear Control **12**, 763–796 (2002)

H. Alwi et al., *Fault Detection and Fault-Tolerant Control Using Sliding Modes*,
Advances in Industrial Control,
DOI 10.1007/978-0-85729-650-4,

17. Balestrino, A., DeMaria, G., Zinober, A.S.I.: Nonlinear adaptive model-following control. Automatica **20**, 559–568 (1984)
18. Barbot, J.P., Boukhobza, T., Djemai, M.: Sliding mode observer for triangular input form. In: Proceedings of the 35th IEEE Conference on Decision and Control, CDC, pp. 1489–1490 (1996)
19. Barbot, J.P., Djemai, M., Boukhobza, T.: Implicit triangular observer form dedicated to a sliding mode observer for systems with unknown inputs. Asian J. Control **5**, 513–527 (2003)
20. Beck, R.E.: Application of control allocation methods to linear systems with four or more objectives. Ph.D. thesis, Virginia Polytechnic Institute and State University, Blacksburg, Virginia (2002)
21. Bejarano, F.J., Fridman, L., Poznyak, A.: Estimation of unknown inputs, with application to fault detection via partial hierarchical observation. In: Proceedings of the European Control Conference, Kos, Greece, pp. 5154–5161 (2007)
22. Bejarano, F.J., Fridman, L., Poznyak, A.: Exact state estimation for linear systems with unknown inputs based on hierarchical super-twisting algorithm. Int. J. Robust Nonlinear Control **17**, 1734–1753 (2007)
23. Bejarano, F.J., Fridman, L., Poznyak, A.: Hierarchical observer for strongly detectable systems via second order sliding mode. In: Proceedings of the Conference on Decision and Control, New Orleans, USA, pp. 487–492 (2007)
24. Blanke, M., Kinnaert, M., Lunze, J., Staroswiecki, M.: Diagnosis and Fault-tolerant Control, 2nd edn. Springer, Berlin (2006)
25. Blanke, M., Staroswiecki, M., Wu, N.E.: Concepts and methods in fault-tolerant control. In: Proceedings of the American Control Conference, Arlington, TX, USA, pp. 2606–2020 (2001)
26. Bocaniala, C.D., Palade, V.: Computational Intelligence Methodologies in Fault Diagnosis: Review and State of the Art. Springer, Berlin (2006)
27. Bokor, J., Balas, G.: Detection filter design for LPV systems—a geometric approach. Automatica **40**, 511–518 (2004)
28. Bokor, J., Szabo, Z., Stikkel, G.: Failure detection for quasi LPV systems. In: IEEE Conference on Decision and Control, Maui, Hawaii, USA (2002)
29. Bordignon, K.A., Durham, W.C.: Closed-form solutions to constrained control allocation problem. J. Guid. Control Dyn. **18**(5), 1000–1007 (1995)
30. Bošković, J.D., Mehra, R.K.: A multiple model-based reconfigurable flight control system design. In: Proceedings of the 37th IEEE Conference on Decision and Control, Tampa, FL, USA, pp. 4503–4508 (1998)
31. Bošković, J.D., Mehra, R.K.: Control allocation in overactuated aircraft under position and rate limiting. In: Proceedings of the American Control Conference, Anchorage, AL, USA, pp. 791–796 (2002)
32. Bošković, J.D., Mehra, R.K.: Failure detection, identification and reconfiguration in flight control. In: Fault Diagnosis and Fault Tolerance for Mechatronic Systems: Recent Advances. Springer, Berlin (2002)
33. Boyd, S.P., Ghaoui, L.E., Feron, E., Balakrishnan, V.: Linear Matrix Inequalities in Systems and Control Theory. SIAM, Philadelphia (1994)
34. Brière, D., Favre, C., Traverse, P.: A family of fault-tolerant systems: electrical flight controls, from Airbus A320/330/340 to future military transport aircraft. Microprocess. Microsyst. **19**, 75–82 (1995)
35. Brière, D., Traverse, P.: Airbus A320/A330/A340 electrical flight controls: A family of fault-tolerant systems. Digest of papers FTCS-23 the twenty-third International Symposium on Fault-Tolerant Computing, Toulouse, France, pp. 616–623 (1993)
36. Bryson, A.E.: Control of Spacecraft and Aircraft. Princeton University Press, Princeton (1994)
37. Buffington, J.: Tailless aircraft control allocation. In: AIAA Guidance, Navigation and Control, pp. 737–747 (1997)
38. Buffington, J., Chandler, P., Pachter, M.: Online system identification for aircraft with distributed control effectors. Int. J. Robust Nonlinear Control **9**, 1033–1049 (1999)

39. Buffington, J.M., Enns, D.F.: Lyapunov stability analysis of daisy chain control allocation. J. Guid. Control Dyn. **19**, 1226–1230 (1996)
40. Burcham, F.W., Burken, J., Maine, T.A., Bull, J.: Emergency flight control using only engine thrust and lateral center-of-gravity offset: A first look. Technical memorandum NASA/TM-4789, NASA (1997)
41. Burcham, F.W., Fullertron, C.G., Maine, T.A.: Manual manipulation of engine throttles for emergency flight control. Technical report NASA/TM-2004-212045, NASA (2004)
42. Burcham, F.W., Maine, T.A., Burken, J., Bull, J.: Using engine thrust for emergency flight control: MD-11 and B-747 results. Technical memorandum NASA/TM-1998-206552, NASA (1998)
43. Burcham, F.W., Maine, T.A., Kaneshinge, J., Bull, J.: Simulator evaluation of simplified propulsion-only emergency flight control system on transport aircraft. Technical report NASA/TM-1999-206578, NASA (1999)
44. Burken, J., Burcham, J.W., Trindel, A.M., Feather, J., Goldthorpe, S., Kahler, J.A.: Flight test of propulsion-based emergency control system on the MD-11 airplane with emphasis on the lateral axis. Technical report NASA/TM-4746, NASA (1996)
45. Burton, J.A., Zinober, A.S.I.: Continuous approximation of variable structure control. Int. J. Syst. Sci. **17**, 876–885 (1986)
46. Butcher, J.C.: Numerical Methods for Ordinary Differential Equations. Wiley, Chichester (2003)
47. Caccavale, F., Villani, L.: Fault Diagnosis and Fault Tolerance for Mechatronic Systems: Recent Advances. Springer, New York (2003)
48. Campo, P.J., Morari, M., Nett, C.N.: Multivariable anti-windup and bumpless-transfer: A general theory. In: Proceedings of the American Control Conference, Pittsburgh, PA, USA, pp. 1706–1711 (1989)
49. Cattarius, J., Inman, D.J.: Experimental verification of intelligent fault detection in rotor blades. Int. J. Syst. Sci. **31**(11), 1375–1379 (2000)
50. Chen, J., Patton, R., Zhang, H.: Design of unknown input observers and robust fault detection filters. Int. J. Control **63**, 85–105 (1996)
51. Chen, J., Patton, R.J.: Robust Model-based Fault Diagnosis for Dynamic Systems. Kluwer Academic, Norwell (1999)
52. Chen, J., Zhang, H.: Robust detection of faulty actuators via unknown input observers. Int. J. Syst. Sci. **22**, 1829–1839 (1991)
53. Chen, W., Jiang, J.: Fault-tolerant control against stuck actuator faults. IEE Proc., Control Theory Appl. **152**, 138–146 (2005)
54. Chen, W., Saif, M.: Actuator fault diagnosis for uncertain linear systems using a high-order sliding mode robust differentiator (HOSMRD). Int. J. Robust Nonlinear Control **18**, 413–426 (2008)
55. Chilali, M., Gahinet, P.: $\mathcal{H}_\infty$ design with pole placement constraints: an LMI approach. IEEE Trans. Autom. Control **41**, 358–367 (1996)
56. Choi, H.H.: A new method for variable structure control system design: a linear matrix inequality approach. Automatica **33**, 2089–2092 (1997)
57. Chu, Q.P., Mulder, J.A., Sridhar, J.K.: Decomposition of aircraft state and parameter estimation problems. In: 10th IFAC Symposium on System Identification, Copenhagen, Denmark (1995)
58. Corradini, M.L., Orlando, G., Parlangeli, G.: A fault tolerant sliding mode controller for accommodating actuator failures. In: 44th IEEE Conference on Decision and Control, Seville, Spain, pp. 3091–3096 (2005)
59. Cunha, J.P.V.S., Costa, R.R., Hsu, L.: Cooperative actuators for fault tolerant model-reference sliding mode control. In: IEEE International Symposium on Industrial Electronics, Rio de Janeiro, Brazil, pp. 690–695 (2003)
60. Cunha, J.P.V.S., Hsu, L., Costa, R.R., Lizarralde, F.: Output-feedback model-reference sliding mode control of uncertain multivariable systems. IEEE Trans. Autom. Control **48**, 2245–2250 (2003)

61. Darouach, M.: On the novel approach to the design of unknown input observers. IEEE Trans. Autom. Control **39**, 698–699 (1994)
62. Darouach, M., Zasadzinski, M., Xu, S.J.: Full-order observers for linear systems with unknown inputs. IEEE Trans. Autom. Control **39**, 606–609 (1994)
63. Davidson, J.B., Lallman, F.J., Bundick, W.T.: Real-time adaptive control allocation applied to a high performance aircraft. In: 5th SIAM Conference on Control & Its Application, San Diego, CA, USA (2001)
64. Davies, R., Spurgeon, S.K.: Robust implementation of sliding mode control schemes. Int. J. Syst. Sci. **24**, 733–743 (1993)
65. De Gaay Fortman, W.F., Van Paassen, M.M., Mulder, M., In't Veld, A.C., Clarke, J.P.: Implementing time-based spacing for decelerating approaches. J. Aircr. **44**, 106–118 (2007)
66. De Prins, J.L., Schippers, F.K.M., Mulder, M., Van Paassen, M.M., In't Veld, A.C., Clarke, J.P.: Enhanced self-spacing algorithm for three-degree decelerating approaches. J. Guid. Control Dyn. **30**, 576–90 (2007)
67. de Jager, B.: Comparison of methods to eliminate chattering and avoid steady state errors in sliding mode digital control. In: Proceedings of the IEEE VSC and Lyapunov Workshop, Sheffield, UK, pp. 37–42 (1992)
68. Deyst Jr., J.J., Harrison, J.V., Gai, E., Daly, K.C.: Fault detection, identification and reconfiguration for spacecraft systems. J. Astronaut. Sci. **29**, 113–26 (1981)
69. Ding, S.X.: Model-based Fault Diagnosis Techniques: Design Schemes, Algorithms and Tools. Springer, Berlin (2008)
70. Dong-Gyun, C., Jong-Hwan, K.: Pitch autopilot design using model-following adaptive sliding mode control. J. Guid. Control Dyn. **25**, 826–829 (2002)
71. Dorling, C.M., Zinober, A.S.I.: A comparative study of the sensitivity of observers. In: Proceedings of the IASTED Symposium on Applied Control and Identification, Copenhagen, Denmark, pp. 6–32638 (1983)
72. Dorling, C.M., Zinober, A.S.I.: Two approaches to hyperplane design in multivariable variable structure control systems. Int. J. Control **44**, 65–82 (1986)
73. Dorling, C.M., Zinober, A.S.I.: Robust hyperplane design in multivariable variable structure control systems. Int. J. Control **48**, 2043–2054 (1988)
74. Drakunov, S., Utkin, V.I.: Sliding mode observers: tutorial. In: Proceedings of the 34th IEEE Conference of Decision and Control, pp. 3376–3378 (1995)
75. Ducard, G.J.J.: Fault-tolerant Flight Control and Guidance Systems: Practical Methods for Small Unmanned Aerial Vehicles, Advances in Industrial Control Series. Springer, Berlin (2009)
76. Dumont, G.A., Huzmezan, M.: Concepts, methods and techniques in adaptive control. In: Proceedings of the American Control Conference, Anchorage, AK, USA, pp. 1137–1150 (2002)
77. Durham, W.C.: Constrained control allocation. J. Guid. Control Dyn. **16**(4), 717–25 (1993)
78. Eberhardt, R.L., Ward, D.G.: Indirect adaptive flight control system interactions. Int. J. Robust Nonlinear Control **9**, 1013–1031 (1999)
79. Edwards, C.: A practical method for the design of sliding mode controllers using linear matrix inequalities. Automatica **40**, 1761–1769 (2004)
80. Edwards, C., Lombaerts, T., Smaili, H.: Fault Tolerant Flight Control: A Benchmark Challenge vol. 399. Springer, Berlin (2010)
81. Edwards, C., Postlethwaite, I.: Anti-windup and bumpless-transfer schemes. Department of engineering report 96-5-Feb, Leicester University (1996)
82. Edwards, C., Spurgeon, S.K.: On the development of discontinuous observers. Int. J. Control **59**, 1211–1229 (1994)
83. Edwards, C., Spurgeon, S.K., Patton, R.J.: Sliding mode observers for fault detection. Automatica **36**, 541–553 (2000)
84. Edwards, C., Spurgeon, S.K.: Sliding mode stabilization of uncertain systems using only output information. Int. J. Control **62**, 1129–1144 (1995)
85. Edwards, C., Spurgeon, S.K.: Sliding Mode Control: Theory and Applications. Taylor & Francis, London (1998)

86. Edwards, C., Spurgeon, S.K.: A sliding mode control observer based FDI scheme for the ship benchmark. Eur. J. Control **6**, 341–356 (2000)
87. Edwards, C., Tan, C.P.: A comparison of sliding mode and unknown input observers for fault reconstruction. Eur. J. Control **16**, 245–260 (2006)
88. El-Ghezawi, O.M.E., Billings, S.A., Zinober, A.S.I.: Variable-structure systems and system zeros. IEE Proc. Part D **130**, 1–5 (1983)
89. Enns, D.: Control allocation approaches. In: AIAA Guidance, Navigation and Control Conference and Exhibit, Boston, MA, pp. 98–108 (1998)
90. Farineau, J.: Lateral electric flight control laws of a civil aircraft based upon eigenstructure assignment technique. In: AIAA Guidance, Navigation and Control Conference, Boston, MA, USA (1989)
91. Field, E.J., Pinney, T.R., (René) van Paassen, M.M., Stroosma, O., Rivers, R.A.: Effects of implementation variations on the results of piloted simulator handling qualities evaluations. In: AIAA Modeling and Simulation Technologies Conference and Exhibit, Providence, RI, USA, pp. 43–58 (2004)
92. Fisher, J.R.: Aircraft control using nonlinear dynamic inversion in conjunction with adaptive robust control. Master of Science Thesis, Texas A&M University (2004)
93. Floquet, T., Barbot, J.P.: An observability form for linear systems with unknown inputs. Int. J. Control **79**, 132–139 (2006)
94. Floquet, T., Barbot, J.P.: Simultaneous robust state observation and unknown input estimation. In: International Workshop on Variable Structure Systems, Barcelona, Spain (2007)
95. Floquet, T., Edwards, C., Spurgeon, S.K.: On sliding mode observers with unknown inputs. In: International Workshop on Variable Structure Systems, Alghero, Italy (2006)
96. Floquet, T., Edwards, C., Spurgeon, S.K.: On sliding mode observers for systems with unknown inputs. Int. J. Adapt. Control Signal Process. **21**, 638–656 (2007)
97. Forssell, L., Nilsson, U.: ADMIRE, the aero-data model in a research environment version 4.0, model description. Technical report FOI-R-1624-SE, Swedish Defence Agency (FOI) (2005)
98. Frank, P.M., Ding, X.: Frequency domain approach to optimally robust residual generation and evaluation for model-based fault diagnosis. Automatica **30**, 789–804 (1994)
99. Franklin, G.F., Powell, J.D., Emami-Naeini, A.: Feedback Control of Dynamic Systems. 4th edn. Prentice Hall, New York (2002)
100. Fridman, L., Davila, J., Levant, A.: High-order sliding mode observation and fault detection. In: Proceedings of the Conference on Decision and Control, New Orleans, USA, pp. 4317–4322 (2007)
101. Fridman, L., Davila, J., Levant, A.: High-order sliding mode observation of linear systems with unknown inputs. In: Proceedings of the IFAC World Congress, Seoul, pp. 4779–4790 (2008)
102. Fridman, L., Shtessel, Y., Edwards, C., Yan, X.G.: Higher-order sliding mode observer for state estimation and input reconstruction in nonlinear systems. Int. J. Robust Nonlinear Control **18**(4–5), 399–412 (2008)
103. Gaeid, K.S., Mohamed, H.A.F.: Diagnosis and fault tolerant control of the induction motors techniques a review. Aust. J. Basic Appl. Sci. **4**(2), 227–246 (2010)
104. Gahinet, P., Nemirovski, A., Laub, A., Chilali, M.: Lmi Control Toolbox, User Guide. The Mathworks, Natick (1995)
105. Ganguli, S., Marcos, A., Balas, G.J.: Reconfigurable LPV control design for Boeing 747-100/200 longitudinal axis. In: American Control Conference, Anchorage, AK, USA, pp. 3612–3617 (2002)
106. Gao, Z., Antsaklis, P.J.: Stability of the pseudo–inverse method for reconfigurable control. Int. J. Control **53**(3), 717–729 (1991)
107. Garcia-Velo, J., Walker, B.K.: Aerodynamic parameter estimation for high-performance aircraft using extended Kalman filtering. J. Guid. Control Dyn. **20**, 1257–9 (1997)
108. Georgie, J., Valasek, J.: Evaluation of longitudinal desired dynamics for dynamic-inversion controlled generic re-entry vehicles. J. Guid. Control Dyn. **26**, 811–819 (2003)

109. Gero, D.: Aviation Disasters: The World's Major Civil Airliner Crashes Since 1950. Patrick Stephens, Sparkford (2006)
110. Gobbo, D.D., Napolitano, M., Famouri, P., Innocenti, M.: Experimental application of extended Kalman filtering for sensor validation. IEEE Trans. Control Syst. Technol. **9**, 376–380 (2001)
111. Gopinathan, M., Bošković, J.D., Mehra, R.K., Rago, C.: A multiple model predictive scheme for fault-tolerant flight control design. In: Proceedings of the 37th IEEE Conference on Decision and Control, Tampa, FL, USA, pp. 1376–1381 (1998)
112. Goupil, P.: AIRBUS state of the art and practices on FDI and FTC. In: Proceedings of the IFAC Symposium SAFEPROCESS '09, Barcelona, Spain, pp. 564–572 (2009)
113. Gouverneur, B., Mulder, J.A.B., van Paassen, M.M.R., Stroosma, O.: Optimization of the SIMONA Research Simulator's Motion Filter Settings for Handling Qualities Experiments. In: AIAA Modeling and Simulation Technologies Conference and Exhibit, Austin, TX, USA (2003)
114. Grenaille, S., Henry, D., Zolghadri, A.: A method for designing fault diagnosis filters for LPV polytopic systems. J. Control Sci. Eng. **2008**, 1–11 (2008). doi:10.1155/2008/231697
115. Gunnarsson, M.: Parameter estimation for fault diagnosis of an automotive engine using extended Kalman filters. Master Thesis, Linköpings University (2001)
116. Hallouzi, R., Verdult, V., Babuska, R., Verhaegen, M.: Fault detection and identification of actuator faults using linear parameter varying models. In: IFAC World Congress, Prague, Czech Republic, pp. 119–124 (2005)
117. Hallouzi, R., Verhaegen, M.: Fault-tolerant subspace predictive control applied to a Boeing 747 model. J. Guid. Control Dyn. **31**(4), 873–883 (2008)
118. Hanke, C.: The simulation of a large jet transport aircraft. Mathematical model, vol. I. Technical report CR-1756, NASA and the Boeing company (1971)
119. Hanke, C., Nordwall, D.: The simulation of a jumbo jet transport aircraft. Modelling data, vol. II. Technical report CR-114494/D6-30643-VOL2, NASA and The Boeing Company (1970)
120. Harefors, M., Bates, D.G.: Integrated propulsion-based flight control system design for a civil transport aircraft. In: Proceedings of the 2002 IEEE International Conference on Control Applications, Glasgow, Scotland, UK, pp. 132–137 (2002)
121. Härkegård, O.: Backstepping and control allocation with applications to flight control. PhD thesis, Division of Automatic Control, Department of Electrical Engineering Linköping University, Sweden (2003)
122. Härkegård, O., Glad, S.T.: Resolving actuator redundancy—optimal control vs. control allocation. Automatica **41**(1), 137–144 (2005)
123. Hasakara, I., Özgüner, U., Utkin, V.I.: On sliding mode observers via equivalent control approach. Int. J. Control **71**, 1051–1067 (1998)
124. Heck, B.S., Yallapragada, S.V., Fan, M.K.H.: Numerical methods to design the reaching phase of output feedback variable structure control. Automatica **31**, 275–279 (1995)
125. Heerspink, H.M., Berkouwer, W.R., Stroosma, O., van Paassen, M.M., Mulder, M., Mulder, J.A.: Evaluation of vestibular thresholds for motion detection in the SIMONA research simulator. In: AIAA Modeling and Simulation Technologies Conference, San Francisco, CA, USA, pp. 1212–1231 (2005)
126. Hennig, A., Balas, G.J.: MPC supervisory flight controller: a case study to flight EL AL 1862. In: AIAA Guidance, Navigation and Control Conference and Exhibit, Honolulu, HI, USA, 2008
127. Hess, R.A., Wells, S.R.: Sliding mode control applied to reconfigurable flight control design. J. Guid. Control Dyn. **26**, 452–462 (2003)
128. Hess, R.A., Wells, S.R., Vetter, T.K.: MIMO sliding mode control as an alternative to reconfigurable flight control designs. In: American Control Conference, pp. 3637–3643 (2002)
129. Horn, R.A., Johnson, C.R.: Matrix Analysis. Cambridge University Press, Cambridge (1990)
130. Hou, M., Patton, R.P.: An LMI approach to $\mathcal{H}_-/\mathcal{H}_\infty$ fault detection observers. In: Proceedings of the UKACC International Conference on Control, Exeter, UK, pp. 305–310 (1996)

131. Huber, R.R., McCulloch, B.: Self-repairing flight control system. Society of automotive engineers technical paper, series 841552, 1–20 (1984)
132. Huzmezan, M., Maciejowski, J.: Reconfigurable flight control methods and related issues—a survey. Technical report prepared for the DERA under the research agreement No. ASF/3455, University of Cambridge (1997)
133. Idan, M., Johnson, M., Calise, A.J.: Intelligent aerodynamic/propulsion flight control for flight safety: a nonlinear adaptive approach. In: Proceedings of the American Control Conference, Arlington, VA, USA, pp. 2918–2923 (2001)
134. Isermann, R.: Model-based fault-detection and diagnosis—status and applications. Annu. Rev. Control **29**, 71–85 (2005)
135. Isermann, R.: Fault-Diagnosis Systems: An Introduction from Fault Detection to Fault Tolerance. Springer, Berlin (2006)
136. Isermann, R., Balle, P.: Trends in the application of model-based fault detection and diagnosis of technical processes. Control Eng. Pract. **5**, 709–719 (1997)
137. Ito, D., Georgie, J., Vasalek, J., Ward, D.T.: Re-entry vehicle flight control design guidelines: dynamic inversion. Technical report NASA/TP-2002-210771, NASA (2002)
138. Ito, D., Vasalek, J., Ward, D.T.: Robust dynamic inversion controller design and analysis for the X-38. In: AIAA Guidance, Navigation and Control Conference and Exhibit (2001)
139. Izadi-Zamanabadi, R., Blanke, M.: A ship propulsion system as a benchmark for fault-tolerant control. In: Proceedings of the IFAC Symposium—SAFEPROCESS '97, Hull, UK, pp. 1074–1081 (1997)
140. Jiang, B., Chowdhury, F.N.: Fault estimation and accommodation for linear MIMO discrete-time systems. IEEE Trans. Control Syst. Technol. **13**, 493–9 (2005)
141. Jiang, B., Staroswiecki, M., Cocquempot, V.: Fault estimation in nonlinear uncertain systems using robust sliding mode observers. IEE Proc., Control Theory Appl. **151**, 29–37 (2004)
142. Jiang, J., Zhang, Y.: Accepting performance degradation in fault-tolerant control system design. IEEE Trans. Control Syst. Technol. **14**, 284–292 (2006)
143. Jones, C.N.: Reconfigurable flight control: First year report. Technical report, Cambridge University Engineering Department (2005)
144. Joosten, D.A., van den Boom, T.J.J., Lombaerts, T.J.J.: Effective control allocation in fault-tolerant flight control using feedback linearization and model predictive control. In: European Control Conference (2007)
145. Juricic, D., Moseler, O., Rakar, A.: Model-based condition monitoring of an actuator system driven by a brushless DC motor. Control Eng. Pract. **9**, 545–554 (2001)
146. Kalman, R.E.: A new approach to linear filtering and prediction problems. Trans. ASME J. Basic Eng. **82**(Series D), 35–45 (1960)
147. Kalman, R.E., Bucy, R.S.: A new approach to linear filtering and prediction problems. Trans. ASME J. Basic Eng. **83**, 95–108 (1961)
148. Kanev, S., Verhaegen, M.: A bank of reconfigurable LQG controllers for linear systems subjected to failures. In: Proceedings of the 39th IEEE Conference on Decision and Control, Sydney, Australia, pp. 3684–3689 (2000)
149. Kanev, S., Verhaegen, M.: Controller reconfiguration for non-linear systems. Control Eng. Pract. **8**, 1223–35 (2000)
150. Keller, J.Y., Darouach, M.: Optimal two-stage Kalman filter in the presence of random bias. Automatica **33**, 1745–8 (1997)
151. Keller, J.Y., Darouach, M.: Two-stage Kalman estimator with unknown exogenous inputs. Automatica **35**, 339–342 (1999)
152. Khalil, H.K.: Nonlinear Systems. Prentice Hall, Englewood Cliffs (1992)
153. Khong, T.H., Shin, J.: Robustness analysis of integrated LPV-FDI filters and LTI-FTC system for a transport aircraft. In: AIAA Guidance, Navigation and Control Conference and Exhibit, Hilton Head, SC, USA, pp. 4166–4187 (2007)
154. Kim, Y.W., Rizzoni, G., Utkin, V.: Automotive engine diagnosis and control via nonlinear estimation. IEEE Control Syst. Mag. **18**, 884–99 (1998)

155. Kim, Y.W., Rizzoni, G., Utkin, V.: Developing a fault tolerant power train system by integrating the design of control and diagnostics. Int. J. Robust Nonlinear Control **11**, 1095–1114 (2001)
156. Kleeman, L.: Understanding and applying Kalman filtering. In: Proceedings of the Second Workshop on Perceptive Systems, Perth, Australia (1996)
157. Kobayashi, T., Simon, D.L.: Application of a bank of Kalman filters for aircraft engine fault diagnostics. Technical report NASA/TM-2003-212526, NASA (2003)
158. Kobayashi, T., Simon, D.L.: Evaluation of an enhanced bank of Kalman filters for in-flight aircraft engine sensor fault diagnostics. J. Eng. Gas Turbine Power **127**, 497–504 (2005)
159. Korbicz, J., Koscielny, J.M., Kowalczuk, Z., Cholewa, W.: Fault Diagnosis: Models, Artificial Intelligence, Applications. Springer, Berlin (2004)
160. Kowal, M., Korbicz, J.: Robust fault detection using neuro-fuzzy networks. In: IFAC World Congress, Prague, Czech Republic, pp. 185–190 (2005)
161. Kudva, P., Viswanadham, N., Ramakrishna, A.: Observers for linear systems with unknown inputs. IEEE Trans. Autom. Control **25**, 113–115 (1980)
162. Kwon, W.H., Kim, P.S., Park, P.G.: A receding horizon Kalman fir filter for linear continuous-time systems. IEEE Trans. Autom. Control **44**, 2115–20 (1999)
163. Lam, T.M., Mulder, M., Van Paassen, M.M., Mulder, J.A.: Comparison of control and display augmentation for perspective flight path displays. J. Guid. Control Dyn. **29**, 564–578 (2006)
164. Landau, I.D.: Adaptive Control: The Model Reference Approach. Marcel Dekker, New York (1979)
165. Leith, D.J., Leithead, W.E.: Survey of gain-scheduling analysis and design. Int. J. Control **73**, 1001–1025 (2000)
166. Levant, A.: Robust exact differentiation via sliding mode technique. Automatica **34**(3), 379–384 (1998)
167. Li, R., Olson, J.H.: Fault detection and diagnosis in a closed-loop nonlinear distillation process. application of extended Kalman filters. Ind. Eng. Chem. Res. **30**, 898–908 (1991)
168. Lira, S.D., Puig, V., Quevedo, J.: PEM fuel cell system robust LPV model-based fault diagnosis. In: The 20th International Workshop on Principles of Diagnosis, Stockholm, Sweden (2009)
169. Liu, G.: Control of robots manipulators with consideration of actuators degradation and failures. In: IEEE Int. Conf. on Robotics and Automation, Seoul, Korea, pp. 2256–2571 (2001)
170. Liu, G.P., Patton, R.J.: Eigenstructure Assignment for Control System Design. Wiley, New York (1998)
171. Ljung, L.: System Identification Toolbox: For use with Matlab. The Mathworks, Natick (1995)
172. Lombaerts, T.J.J., Chu, Q.P., Mulder, J.A., Joosten, D.A.: Real time damaged aircraft model identification for reconfiguring flight control. In: AIAA Atmospheric Flight Mechanics Conference and Exhibit, Hilton Head, SC, USA, pp. 1207–1231 (2007)
173. Lombaerts, T.J.J., Huisman, H.O., Chu, Q.P., Mulder, J.A., Joosten, D.A.: Flight control reconfiguration based on online physical model identification and nonlinear dynamic inversion. In: AIAA Guidance, Navigation and Control Conference and Exhibit, Honolulu, HI, USA (2008)
174. Luenberger, D.G.: An introduction to observers. IEEE Trans. Autom. Control **16**, 596–602 (1971)
175. Maciejowski, J.M.: Multivariable Feedback Design. Addison-Wesley, Reading (1989)
176. Maciejowski, J.M.: Predictive Control with Constraints. Prentice Hall, New York (2002)
177. Maciejowski, J.M., Jones, C.N.: MPC fault-tolerant control case study: flight 1862. In: Proceedings of the IFAC Symposium SAFEPROCESS '03, WA, USA, pp. 119–124 (2003)
178. Magni, J.F., Bennani, S., Terlouw, J.: Robust Flight Control: A Design Challenge. Springer, Berlin (1997)
179. Mahmoud, M., Jiang, J., Zhang, Y.: Active Fault Tolerant Control Systems Stochastic Analysis and Synthesis. Springer, Berlin (2003)

180. Maine, R.E., Iliff, K.W.: Formulation of a practical algorithm for parameter estimation with process and measurement noise. In: Identification and System Parameter Estimation 1982. Proceedings of the Sixth IFAC Symposium, Washington, USA (1983)
181. Maine, R.E., Iliff, K.W.: Agardograph 300 flight test technique series: vol. 3 on identification of dynamic systems—applications to aircraft. Part 1: The output error approach. Technical report AGARD-AG-300 vol. 3 Part 1, NATO (1986)
182. Marcos, A.: A linear parameter varying model of the BOEING 747-100/200 longitudinal motion. Master of science thesis, University of Minnesota (2001)
183. Marcos, A., Balas, G.J.: A Boeing 747-100/200 aircraft fault tolerant and diagnostic benchmark. Technical report AEM-UoM-2003-1, Department of Aerospace and Engineering Mechanics, University of Minnesota (2003)
184. Marcos, A., Balas, G.J.: A robust integrated controller/diagnosis aircraft application. Int. J. Robust Nonlinear Control **15**(12), 531–551 (2005)
185. Marcos, A., Ganguli, S., Balas, G.J.: An application of H_∞ fault detection and isolation to a transport aircraft. Control Eng. Pract. **13**(1), 105–119 (2005)
186. Maybeck, P.S.: Stochastic models, estimation, and control. In: Mathematics in Science and Engineering, vol. 141. Academic Press, New York (1979)
187. McLean, D., Aslam-Mir, S.: Reconfigurable flight control systems. In: International Conference on Control '91 (Conf. Publ. No. 332), pp. 234–242 (1991)
188. Mirea, L., Patton, R.J.: Component fault diagnosis using wavelet neural networks with local recurrent structure. In: Proceedings of the IFAC Symposium SAFEPROCESS '06, Beijing, China, pp. 91–96 (2006)
189. Mirea, L., Patton, R.J.: A new dynamic neuro-fuzzy system applied to fault diagnosis of an evaporation station. In: Proceedings of the IFAC Symposium SAFEPROCESS '06, Beijing, China, pp. 253–258 (2006)
190. Moreno, J.A., Osorio, M.: A Lyapunov approach to second-order sliding mode controllers and observers. In: 47th IEEE Conference on Decision and Control, Cancun, Mexico, pp. 2856–2861 (2008)
191. Moseler, O., Isermann, R.: Application of model-based fault detection to a brushless DC motor. IEEE Trans. Ind. Electron. **47**(5), 1015–1020 (2000)
192. Mulder, J.A., Chu, Q.P., Sridhar, J.K., Breeman, J.H., Laban, M.: Non-linear aircraft flight path reconstruction review and new advances. Prog. Aerosp. Sci. **35**, 673–726 (1999)
193. Mulder, J.A., Sridhar, J.K., Breeman, J.H.: Agardograph 300 flight test technique series: vol. 3 on identification of dynamic systems—applications to aircraft. Part 2: Nonlinear analysis and manoeuvre design. Technical report AGARD-AG-300 vol. 3 Part 2, NATO (1994)
194. Mulder, M., Veldhuijzen, A.R., Van Paassen, M.M., Mulder, J.A.: Integrating fly-by-wire controls with perspective flight path displays. J. Guid. Control Dyn. **28**, 1263–1274 (2005)
195. Narendra, K.S., Balakrishnan, J.: Adaptive control using multiple models. IEEE Trans. Autom. Control **42**, 171–187 (1997)
196. Narendra, K.S., Driollet, O.A., Feiler, M., George, K.: Adaptive control using multiple models. Int. J. Adapt. Control Signal Process. **17**, 87–102 (2003)
197. Ng, K.Y., Tan, C.P., Edwards, C., Kuang, Y.C.: New results in robust actuator fault reconstruction in linear uncertain systems. Int. J. Robust Nonlinear Control **17**, 1294–1319 (2007)
198. Ni, L., Fuller, C.R.: Control reconfiguration based on hierarchical fault detection and identification for unmanned underwater vehicle. J. Vib. Control **9**, 753–748 (2003)
199. Noura, H., Theilliol, D., Ponsart, J.C., Chamseddine, A.: Fault-tolerant Control Systems: Design and Practical Applications. Mathematics in Science and Engineering. Springer, Berlin (2009)
200. de Oca, S.M., Puig, V., Theilliol, D., Tornil-Sin, S.: Fault-tolerant control design using LPV admissible model matching: application to a two-degree of freedom helicopter. In: 17th Mediterranean Conference on Control & Automation, Thessaloniki, Greece, pp. 522–527 (2009)
201. Oliveira, J., Chu, Q.P., Mulder, J.A., Balini, H.M.N.K., Vos, W.G.M.: Multi-input design for aerodynamic parameter estimation. In: AIAA Guidance, Navigation, and Control Conference (2005)

202. Omerdic, E., Toal, D.: Control allocation of over-actuated thruster-propelled underwater vehicles. In: 1st Workshop on Networked Control System and Fault Tolerant Control, Ajaccio, France, pp. 163–171 (2005)
203. Oppenheimer, M.W., Doman, D.B.: Control allocation for overactuated systems. Technical report AFRL-VA-WP-TP-2006-321, Air Force Research Laboratory (2006)
204. Papageorgiou, G., Glover, K., D'Mello, D., Patel, Y.: Taking robust LPV control into flight on the VAAC harrier. In: 39th IEEE Conference on Decision and Control, Sydney, Australia, pp. 4558–4564 (2000)
205. Patton, R.J., Frank, P.M., Clark, R.N.: Fault Diagnosis in Dynamic Systems: Theory and Application. Prentice Hall, New York (1989)
206. Patton, R.J.: Fault tolerant control: the 1997 situation. In: Proceedings of the IFAC Symposium—SAFEPROCESS '97, Hull, UK, pp. 1035–1055 (1997)
207. Patton, R.J., Chen, J.: Optimal unknown input distribution matrix selection for robust fault diagnosis. Automatica **29**, 837–841 (1993)
208. Patton, R.J., Chen, J., Benkhedda, H.: A study on neuro-fuzzy systems for fault diagnosis. Int. J. Syst. Sci. **31**(11), 1441–1448 (2000)
209. Patton, R.J., Korbicz, J.: Advances in computational intelligence for fault diagnosis systems. Special issue of International Journal of Applied Mathematics and Computer Science **9**(3) (1999)
210. Patton, R.J., Uppal, F.J., Lopez-Toribio, C.J.: soft computing approaches to fault diagnosis for dynamic systems: a survey. In: Proceedings of the IFAC Symposium SAFEPROCESS '00, Budapest, Hungary, pp. 298–311 (2000)
211. Perruquetti, W., Barbot, J.P.: Sliding Mode Control in Engineering. Dekker, New York (2002)
212. Polycarpou, M.M., Vemuri, A.T.: Learning methodology for failure detection and accommodation. IEEE Control Syst. Mag. **15**(3), 16–24 (1995)
213. Puig, V., Witczak, M., Nejjari, F., Quevedo, J., Korbicz, J.: A GMDH neural network-based approach to passive robust fault detection using a constraint satisfaction backward test. Eng. Appl. Artif. Intell. **20**(7), 886–897 (2007)
214. Rago, C., Prasanth, R., Mehra, R.K., Fortenbaugh, R.: Failure detection and identification and fault tolerant control using the IMM-KF with applications to the eagle-eye UAV. In: Proceedings of the 37th IEEE Conference on Decision and Control, Tampa, FL, USA, pp. 4208–4213 (1998)
215. Rodrigues, M., Theilliol, D., Aberkane, S., Sauter, D.: Fault tolerant control design for polytopic LPV systems. Int. J. Appl. Math. Comput. Sci. **17**, 27–37 (2007)
216. Rosenbrock, H.H.: State Space and Multivariable Theory. Wiley, New York (1970)
217. Ryan, E.P., Corless, M.: Ultimate boundedness and asymptotic stability of a class of uncertain dynamical systems via continuous and discontinuous control. IMA J. Math. Control Inf. **1**, 223–242 (1984)
218. Saif, M., Chen, W., Wu, Q.: High order sliding mode observers and differentiators-application to fault diagnosis problem. In: Bartolini, G., Fridman, L., Pisano, A., Usai, E. (eds.) Modern Sliding Mode Control Theory. Lecture Notes in Control and Information Sciences, vol. 375, pp. 321–344. Springer, Berlin (2008)
219. Saif, M., Guan, Y.: A new approach to robust fault detection and identification. IEEE Trans. Aerosp. Electron. Syst. **29**, 685–695 (1993)
220. Sánchez, J.M.M., Rodellar, J.: Adaptive Predictive Control: From the Concepts to Plant Optimization. Prentice Hall, New York (1996)
221. Shamma, J.S., Cloutier, J.R.: Gain-scheduled missile autopilot design using linear parameter varying transformations. J. Guid. Control Dyn. **16**(2), 256–261 (1993)
222. Shin, D., Moon, G., Kim, Y.: Design of reconfigurable flight control system using adaptive sliding mode control: actuator fault. Proc. Inst. Mech. Eng., G J. Aerosp. Eng. **219**(4), 321–328 (2005),
223. Shin, J., Belcastro, C.M.: Performance analysis on fault tolerant control system. IEEE Trans. Control Syst. Technol. **14**, 920–925 (2006)

224. Shin, J., Wu, N.E., Belcastro, C.: Adaptive linear parameter varying control synthesis for actuator failure. J. Guid. Control Dyn. **27**, 787–794 (2004)
225. Shtessel, Y., Buffington, J., Banda, S.: Multiple time scale flight control using re-configurable sliding modes. J. Guid. Control Dyn. **22**, 873–883 (1999)
226. Shtessel, Y., Buffington, J., Banda, S.: Tailless aircraft flight control using multiple time scale re-configurable sliding modes. IEEE Trans. Control Syst. Technol. **10**(2), 288–296 (2002)
227. Sira-Ramirez, H., Spurgeon, S.K.: On the robust design of sliding observers for linear systems. Syst. Control Lett. **23**, 9–14 (1994)
228. Skogestad, S., Postlethwaite, I.: Multivariable Feedback Control: Analysis and Design. Wiley, New York (1996)
229. Slotine, J.J.E., Hedrick, J.K., Misawa, E.A.: On sliding observers for nonlinear systems. Trans. ASME J. Dyn. Syst. Meas. Control **109**, 245–252 (1987)
230. Slotine, J.J.E., Li, W.: Applied Nonlinear Control. Prentice Hall, Englewood Cliffs (1991)
231. Smaili, H., Breeman, J., Lombaerts, T., Joosten, D.: A benchmark for integrated fault tolerant flight control evaluation. In: Fault Tolerant Flight Control: A Benchmark Challenge, vol. 399, pp. 168–221. Springer, Berlin (2010)
232. Smaili, M.H.: Flight data reconstruction and simulation of EL AL Flight 1862. Graduation report, Delft University of Technology, 1997
233. Smaili, M.H.: Flightlab 747: Benchmark for advance flight control engineering. Technical report, Technical University Delft, The Netherlands, 1999
234. Smaili, M.H., Breeman, J., Lombaerts, T.J.J., Joosten, D.A.: A simulation benchmark for integrated fault tolerant flight control evaluation. In: AIAA Modeling and Simulation Technologies Conference and Exhibit, Keystone, CO, USA, pp. 563–585 (2006)
235. Smaili, M.H., Mulder, J.A.: Flight data reconstruction and simulation of the 1992 Amsterdam Bijlmermeer airplane accident. In: AIAA Modeling and Simulation Technologies Conference, Denver, CO, USA, 2000
236. Snell, S.A., Enns, D.F., Garrard Jr., W.L.: Nonlinear inversion flight control for a supermaneuverable aircraft. J. Guid. Control Dyn. **15**, 976–984 (1992)
237. Sorenson, H.W.: Kalman Filtering: Theory and Application. IEEE Press, New York (1985)
238. Spirkovska, L., Iverson, D.L., Poll, S., Pryor, A.: Inductive learning approaches for improving pilot awareness of aircraft faults. Technical report 20060017823, NASA (2005)
239. Spurgeon, S.K., Patton, R.J.: Robust variable structure control of model reference systems. Proc. IEE D **137**, 341–348 (1990)
240. Stengel, R.F.: Flight Dynamics. Princeton University Press, Princeton (2004)
241. Stewart, G.W.: On scaled projections and pseudoinverses. Linear Algebra Appl. **112**, 189–193 (1989)
242. Stroosma, O., Smaili, H., Lombaerts, T., Mulder, J.A.: Piloted simulator evaluation of new fault-tolerant flight control algorithms for reconstructed accident scenarios. In: AIAA Modeling and Simulation Technologies Conference and Exhibit, Honolulu, HI, USA (2008)
243. Stroosma, O., van Paassen, M.M., Mulder, M.: Using the SIMONA research simulator for human-machine interaction research. In: AIAA Modeling and Simulation Technologies Conference, WA, USA (2003)
244. Szaszi, I., Ganguli, S., Marcos, A., Balas, G.J., Bokor, J.: Application of FDI to a nonlinear Boeing-747 aircraft. In: Mediterranean Conference on Control and Automation, Lisbon, Portugal (2002)
245. Szaszi, I., Marcos, A., Balas, G.J., Bokor, J.: Linear parameter-varying detection filter design for a Boeing 747-100/200 aircraft. J. Guid. Control Dyn. **28**(3), 461–470 (2005)
246. Takahashi, R.H.C., Peres, P.L.D.: Unknown input observers for uncertain systems: a unifying approach. Eur. J. Control **5**, 261–275 (1995)
247. Tan, C.P.: Sliding mode observers for fault detection and isolation. Ph.D. thesis, University of Leicester, 2002
248. Tan, C.P., Edwards, C.: Sliding mode observers for robust detection and reconstruction of actuator and sensor faults. Int. J. Robust Nonlinear Control **13**, 443–463 (2003)
249. Tandale, M.D., Valasek, J.: Adaptive dynamic inversion control of a linear scalar plant with constrained control inputs. In: Proceedings of the 2005 American Control Conference (2005)

250. Tandale, M.D., Valasek, J.: Fault-tolerant structured adaptive model inversion control. J. Guid. Control Dyn. **29**, 635–642 (2006)
251. Tao, G., Joshi, S.M., Ma, X.: Adaptive state feedback and tracking control of systems with actuator failures. IEEE Trans. Autom. Control **46**(1), 78–95 (2001)
252. Tournes, C., Landrum, D.B., Shtessel, Y., Hawk, C.W.: Ramjet-powered re-usable vehicle control by sliding modes. J. Guid. Control Dyn. **21**(3), 409–415 (1998)
253. Tucker, T.: Touchdown: The Development of Propulsion Controlled Aircraft at Nasa Dryden. Monographs in Aerospace History, vol. 16 (1999)
254. Uppal, F.J., Patton, R.J., Witczak, M.: A neuro-fuzzy multiple-model observer approach to robust fault diagnosis based on the DAMADICS benchmark problem. Control Eng. Pract. **14**, 699–717 (2006)
255. Utkin, V., Guldner, J., Shi, J.: Sliding Mode Control in Electromechanical Systems. Taylor & Francis, London (1999)
256. Utkin, V.I.: Sliding Modes in Control Optimization. Springer, Berlin (1992)
257. Utkin, V.I., Young, K.K.D.: Methods for constructing discontinuity planes in multidimensional variable structure systems. Autom. Remote Control **39**, 1466–1470 (1978)
258. Valente Pais, A.R., Mulder, M., Van Paassen, M.M., Wentink, M., Groen, E.: Modeling human perceptual thresholds in self-motion perception. In: AIAA Modeling and Simulation Technologies Conference, Keystone, CO, USA, pp. 900–914 (2006)
259. van der Linden, C.A.A.M.: DASMAT: Delft University aircraft simulation model and analysis tool. Technical report LR-781, Technical University of Delft, the Netherlands, 1996
260. van der Linden, C.A.A.M., Sridhar, J.K., Mulder, J.A.: Multi-input design for aerodynamic parameter estimation. In: Proceedings of the American Control Conference, Seattle, WA, USA, pp. 703–707 (1995)
261. Varga, A.: Design of least order residual generators for fault detection and isolation with application to monitoring actuator/surface faults for a Boeing 747 100/200 aircraft. Technical report DLR IB 515-08-28, DLR, 2008
262. Verhaegen, M., Kanev, S., Hallouzi, R., Jones, C., Maciejowski, J., Smail, H.: Fault tolerant flight control—a survey. In: Fault Tolerant Flight Control: A Benchmark Challenge, vol. 399, pp. 46–89. Springer, Berlin (2010)
263. Vetter, T.K., Wells, S.R., Hess, R.A.: Designing for damage-robust flight control design using sliding mode techniques. Proc. Inst. Mech. Eng., G J. Aerosp. Eng. **217**, 245–261 (2003)
264. Vormer, F.J., Mulder, M., Van Paassen, M.M., Mulder, J.A.: Optimization of flexible approach trajectories using a genetic algorithm. J. Aircr. **43**, 941–952 (2006)
265. Walcott, B.L., Corless, M.J., Żak, S.H.: Comparative study of nonlinear state observation techniques. Int. J. Control **45**, 2109–2132 (1987)
266. Walcott, B.L., Żak, S.H.: State observation of nonlinear uncertain dynamical systems. IEEE Trans. Autom. Control **32**, 166–170 (1987)
267. Wei, X., Verhaegen, M.: Mixed $\mathcal{H}_-/\mathcal{H}_\infty$ fault detection observer design for LPV systems. In: IEEE Conference on Decision and Control, Cancun, Mexico, pp. 1073–1078 (2008)
268. Welch, G., Bishop, G.: An introduction to the Kalman filter. Technical report TR 95-041, University of North Carolina, 2006
269. Wells, S.R., Hess, R.A.: Multi-input/multi-output sliding mode control for a tailless fighter aircraft. J. Guid. Control Dyn. **26**(3), 463–473 (2003)
270. Wheeler, G., Su, C., Stepanenko, Y.: Sliding mode controller with improved adaptation laws for the upper bounds on the norm of uncertainties. Automatica **34**, 1657–1661 (1998)
271. Willems, J.C.: Least squares optimal control and the algebraic Ricatti equation. IEEE Trans. Autom. Control **16**, 621–634 (1971)
272. Witczak, M.: Advances in model-based fault diagnosis with evolutionary algorithms and neural networks. Int. J. Appl. Math. Comput. Sci. **16**(1), 85–99 (2006)
273. Wu, F.: A generalized LPV system analysis and control synthesis framework. Int. J. Control **74**, 745–759 (2001)
274. Wu, N.E., Zhang, Y., Zhou, K.: Control effectiveness estimation using an adaptive Kalman estimator. In: Proceedings of the 1998 IEEE International Symposium on Intelligent Control

(ISIC) Held Jointly with IEEE International Symposium on Computational Intelligence in Robotics and Automation (CIRA) Intelligent Systems and Semiotics (ISAS), Gaithersburg, MD, USA, pp. 181–186 (1998)
275. Wu, N.E., Zhang, Y., Zhou, K.: Detection, estimation, and accommodation of loss of control effectiveness. Int. J. Adapt. Control Signal Process. **14**, 775–95 (2000)
276. Xiong, Y., Saif, M.: A novel design for robust fault diagnostic observer. In: IEEE Conference on Desicion and Control, pp. 952–957 (1998)
277. Xiong, Y., Saif, M.: Sliding mode observer for nonlinear uncertain systems. IEEE Trans. Autom. Control **46**, 2012–2017 (2001)
278. Xu, J.X., Jia, Q.W., Lee, T.H.: On the design of a nonlinear adaptive variable structure derivative estimator. IEEE Trans. Autom. Control **45**, 1028–1033 (2000)
279. Yan, X.G., Edwards, C.: Sensor fault detection and isolation for nonlinear systems based on sliding mode observers. Int. J. Adapt. Control Signal Process. **21**, 657–673 (2007)
280. Yan, X.G., Edwards, C.: Adaptive sliding mode observer-based fault reconstruction for nonlinear systems with parametric uncertainties. IEEE Trans. Ind. Electron. **55**, 4029–4036 (2008)
281. Yan, X.G., Edwards, C.: Fault estimation for single output nonlinear systems using an adaptive sliding mode observer. IET Control Theory Appl. **2**, 841–850 (2008)
282. Yan, X.G., Edwards, C.: Nonlinear robust fault reconstruction and estimation using a sliding mode observer. Automatica **43**(9), 1605–1614 (2007)
283. Yang, S.K.: An experiment of state estimation for predictive maintenance using kalman filter on a DC motor. Reliab. Eng. Syst. Saf. **75**(1), 103–111 (2002)
284. Yang, Z., Blanke, M., Verhaegen, M.: Robust control mixer method for reconfigurable control design using model matching. IET Control Theory Appl. **1**, 349–357 (2007)
285. Young, K.-K.D.: Asymptotic stability of model reference systems with variable structure control. IEEE Trans. Autom. Control **22**, 279–281 (1977)
286. Young, K.K.D.: Design of variable structure model-following control systems. IEEE Trans. Autom. Control **23**, 1079–1085 (1978)
287. Yu, D.L., Gomm, J.B.: Implementation of neural network predictive control to a multivariable chemical reactor. Control Eng. Pract. **11**(11), 1315–1323 (2003)
288. Zaal, P.M.T., Nieuwenhuizen, F.M., Mulder, M., Van Paassen, M.M.: Perception of visual and motion cues during control of self-motion in optic flow environments. In: AIAA Modeling and Simulation Technologies Conference, Keystone, CO, USA, pp. 915–932 (2006)
289. Zhang, Y., Jiang, J.: Design of integrated fault detection, diagnosis and reconfigurable control systems. In: Proceedings of the IEEE Conference on Decision and Control, Phoenix, AZ, USA, pp. 3587–3592 (1999)
290. Zhang, Y., Jiang, J.: Integrated active fault-tolerant control using IMM approach. IEEE Trans. Aerosp. Electron. Syst. **37**, 1221–1235 (2001)
291. Zhang, Y., Jiang, J.: Bibliographical review on reconfigurable fault tolerant control systems. In: Proceedings of the IFAC Symposium SAFEPROCESS '03, WA, USA, pp. 265–276 (2003)
292. Zhang, Y., Jiang, J.: Fault tolerant control system design with explicit consideration of performance degradation. IEEE Trans. Aerosp. Electron. Syst. **39**, 838–848 (2003)
293. Zhang, Y., Jiang, J.: Issues on integration of fault diagnosis and reconfigurable control in active fault-tolerant control systems. In: Proceedings of the IFAC Symposium SAFEPROCESS '06, Beijing, China, pp. 1437–1448 (2006)
294. Zhang, Y., Jiang, J.: Bibliographical review on reconfigurable fault-tolerant control systems. Annu. Rev. Control **32**, 229–252 (2008)
295. Zhang, Y., Suresh, V.S., Jiang, B., Theilliol, D.: Reconfigurable control allocation against aircraft control effector failures. In: IEEE International Conference on Control Applications, pp. 1197–1202 (2007)
296. Zhang, Y.M., Jiang, J.: Active fault-tolerant control system against partial actuator failures. IEE Proc., Control Theory Appl. **149**, 95–104 (2002)
297. Zhou, K., Doyle, J.C., Glover, K.: Robust and Optimal Control. Prentice Hall, New Jersey (1996)

298. Zhou, K., Rachinayani, P.K., Liu, N., Ren, Z., Aravna, J.: Fault diagnosis and reconfigurable control for flight control systems with actuator failures. In: 43rd IEEE Conference on Decision and Control, Bahamas, pp. 5266–5271 (2004)
299. Zinober, A.S.I.: An introduction to variable structure control. In: Zinober, A.S.I. (ed.) Deterministic Control of Uncertain Systems, pp. 1–26. Peter Peregrinus, Stevenage (1990)
300. Zinober, A.S.I.: An introduction to sliding mode variable structure control. In: Zinober, A.S.I. (ed.) Variable Structure and Lyapunov Control, pp. 1–22. Springer, Berlin (1994)
301. Zinober, A.S.I., El-Ghezawi, O.M.E., Billings, S.A.: Multivariable variable-structure adaptive model-following control systems. Proc. IEE, Part D **129**, 6–12 (1982)

Index

H. Alwi et al., *Fault Detection and Fault-Tolerant Control Using Sliding Modes*, Advances in Industrial Control,
DOI 10.1007/978-0-85729-650-4,

Other titles published in this series (continued):

Soft Sensors for Monitoring and Control of Industrial Processes
Luigi Fortuna, Salvatore Graziani, Alessandro Rizzo and Maria G. Xibilia

Adaptive Voltage Control in Power Systems
Giuseppe Fusco and Mario Russo

Advanced Control of Industrial Processes
Piotr Tatjewski

Process Control Performance Assessment
Andrzej W. Ordys, Damien Uduehi and Michael A. Johnson (Eds.)

Modelling and Analysis of Hybrid Supervisory Systems
Emilia Villani, Paulo E. Miyagi and Robert Valette

Process Control
Jie Bao and Peter L. Lee

Distributed Embedded Control Systems
Matjaž Colnarič, Domen Verber and Wolfgang A. Halang

Precision Motion Control (2nd Ed.)
Tan Kok Kiong, Lee Tong Heng and Huang Sunan

Optimal Control of Wind Energy Systems
Julian Munteanu, Antoneta luliana Bratcu, Nicolaos-Antonio Cutululis and Emil Ceangă

Identification of Continuous-time Models from Sampled Data
Hugues Gamier and Liuping Wang (Eds.)

Model-based Process Supervision
Arun K. Samantaray and Belkacem Bouamama

Diagnosis of Process Nonlinearities and Valve Stiction
M.A.A. Shoukat Choudhury, Sirish L. Shah and Nina F. Thornhill

Magnetic Control of Tokamak Plasmas
Marco Ariola and Alfredo Pironti

Real-time Iterative Learning Control
Jian-Xin Xu, Sanjib K. Panda and Tong H. Lee

Deadlock Resolution in Automated Manufacturing Systems
ZhiWu Li and MengChu Zhou

Model Predictive Control Design and Implementation Using MATLAB®
Liuping Wang

Predictive Functional Control
Jacques Richalet and Donal O'Donovan

Fault-tolerant Flight Control and Guidance Systems
Guillaume Ducard

Fault-tolerant Control Systems
Hassan Noura, Didier Theilliol, Jean-Christophe Ponsart and Abbas Chamseddine

Detection and Diagnosis of Stiction in Control Loops
Mohieddine Jelali and Biao Huans (Eds.)

Stochastic Distribution Control System Design
Lei Guo and Hong Wang

Dry Clutch Control for Automotive Applications
Pietro J. Dolcini, Carlos Canudas-de-Wit and Hubert Béchart

Advanced Control and Supervision of Mineral Processing Plants
Daniel Sbárbaro and René del Villar (Eds.)

Active Braking Control Design for Road Vehicles
Sergio M. Savaresi and Mara Tanelli

Active Control of Flexible Structures
Alberto Cavallo, Giuseppe de Maria, Ciro Natale and Salvatore Pirozzi

Induction Motor Control Design
Riccardo Marino, Patrizio Tomei
and Cristiano M. Verrelli

Fractional-order Systems and Controls
Concepcion A. Monje, YangQuan Chen,
Blas M. Vinagre, Dingyu Xue and Vincente
Feliu

Model Predictive Control of Wastewater Systems
Carlos Ocampo-Martinez

Wastewater Systems
Carlos Ocampo-Martinez

Tandem Cold Metal Rolling Mill Control
John Pitter and Marwan A. Simaan